Adaptive Management of Social-Ecological Systems

Craig R. Allen • Ahjond S. Garmestani
Editors

Adaptive Management of Social-Ecological Systems

Springer

Editors
Craig R. Allen
U.S. Geological Survey
Nebraska Cooperative Fish and Wildlife Research Unit
University of Nebraska
Lincoln
Nebraska
USA

Ahjond S. Garmestani
Office of Research and Development
U.S. Environmental Protection Agency
Cincinnati
Ohio
USA

ISBN 978-94-017-9681-1 ISBN 978-94-017-9682-8 (eBook)
DOI 10.1007/978-94-017-9682-8

Library of Congress Control Number: 2014957649

Springer Dordrecht Heidelberg New York London

Printed on acid-free paper

Springer Netherlands is part of Springer Science+Business Media (www.springer.com)

Contents

Contributors

Steven Alexander Environmental Change and Governance Group, Faculty of Environment, University of Waterloo, Waterloo, ON, Canada

Craig R. Allen U.S. Geological Survey, Nebraska Cooperative Fish and Wildlife Research Unit, School of Natural Resources, University of Nebraska, Lincoln, USA

Mark Andrachuk Environmental Change and Governance Group, Faculty of Environment, University of Waterloo, Waterloo, ON, Canada

Derek Armitage Environmental Change and Governance Group, Faculty of Environment, University of Waterloo, Waterloo, ON, Canada

Melinda Harm Benson Department of Geography & Environmental Studies, University of New Mexico, Albuquerque, NM, USA

Samantha Berdej Environmental Change and Governance Group, Faculty of Environment, University of Waterloo, Waterloo, ON, Canada

Jim Berkley U.S. Environmental Protection Agency, Denver, CO, USA

Brian C. Chaffin Oregon State University, Corvallis, OR, USA

Bianca Currie Sustainability Research Unit, Nelson Mandela Metropolitan University, George, South Africa

Thomas Dyck Environmental Change and Governance Group, Faculty of Environment, University of Waterloo, Waterloo, ON, Canada

Geography and Environmental Studies, Faculty of Arts, Wilfrid Laurier University, Waterloo, ON, Canada

Christo Fabricius Sustainability Research Unit, Nelson Mandela Metropolitan University, George, South Africa

Ahjond S. Garmestani Office of Research and Development, U.S. Environmental Protection Agency, Cincinnati, OH, USA

Hannah Gosnell Oregon State University, Corvallis, OR, USA

Nicholas A. J. Graham ARC Centre of Excellence for Coral Reef Studies, James Cook University, Townsville, QLD, Australia

Lorne A. Greig ESSA Technologies Ltd., Vancouver, BC, Canada

Lance Gunderson Department of Environmental Sciences, Emory University, Atlanta, GA, USA

Christina C. Hicks ARC Centre of Excellence for Coral Reef Studies, James Cook University, Townsville, QLD, Australia

Center for Ocean Solutions, Stanford Woods Institute for the Environment, Stanford University, Monterey, CA, USA

C. S. Holling Resilience Center, Nanaimo, BC, Canada

Fred A. Johnson U.S. Geological Survey, Southeast Ecological Science Center, Gainesville, FL, USA

David R. Marmorek ESSA Technologies Ltd., Vancouver, BC, Canada

Carol L. Murray ESSA Technologies Ltd., Vancouver, BC, Canada

Prateep Kumar Nayak Environmental Change and Governance Group, Faculty of Environment, University of Waterloo, Waterloo, ON, Canada

School of Environment, Enterprise and Development, Faculty of Environment, University of Waterloo, Waterloo, ON, Canada

Marmar Nekoro Stockholm University, Stockholm, Sweden

Olivia Odom Green U.S. Environmental Protection Agency, Cincinnati, Ohio, USA

Jeremy Pittman Environmental Change and Governance Group, Faculty of Environment, University of Waterloo, Waterloo, ON, Canada

Kaitlyn Rathwell Environmental Change and Governance Group, Faculty of Environment, University of Waterloo, Waterloo, ON, Canada

Courtney Schultz Department of Forest and Rangeland Stewardship, Colorado State University, Fort Collins, CO, USA

Lisen Schultz Stockholm Resilience Centre, Stockholm University, Stockholm, Sweden

Shana M. Sundstrom Nebraska Cooperative Fish and Wildlife Research Unit, School of Natural Resources, University of Nebraska, Lincoln, NE, USA

Byron K. Williams The Wildlife Society, Bethesda, MD, USA

List of Figures

List of Tables

Chapter 1
Adaptive Management

Craig R. Allen and Ahjond S. Garmestani

Keywords Adaptive management · Ecology · Uncertainty · Resilience · Natural resource management

Introduction

Management is complicated by social pressures, which are often poorly understood and affect the application of science to management problems. Often natural resource managers are under pressure from stakeholders and mandates from central offices to promote a narrow focus on single-species management (e.g., game species). Thus, managers are often forced to select between particular kinds of resource use, weighing ecosystem services against one another (Millennium Ecosystem Assessment 2005, Rodríguez et al. 2005). If tradeoffs between services are ignored, future problems may be created that can result in expensive remedial actions to restore previously available ecosystem services.

An additional problem in developing effective management for complex systems is that social and ecological components may not be aligned at the appropriate scales to achieve consistent regional and local management (Conroy et al. 2003, Cumming et al. 2006). It is not uncommon for agency administrators to demand local management actions that are impossible or inappropriate. For example, statewide hunting regulations may be inappropriate where local wildlife populations are overabundant or on the verge of extinction. Conversely, many ecosystem processes

C. R. Allen (✉)
U.S. Geological Survey, Nebraska Cooperative Fish and Wildlife Research Unit,
University of Nebraska, Lincoln 68583, USA
e-mail: allencr@unl.edu

A. S. Garmestani
Office of Research and Development, U.S. Environmental Protection Agency,
26 West Martin Luther King Drive, Cincinnati, OH 45268, USA
e-mail: garmestani.ahjond@epa.gov

C. R. Allen, A. S. Garmestani (eds.), *Adaptive Management of Social-Ecological Systems*,
DOI 10.1007/978-94-017-9682-8_1

are difficult to manage at the local scale, and appropriate regional authorities and mandates may not exist.

Paradigms for multi-species and ecosystem management have existed for decades, but their implementation within management agencies lags in acceptance, despite compelling arguments for their usefulness (Barrows et al. 2005). The failure of many federal and state management agencies to embrace ecosystem management may be attributable to restrictive institutional mandates and agendas, inflexibility in their ability to adopt new approaches and avoidance of risk taking, and lack of funding (particularly for long-term monitoring and intensive schemes that ecosystem management often demands). Additionally there are real and perceived shortcomings in the associated science, especially in basic understanding of social-ecological systems and in translating theory-derived guidelines into practical, unambiguous recommendations for managers.

Changes in natural resource management through time have been driven by changes in scientific understanding, as well as by a wide range of changes in society and politics. The usual goal of management is to ensure that one or more properties of a system of interest are maintained through time. This is often interpreted as a need for managers to either seek to maintain system stability, or to maintain particular system components and relationships while allowing or encouraging the system to change. In considering the dynamics of management and system change, an understanding of resilience is particularly relevant.

Attempts to optimize economic returns, physical connectivity, or other single system properties are typically doomed in the long-term to failure because related, critical variables are often negatively affected by such management (Holling and Meffe 1996). Available evidence suggests that managing for single variables usually fails because such approaches do not account for potential feedbacks, thresholds, or surprises arising from interactions with other components of the system (Holling and Meffe 1996). Optimization or efficiency approaches applied to single variables or to maximize output over short time frames often fail because of the complex interactions between social and ecological components of the system (Mascia et al. 2003). Unfortunately, strategies for managing multiple variables are seldom applied, and if they are, appropriate factors for maintaining resilience are rarely identified, monitored, and enforced.

Adaptive management, applied in an appropriate way to an appropriate problem, can speed the process of learning about complex natural resource problems. Adaptive management is an approach to natural resource management that emphasizes learning through management where knowledge is incomplete, and when, despite inherent uncertainty, managers and policymakers must act (Walters 1986). Although the concept of adaptive management has resonated with resource management scientists and practitioners following its formal description (Holling 1978), it has been and continues to remain frequently misapplied and misunderstood. Misunderstanding is largely based upon the belief that adaptive management is a trial and error attempt to improve management outcomes, that is, the adaptive component is interpreted as a willingness to try something new when current approaches fail, rather than a structured approach focused on learning. Unlike a traditional trial and error approach, adaptive management has explicit structure, including a careful elucidation of goals, identification of alternative management objectives and hypotheses of causation, and procedures for the collection of data followed by evaluation and reiteration. Regardless of the particular definition of adaptive management used, and there are many, adaptive management emphasizes learning and subsequent adaptation of management based upon that learning. The process is iterative, and serves to reduce uncertainty, build knowledge and improve management over time in a goal-oriented and structured process. Adaptive management is a poor fit for solving problems of intricate complexity, high external influences, long time spans, high structural uncertainty and with low confidence in assessments (Gregory et al. 2006) (e.g., climate change). However, even in such situations, adaptive management may be the preferred alternative, and can be utilized to resolve or reduce structural uncertainty.

Adaptive management is now common to a variety of resource management issues, and represents an evolving approach to natural resource management in particular, and structured decision making in general. Founded in the decision approaches of other fields (Williams 2010) including business (Senge 1990), experimental science (Popper 1968), systems theory (Ashworth 1982) and industrial ecology (Allenby and Richards 1994), the first reference to adaptive management philosophies in natural resource management may be traced back to the work of Beverton and Holt (1957) in fisheries management, though the term adaptive management was yet to be used (reviewed in Williams 2010). The term adaptive management would not become common vernacular until C.S. Holling, widely recognized as the "father" of adaptive management, edited "Adaptive Environmental Assessment and Management" in 1978 (Holling 1978). The work was spawned by the experiences of Holling and colleagues at the University of British Columbia following from the development of resilience theory (Holling 1973). The concept of resilience, predicated upon the existence of more than one alternative stable state for ecosystems, had several ramifications. For one, it meant that managers should be very careful not to exceed a threshold that might change the state of the system being managed; and the location of those thresholds is unknown. Second, for ecological systems in a favorable state, management should focus on maintaining that state, and its resilience. Adaptive management then, was a method to probe the dynamics and resilience of systems while continuing with 'management' via management experiments developed to enhance learning and reduce uncertainty.

Eventually Carl Walters (1986) built upon Holling's original book (1978) and further developed the ideas, especially in the realm of mathematical modeling. Whereas Holling's original emphasis was in bridging the gap between science and practice, Walters emphasized treating management activities as designed experiments to reduce uncertainty. Both scientists sought an approach that allowed resource management and exploitation to continue while explicitly embracing uncertainties and seeking to reduce them through that management. Walters (1986) described the process of adaptive management as beginning "with the central tenet that management involves a continual learning process that cannot conveniently be separated into functions like research, ongoing regulatory activities, and probably never converges to a state of blissful equilibrium involving full knowledge and optimum productivity." He characterized adaptive management as the process of defining and bounding the management problem, identifying and representing what we know through models of dynamics that identify assumptions and predictions so experience can further learning, identifying possible sources of uncertainty and alternate hypotheses, and designing policies to allow continued resource management while enhancing learning.

The confusion over the term "adaptive management" may stem from the flexibility inherent in the approach which has resulted in multiple interpretations of "adaptive management" that fall upon a continuum of complexity and *a priori* design, starting from the simple (e.g., "learning by doing") and progressing to the more explicit (e.g., "a rigorous process that should include sound planning and experimental design with a systematic evaluation process that links monitoring to management") (Wilhere 2002, Aldridge et al. 2004). Obviously there is a clear distinction in intent, investment and success between approaches that propose to learn from prior management decisions and those that outline a concise feedback mechanism dependent upon sound scientific principles on which future management decisions will be made. Central to the success of the structured decision making process is the requirement to clearly articulate fundamental objectives, explicitly acknowledge uncertainty, and respond transparently to stakeholder interests in the decision process. The conceptual simplicity inherent in structured decision making makes the process useful for all decisions from minor decisions to complex problems involving multiple stakeholders.

A key component of any management approach, whether it is adaptive or not, is deciding on the objectives, goals, and management options that may best achieve the desired goals. Unfortunately, as with many decisions, deciding upon a proper set of objectives and the means to reach those objectives can prove challenging. Resource management decisions are further complicated because social-ecological systems are complex (e.g., multiple objectives and stakeholders, overlapping jurisdictions, short and long term effects) and are characterized by a high degree of uncertainty (e.g., appropriate management action or monitoring protocols, future economic or ecological conditions) and therefore present decision makers with challenging judgments (e.g., predicted consequences of proposed alternatives, value-based judgments about priorities, preferences and risk tolerances) often under enormous pressure (economic, environmental, social and political) and with limited resources to ensure success. The resulting outcome of such conditions too often leads to management paralysis, or continuation of the status quo, as managers and policy makers become overwhelmed by the process of the decision and lose track of the desired social-ecological conditions they are charged with achieving. Indeed, the process of resource management can be arduous and even controversial, particularly if there are a variety of stakeholders vying to push an agenda. Fortunately, there are options to overcome these pitfalls and maximize the potential for success.

A method to overcome management paralysis and mediate multiple stakeholder interests is structured decision making. Structured decision making is a term often used in conjunction with or as a synonym for adaptive management, but in actuality it is a problem solving approach borrowed from the sociological fields, and best used to identify and evaluate alternative resource management options by engaging stakeholders, experts and decision makers in the decision process and addressing the complexity and uncertainty inherent in resource management in a proactive and transparent manner. As such, the framework of structured decision making is an ideal template to facilitate the iterative decision making and learning that defines adaptive management. To achieve this goal, structured decision making uses a simple set of steps to evaluate a problem and integrate planning, analysis and management into a transparent process that provides a roadmap focused on achieving the fundamental objectives of the program.

Adaptive governance is a form of governance that incorporates formal institutions, intermediaries (e.g., bridging organizations and networks) and individuals at multiple scales for purposes of collaborative environmental management (Folke et al. 2005). Intermediaries play a critical role in facilitating adaptive co-management and governance, and are essential to managing for resilience in social-ecological systems (Olsson et al. 2007). Intermediaries have the capacity to allow for development of new ideas, to facilitate communication between entities, and create the flexibility necessary for the interplay of the fluid (e.g., ecological systems) and the rigid (e.g., institutions) to be successful for environmental management (Folke et al. 2005). Leadership has also been established as a critical factor in facilitating good environmental governance. Leaders can develop and facilitate a vision for environmental governance, incorporating local knowledge and information from intermediaries (Folke et al. 2005). Enabling legislation and government policies can also contribute to the success of an adaptive governance framework, whereby governance creates a vision and management actualizes the vision (Folke et al. 2005).

The reality of non-stationary ecological systems, and uncertainty associated with most all environmental management, clashes with the certainty the legal system is predicated upon. Alternative regimes, non-linear responses and surprise (Folke et al. 2007) are difficult for law to cope with. Required legal enforceability, and the associated rigidity of institutions limit adaptive experimentation and management (Garmestani et al. 2009). The two-step legal process of administrative law in the United States, of public comment on draft documents and alternatives followed by agency action, is based on the assumption that agencies have the capacity to predict the consequences of the identified final actions (Ruhl and Fischman 2010). This creates a basic tension between linear legal processes, fundamentally uncertain ecological systems and the iterative process of adaptive management (Karkkainen 2005). This clash makes adaptive management very difficult under the current administrative law framework (Ruhl 2008). Because of this, many agencies conduct adaptive management "lite" (Ruhl and Fischman 2010). Courts in the United States have allowed adaptive management projects, if they include components that are legally enforceable. This approach falls considerably short of ideal in terms of adaptive management (Holling 1978). Ruhl and Fischman (2010) profess that we will not see adaptive management as idealized by Holling until the U.S. Congress provides sufficient funding and clear standards. Unfortunately, this means that environmental law is at odds with science, because the certainty required by law is largely incompatible with adaptive management. Thus, since adaptive management is a superior approach for a subset of environmental problems, environmental law itself will need to be transformed (Ruhl 2008).

The twelve contributed chapters that comprise the majority of this volume span a breadth of approaches to adaptive management, including some legal aspects.

Other aspects of the relationship between law and the management of complex social-ecological systems may be found in Garmestani and Allen (2014). The approaches described here represent a number of trends related to adaptive management, with differing emphases on collaboration, governance, structured decision making and more, all which reflect the current areas of emphasis within the field. We begin with a personal history of adaptive management by C.S. (Buzz) Holling (Chap. 2). Holling's contribution was expanded by Sundstrom, who probed Holling's recollections with a series of interview questions. Their combined contribution offers insight into the creation of adaptive management, which was intricately linked with the development of resilience theory. Holling's research career included frequent shifts between theory and practice, evident in adaptive management.

Gunderson (Chap. 3) provides an overview of adaptive management, from a collaborative and resilience perspective. He provides three key lessons derived from the theory and practice of adaptive management. The first is that adaptive assessments have led to creative syntheses of scientific understanding while catalyzing management. The second is that experimentation has resulted in organizational learning. The third outcome is that obstacles to adaptive management are a result of institutional and governance resistance. Therefore, Gunderson states that adaptive governance should complement adaptive management to overcome barriers.

Benson and Schultz (Chap. 4) explore the integration of adaptive management into current legal and regulatory frameworks in the United States. They recommend the provision of adequate funding for adaptive management and suggest that Congress should explicitly require adaptive management plans to articulate measurable goals, identify testable hypotheses, and state criteria for evaluation. Benson and Schultz further suggest that a federal agency be in charge of implementation, that stakeholders should be included in committees to oversee and recommend changes to adaptive management efforts, and that there be congressionally specified procedures for carrying out adaptive management.

Johnson and Williams (Chap. 5) offer hope that adaptive management may yet live up to its promise. They, as does Gunderson, suggest that the basic contribution of adaptive management is in helping to change the culture of resource management. Through this transformation adaptive management may have an impact broader than any technical improvements in approach or resource condition. Johnson and Williams approach adaptive management through decision analysis and suggest that the expanding use of decision-theoretic approaches is contributing to an increased capacity for rational decision making and adaptation in resource use and conservation. The structuring of a decision-making process—i.e., bounding and focusing the debate over choices, outcomes, and values—is a key element that has been lacking in many complex and contentious natural-resource issues.

Chaffin and Gosnell (Chap. 6) describe a framework for measuring success in adaptive management, and offer metrics for measuring the efficacy of adaptive management. They describe their proposed metrics and demonstrate their utility by applying them to the Glen Canyon Dam Adaptive Management Program. They suggest that adaptive management offers a more explicit, inclusive, and rigorous approach to managing a social-ecological system surrounded by uncertainty.

Green et al. (Chap. 7) focus on the role of bridging organizations in adaptive management success. They suggest that the nested nature of complex systems requires a multi-scale approach, but that in many cases this is hindered by hierarchical organizational structures and bureaucratic procedures. Bridging organizations that facilitate collaboration and learning across sectors and scales are key to adaptive governance, according to Green et al. Bridging organizations can facilitate cross-scale linkages, enabling formal management entities operating at discrete scales to improve communication channels, create opportunities for collaboration, and foster the resilience of social-ecological systems and the provisioning of ecosystem services.

Graham and Hicks (Chap. 8) focus on the application of adaptive management to novel ecosystems. They suggest that changes in systems following perturbation can lead to persistent new ecosystem configurations which may still provide valuable goods and services to society. Consequently, there is a need to understand the properties of emergent novel ecosystems and determine the most appropriate management in the new ecosystem contexts. Because adaptive management is designed to employ a diversity of testable actions, it affords a system greater resilience than the current tendency to employ single policies across ecological space or time that could potentially lead to disaster. Rather than presenting an obstacle to management, adaptive management turns a lack of evidence into an opportunity to simultaneously manage, monitor, and learn about a system.

Fabricius and Currie (Chap. 9) address adaptive co-management, a process that allows stakeholders to share responsibility within a system where stakeholders can explore their objectives, find common ground, learn from their institutions and practices, and adapt and modify them for subsequent cycles. Like adaptive management, the focus remains learning by doing, and takes into account a diversity of knowledge systems. This allows for the inclusion of informal, local and traditional knowledge, formal scientific knowledge and the sharing of rights, responsibilities and power among the diverse range of relevant stakeholders. Adaptive co-management is no panacea and may be inappropriate in contexts where stakeholder capacity is lacking, governance is weak, problems are 'tame', solutions are urgent and trust is low. It is, however, one of the few workable options where co-management is a necessity and current knowledge is incomplete.

Murray et al. (Chap. 10) describe three dimensions to the application of adaptive management. The first pertains to the nature of practice, the second to the nature of the problem, and the last to the social/organizational environment. For each of these dimensions the authors describe the set of attributes or characteristics that would lead to the successful implementation of adaptive management. They suggest that adaptive management is more likely to be successful at smaller spatial scales where treatments can be more readily replicated and controlled, and if the time required to test hypotheses can be measured in years rather than decades.

Berkley and Gunderson (Chap. 11) describe adaptive management networks, and organize their contribution around a series of interviews of key decision makers.

The participants were part of the Collaborative Adaptive Management Network, and experienced in the application of adaptive management. Responses to the interview questions indicate that the involvement of non-governmental organizations in adaptive management arose because they filled a gap in adaptive management education and implementation.

Williams and Johnson (Chap. 12) consider the tradeoff between optimization and resilience in the management of natural resources. They argue that under the conditions of uncertainty that characterize many resource systems, an optimal decision process that focuses on robustness does not automatically induce a loss of resilience. They argue that when a system is appropriately represented, its stochastic behavior is incorporated, and uncertainty accounted for, there is no reason to believe that optimization will automatically lead to the loss of resilience. Indeed, long-term sustainability can itself be the focus of management, with the idea of re-orienting management to ensure that decision making sustains productive capacity. Williams and Johnson conclude that optimal decision making should be an ally rather than an adversary of resilience.

Armitage et al. (Chap. 13) examine factors that are central to advance the theory and practice of adaptive management of natural resources. The authors suggest that the promise and elegance of adaptive management is more likely to emerge if practitioners and researchers situate their thinking in a transdisciplinary context of linked systems of people and nature, with reference to the issues of governance, power and knowledge, and as a strategy to encourage broader reflection on societies' interaction with natural resources.

Conclusion

Adaptive management is not new, but its application is evolving along many fronts. This volume illustrates some of those fronts, and serves to advance the use of adaptive management in appropriate settings. The conceptual underpinnings for adaptive management are simple; there will always be inherent uncertainty and unpredictability in the dynamics and behavior of ecosystems as a result of non-linear interactions among components and emergence, yet management decisions must still be made. The strength of adaptive management is in the recognition and confrontation of such uncertainty. Rather than ignore uncertainty, or use it to preclude management actions, adaptive management can foster resilience and flexibility to cope with an uncertain future, and develop safe to fail management approaches that acknowledge inevitable changes and surprises. Since its initial introduction, adaptive management has been hailed as a solution to endless trial and error approaches to complex natural resource management challenges. Adaptive management does not produce easy answers, and is only appropriate in a subset of natural resource management problems, but has great potential for a turbulent future.

References

Aldridge, C. L., Boyce, M. S., & Baydack, R. K. (2004). Adaptive management of prairie grouse: How do we get there? *Wildlife Society Bulletin, 32,* 92–103.

Allenby, B. R., & Richards, D. J. (1994). *The greening of industrial ecosystems*. Washington, D.C.: National Academy.

Ashworth, M. J. (1982). *Feedback design of systems with significant uncertainty*. Chichester: Research Studies.

Barrows, C. W., Swartz, M. S., Hodges, W. L., Allen, M. F., Rotenberry, J. T., Li, B., et al. (2005). A framework for monitoring multiple species conservation plans. *Journal of Wildlife Management, 69,* 1333–1345.

Beverton, R. J. H., & Holt, S. J. (1957). *On the dynamics of exploited fish populations*. London: Her Majesty's Stationery Office.

Conroy, M. J., Allen, C. R., Peterson, J. T., Pritchard, L., Jr., & Moore C. T. (2003). Landscape change in the southern piedmont: Challenges, solutions, and uncertainty across scales. *Conservation Ecology, 8*(2), 3. http://www.consecol.org/vol8/iss2/art3.

Cumming, G. S., Cumming, D. H. M., & Redman, C. L. (2006). Scale mismatches in social-ecological systems: Causes, consequences, and solutions. *Ecology and Society, 11*(1), 14. http://www.ecologyandsociety.org/vol11/iss1/art14/.

Folke, C., Hahn, T., Olsson, P., & Norberg, J. (2005). Adaptive governance of social-ecological systems. *Annual Review of Environment and Resources, 30,* 441–473.

Folke, C., Pritchard, L., Berkes, F., Colding, J., & Svedin, U. (2007). The problem of fit between ecosystems and institutions: Ten years later. *Ecology and Society, 12*(1), 30. http://www.ecologyandsociety.org/vol12/iss1/art30/.

Garmestani, A. S., & Allen, C. R. (2014). *Social-ecological resilience and law*. New York: Columbia University Press

Garmestani, A. S., Allen, C. R., & Cabezas, H. (2009). Panarchy, adaptive management and governance: Policy options for building resilience. *Nebraska Law Review, 87,* 1036–1054.

Gregory, R., Ohlson, D., & Arvai, J. (2006). Deconstructing adaptive management: Criteria for applications to environmental management. *Ecological Applications, 16,* 2411–2425.

Holling, C. S. (1973). Resilience and stability of ecological systems. *Annual Review of Ecology and Systematics, 4,* 1–23.

Holling, C. S. (Ed.). (1978). *Adaptive environmental assessment and management*. Chichester: Wiley.

Holling, C. S., & Meffe, G. K. (1996). Command and control and the pathology of natural resource management. *Conservation Biology, 10,* 328–337.

Karkkainen, B. C. (2005). Panarchy and adaptive change: Around the loop and back again. *Minnesota Journal of Law, Science & Technology, 7,* 59–77.

Mascia, M., Brosius, J. P., Dobson, T., Forbes, B. C., Nabhan, G., & Tomforde, M. (2003). Conservation and the social sciences. *Conservation Biology, 17,* 649–650.

Millennium Ecosystem Assessment. (2005). *Ecosystem and human well-being: General synthesis*. Washington, D.C.: Island Press.

Olsson, P., Gunderson, L. H., Carpenter, S. R., Ryan, P., Lebel, L., Folke, C., et al. (2006). Shooting the rapids: Navigating transitions to adaptive governance of social-ecological systems. *Ecology and Society, 11*(1), 18. http://www.ecologyandsociety.org/vol11/iss1/art18/.

Olsson, P., Folke, C., Galaz, V., Hahn, T., & Schultz, L. (2007). Enhancing the fit through adaptive co-management: Creating and maintaining bridging functions for matching scales in the Kristianstads Vattenrike Biosphere Reserve Sweden. *Ecology and Society, 12*(1), 28. http://www.ecologyandsociety.org/vol12/iss1/art28/.

Popper, K. R. (1968). *The logic of scientific discovery* (2nd ed.). New York: Harper and Row.

Rodríguez, J. P., Beard, T. D., Jr., Agard, J., Bennett, E., Cork, S., Cumming, G., et al. (2005). Interactions among ecosystem services. In S. R Carpenter, P. L. Pingali, E. M. Bennett, & M. B. Zurek (Eds.), *Ecosystems and human well-being: Scenarios. Volume 2. Findings of the scenarios working group, millennium ecosystem assessment* (pp. 431–448). Washington, D.C.: Island Press.

Ruhl, J. B. (2008). Adaptive management for natural resources—inevitable, impossible, or both? *Rocky Mountain Law Institute, 54*(11), 01–06.
Ruhl, J. B., & Fischman, R. (2010). Adaptive management in the Courts. Minnesota Law Review *95*, 424–484.
Senge, P. M. (1990). *The fifth discipline: The art and practice of the learning organization*. New York: Currency Doubleday.
Walters, C. J. (1986). *Adaptive management of renewable resources*. New York: McMillan.
Wilhere, G. F. (2002). Adaptive management in habitat conservation plans. *Conservation Biology, 16,* 20–29.
Williams, B. K. (2010). Adaptive management of natural resources—framework and issues. *Journal of Environmental Management, 92,* 1346–1353.

Chapter 2
Adaptive Management, a Personal History

C. S. Holling and Shana M. Sundstrom

Keywords Resilience · Adaptive management · Panarchy · Ecology · Uncertainty

Introduction: How it All Began

The choice was made at 30,000 feet, flying over Managua, Nicaragua on the way home from a workshop in Venezuela. The workshop was about comparing different disturbed regional systems, by exploring both theory and possible management actions. It was a workshop that brought together a Russian (a grand and wonderful mathematician who, sadly, later got into trouble with Soviet authorities), Canadians, Americans, Venezuelans, Argentineans, and Europeans from different countries. We had a great meeting but were still groping for an appropriate name for the planned book.

The news was full of the recent Managua earthquake and I had just read a review of Bob Kate's work that emphasized how the poor formal facilities in the city provided little help to survivors of the quake. Instead, they drew upon their extended families, and thereby mobilized the key help needed for nurture and recovery. It became help at the scale of a neighborhood. And the image of that kind of crisis and that kind of recovery acted as a metaphor for the way ecosystems and

C. S. Holling (✉)
Resilience Center, Vancouver Island, Nanaimo, BC, Canada
e-mail: holling@me.com

S. M. Sundstrom
Nebraska Cooperative Fish and Wildlife Research Unit, School of Natural Resources, University of Nebraska, Lincoln, NE 68583, USA
e-mail: sundstrom.shana@gmail.com

C. R. Allen, A. S. Garmestani (eds.), *Adaptive Management of Social-Ecological Systems*,
DOI 10.1007/978-94-017-9682-8_2

regional systems function. Suddenly the word "adaptive" popped in my mind, and the name became the one that was beautifully suited for our work- both applied and theoretical.

We used the term adaptive management for the applied aspects of these new ideas, the adaptive cycle to describe the fundamental structure and dynamics of systems and adaptive assessment for the methods. And resilience, along with panarchy (though developed much later) were the words we used that eventually captured the theoretical foundations of the core ideas—non-linearity, surprise, alternative stable states and cross-scale dynamics in space and time.

The Theory Behind it All

Theory shaped the emergence of adaptive management. The 1973 'resilience' paper (Holling 1973) really launched the adaptive management work we subsequently developed at the University of British Columbia. Resilience is the ability of a system to experience disturbances, to be changed thereby and then to re-organize and still retain the same basic structure and ways of functioning. It includes the ability to learn from disturbance. Flexibility and break points are at its heart. The precepts of adaptive management were developed as a response to defining an ecosystem in terms of resilience.

A resilient system is forgiving of external shocks. If resilience declines because of resource exploitation and loss of diversity, the magnitude of a shock from which it cannot recover gets smaller and smaller. Resilience shifts attention from growth and efficiency to recovery and flexibility. Growth and efficiency alone can often lead ecological systems, businesses and societies into fragile rigidities, exposing them to completely unexpected turbulent transformation. Resilience, in contrast, adds learning, recovery and flexibility as inherent properties of complex systems. It opens eyes to novelty and new worlds of opportunity.

Growth, as economists see it, is important, but equally so are the resilience forces in a healthy system that dominate during infrequent crises and collapses. And systems are healthy when they can grow for periods but can also generate creative collapses and can renew after collapses. During episodes when growth is halted or reversed, deep uncertainty appears, and alternative futures are unexpectedly perceived. Suddenly, the resulting unpredictability stifles informed action or triggers ignorant and fearful reaction, and there is a search for certainty.

That search for certainty smothers opportunity. Alternatives are suppressed and rigidity increases. Security is what is being sought, independent of evidence to the contrary, and often, when possible, such evidence is masked or hidden. In contrast, adaptive management seeks ways for the system itself to provide clues about opportunities and their consequences by setting up policies that in part provide products and in part are experiments that test causes of uncertainty and suggest solutions. For adaptive management the unknown is ever alive and present, with monitoring a constant need that can always be launched, but is difficult to sustain.

Applied Developments

Once some of the relevant theory was developed, it led to more applied phases of investigation, where Carl Walters became a central partner. He was and is a truly brilliant, maverick scientist who walks a non-traditional path that creates new traditions. His work on adaptive management methods has been a classic contribution to the field (Walters 1986), and more recently, he has advanced our understanding of ecosystem dynamics (Walters and Martell 2004).

The resilience research led a group, largely at British Columbia, to mobilize a series of studies of large-scale ecosystems subject to management, including terrestrial, fresh water and marine. Each study was coordinated with the key scientists involved in the ecosystem, and, in some cases, policy people who "owned" the systems and the data. Typically, several organizations were involved because of the different home bases of participants. The process encouraged two major advances.

One advance was that the set of deep studies allowed a comparative analysis of the theoretical foundations of ecosystem behavior and ecosystem management that was ecological, social and economic. That was the part that was particularly interesting, and it led to the book where the term 'adaptive management' was used for the first time (Holling 1978). The second advance was that in the course of conducting these ecosystem studies and comparisons, we developed a sequence of workshop techniques for working with experts in order to develop alternative explanatory models and suggestive policies. In the models, several scales were chosen, based on where we thought the causes lay, and we posed alternative hypotheses for the unknown relationships. Subsequent simulations then showed which, if any, of those alternatives were important in affecting behavior of the integrated system. If they were unimportant they were forgotten; if important they became a focus for further research. The models were then used in a second phase of the workshops to search for effective alternative policies. Three or four extreme policies with contrasting objectives were tested, and then a sustained policy was discovered that balanced economic, social and ecological objectives.

An immense amount was learned from our first experiments, which focused on the beautiful Gulf Islands, an archipelago off the coast of southern British Columbia. We chose to develop a simulation model of recreational property development. I knew little about land speculation, but we made up a marvelous scheme that used my earlier predation equations as the foundation of our modeling exercise—the land of various classes were the "prey", speculators were the "predators" and a highest bidder auction cleared the market each year. The equations were modifications of the general predation equations (Holling 1988). The predictions were astonishingly effective and persisted for at least two decades. As much as anything, it reinforced the earlier conclusion that these equations were powerful and general. But the important conclusion concerned the workshop process and the people.

The essence of those workshop methods were fun to present in a critical paper where the workshop processes were described and where key personalities were represented in delightful cartoons drawn by Roy Peterson, a cartoonist in Vancouver, and methods were expressed as a game (Holling and Chambers 1973). It was fun to reveal the truth about characters like Snively Whiplash, The Blunt Scot, The Utopians, the Compleat Amanuensis and The Peerless Leaders in this way. But a reviewer in *Ecology* turned the manuscript down by saying "no one wants to know about the games people in British Columbia play!" Bioscience reviewers were more enlightened so I happily published there.

I learned that the key design feature for these workshops was to start with two goals. The first goal was long term: to identify large, unattainable goals that can be approached, but never achieved, that relate to fundamental values of freedom, equity, tolerance and education. And then, for the second goal, to add a tough design for the first step, in a way that highlights or creates options to design, later, a second step—and then a third and so on. We found that the results were steps that rapidly covered more ground than could ever be designed at the start. At the heart, that is adaptive design, where the unknown is great, learning is continual and actions evolve. But it is tough for staff of a granting agency, who when they ask what specifics we expected, blanch at being told "wait and see".

Theory and Practice Both Trigger Institutional Needs

My work always shifted between fundamental theory and applied research. Surprises occurring at one stage became explored in the second, so the old categorization into basic or applied research had no meaning where my work was concerned. Each was intertwined with the other, and each benefited thereby. I found so much of existing theory seriously irrelevant at that time in the 1970's. It was too simplistic, too static, too uniform in scale, too linear and perceived by the originators as too certain. Traditional ecological practice based on such ideas was therefore grossly ineffective. As an example, it was no wonder that cod collapsed on the east coast of Canada, and since 1992 still shows only a weak sign of recovery even with fishing banned.

We had no difficulty in facing and discussing these issues when people were in workshops. The majority of participants grasped the non-linearity, the thresholds separating different quasi-stable states, the varied spatial patterns at different scales, the inherent uncertainty, the unknown and the necessary complexity of social-ecological systems. When these concepts are understood, the fixed world of standard environmental protection is recognized as being rigid and wrong. Those who got it became the subset of folks in ecology, economics, social science, political science, etc., the ones who could work together to design different solutions as acts of mutual discovery.

That rationale of mutual discovery, developed over 30 years in workshops and in theoretical studies from fish to forests, led us to form an internet organization in 1999 called the Resilience Alliance (http://www.resalliance.org), that combined several groups around the world in collaborative research and collaborative publishing. Fundamentally it was meant to sustain the international cooperation that had emerged in several of these earlier projects, and assist in the continued search for a deeper understanding and ever-broader examples of complexity in nature.

However, when such groups attempt to encourage implementation of adaptive management, the success is often only partial. Carl Walters described the failures well in a paper (Walters 1997). It is a failure of implementation, not of analysis, evaluation, understanding or policy. I found that two projects I got deeply involved with provided beautiful examples of successes and failures. Both moved to a phase of implementation with variable success, but prior to that, the models, understandings and alternative policies coming from the workshops were central to the initial suspension of regional conflicts in each case.

The first project concerned the Florida Everglades, with Lance Gunderson contributing deep insight and personal experience in the process. The Everglades project was undergoing one of its crises of transformation. That project succeeded in developing an understanding of a system functioning at three different spatial and temporal scales, from sawgrass and tree islands, to slough structures and sugar plantations, to topography. But it was also the example that failed on implementation, because the adaptive experiments were lost and the system became locked into an enormously expensive effort of ecosystem restoration. There was no respected, responsible leader who could survive the political games among the four jurisdictions involved—municipal, regional district, State, and Nation. No one, therefore, could continue with the responsibility to manage a transition. There were just committees of local, state and national government, combined with a good NGO, which became politically active and politically rigid.

The other project concerned forest growth, forest crises and harvesting in the eastern Boreal Forest in the face of spruce budworm outbreaks. This project was housed at the International Institute of Applied Systems Analysis (IIASA), with links to a team in New Brunswick led by Gordon Baskerville. Bill Clark, Dixon Jones, Mike Fiering and I led the effort at IIASA—a wonderful group with a remarkable ability to blend different experiences. Over the centuries, spruce budworm outbreaks periodically swept from Manitoba, through Ontario and Quebec, into New Brunswick and Nova Scotia and still further east to Newfoundland and the eastern U.S. We focused on New Brunswick, Canada, an area just big enough to contain the essential scales and to connect to still larger areas. The project depended on the deep understanding of ecological dynamics that had earlier come from Frank Morris' classic study of Budworm outbreak dynamics and forest growth (Morris 1963). It succeeded in the sense that non-linearities and cross-scale dynamics were discovered, and we also developed a three-scale simulation model (Holling et al. 1977). Simplified versions of the model (Ludwig et al. 1978) were developed, adaptive experiments were identified and eventually flourished. Three scales of management evolved, monitoring and forest inventory was transformed, and the partner on our project, Gordon Baskerville, became continuing leader of the implementation (Clark et al. 1979).

Some of the leaders in those fields were part of IIASA's wonderfully innovative early days and found that the small budworm and its up-scale effects presented a rich set of data at a large scale. Howard Raiffa, George Dantzig and Tschalling Koopmans became our partners in this evaluation of the usefulness of those methods for multi-scale ecological/economic systems. The conclusions are reviewed in Bell (1972, 1977).

This leads us to a third revealing example of implementing adaptive management from Carl Folke, now Director of the Beijer Institute of Ecological Economics, and his colleagues. This is the example of a cycle of the continuing transformation of Kristianstads, a small city in southern Sweden, and its wetland landscapes (Olsson et al. 2004). After a major flooding crisis the traditional response of dike building and land draining was rejected and replaced by a vision of land and waterscapes integrated with a variety of peoples' activities. Solutions, when discovered, often "stayed in the back pocket" until public understanding emerged—a very real process of bottom-up design and implementation. Experiments and education were deeply involved in the transformation, which became a continuing effort with deep and extensive public involvement. A senior leader, Karl Magnusson, orchestrated the process from its beginnings. Vision, practicality, and leadership came together.

Each of these three cases succeeded in having participants from a broad range of organizations. But one, the Everglades, failed in implementation because for political reasons, options or experiments to test uncertainties were not explored and tested as part of the management design. The other two succeeded in opening options for the future through experiments in policy that were tested, allowing some options to be rejected, and some to be adapted.

Complexity

The complexity of these examples is considerable even when you look only at the behavior of people as they create and modify the associated social, economic and ecological processes that are part of systems. But when you examine the core of the causation of each problem, the examples look simpler. Most practical people deal with explanations created by the supposed actions of one or two variables. That is too simple. But we find we do not need more than about 5 variables at different scales to capture the full range of possible qualitatively different behaviors for the core part of the problem. We call that the Rule of Hand!

It is, however, very hard for people to think through explanations and devise policies or strategies by thinking of five things. So the Rule of Hand challenges the conceptual ability of people. That is why models are so essential at the beginning of adaptive management projects, and extensive and persistent monitoring needed at the end and thereafter. Without modeling, practitioners feel they are operating in a barren world of inadequate knowledge and conflicting explanations.

The other reason why the systems discussed seem so intractably complex is because many practitioners and scientists are not truly integrative. Some act purely as environmentalists, or industry advocates, or developers, or citizen helpers, in a society where the environment, economy, society and politics are all in a turbulent relationship with each other. The result is that so often too many become narrow "lobbyists", pushing simplistic explanations, avoiding shared discovery, ignoring uncertainty and the unknown, and are hostile to or fearful of adaptive experiments.

Hence that narrowness and avoidance of the unknown is true of all vested interests, so what is needed is to add an integrative overarching synthesis. That was the goal behind adaptive management.

Key Features of Adaptive Management

Jim Lovelock, the noted chemist, atmospheric scientist and innovative thinker, once asked me, "Why don't ecologists consider the environment?" It left me speechless, since ecology is meant to be focused on the INTER-action between the biota, physical structure and the environment. What he meant was that population and community ecologists at that time ignored the two-way dynamic nature of the interaction of the biota with the physical attributes of the environment. They did not recognize that the interaction between the living world and its physical structure modified and created attributes of both. That is what his Gaia is all about (Lovelock 1988). The biota in Lovelock's Daisy World regulate and manage the atmospheric chemistry to sustain a consistent temperature for life.

But population and community ecologists saw the environment and slow physical structures as a fixed (including stochastic) backdrop for the biota to apply exclusively biotic variables to understand ecological dynamics. Animals and plants can affect each other, so that paradigm goes, but animals and plants cannot develop interactive impacts on the physical structure and environment of their world. But we know, for example, that the savannas of Africa and the forests of Canada have physical spatial patterns caused by the action of animals from the small grazer to elephants, and from the small budworm defoliator to the moose, and that those patterns in turn facilitate the organisms that created them.

These notions created a new paradigm that characterizes ecosystem science. Such interactions between the biota and the physical world generate self-organized patterns that re-enforce the very processes creating them. The consequence is interrelated variables whose relationships are astonishingly robust and resilient (though not infinitely so—hence abrupt surprise). Jim Lovelock was right! Slow variables and fast variables set the interaction across scales, and the slow variables determine the resilience of the system. When you add evolution and natural selection, you get panarchy —clumps of function and structure across scales from centimeters and days of leaves and needles, in a series of steps to the hundreds of kilometers and millennia of forests and savannas (Gunderson and Holling 2002).

Where it Went

Since that flight back from Venezuela, a series of books emerged on Adaptive Management. Adaptive Environmental Assessment and Management (Holling 1978) was the first major book. Work on it started at IIASA and the resulting manuscript was finally reviewed at a meeting of senior international environmental people who offered critiques that were then incorporated into a final version. Carl Walters' book on methods was the next feature (Walters 1986) of significance that laid out the methods to deal with complex non-linear resource systems. Then came Compass and Gyroscope by Kai Lee (1994) and Barriers & Bridges (Gunderson et al. 1995).

The experiences of those early workshops helped shape the essential design and maintain the flexibility of the Resilience Alliance project that began about two decades later. The Resilience Alliance project was the third of a new program run by the Beijer International Institute of Ecological Economics. It produced a turbulent, broad and delightful process of mutual discovery for those who chose to be part of it. Hundreds of people, natural and social scientists, mathematicians, economists and ecologists from many countries attended one or more workshops over 5 years. All or most were held on islands around the world where deep differences could be discussed, resolved and highlighted. Adaptive management and resilience are intricately interrelated. Adaptive management is the process that allows safe to fail experiments of complex, large systems where there is uncertainty but a need to manage.

Collaboration was typical of my work since the discovery of multi-stable states in systems. That was when I realized that my knowledge only covered a part of the full story. Anything that dealt with regional scales and with ecosystems, economies and societies, required partnering with those who were deeply knowledgeable of each particular subject. Comparing different systems needed experts in each, ones who could think and search for commonalities.

The Resilience Alliance is a network of international scholars from many disciplines that continue to collaborate. The internet has provided an alternative means to develop an integrative and adaptive organization at low cost, and a journal, Ecology and Society, that is fully internet-based (arguably the first of this type). The Alliance is formed by about 20 groups from around the world, people who all share the same enthusiasms and flexible desires for novel and relevant work on resilience, complex systems and case studies. They each provide a modest annual membership fee to publish the journal and maintain the organization. Committed people are the key and grants do the rest. Integrative workshops interspersed with integrative research, integrative educational material and programs, and novel modes of communication provide a foundation for both fundamental integrative science and policy research.

A core part of the Resilience Alliance project was the design and preparation of four books. One was the integrative book Panarchy (Gunderson and Holling 2002), which was meant to show what we developed to test and integrate the separate theories and knowledge in ecosystem science, economics and aspects of the social sciences. The other books were designed to separately address the ecosystem, social and economic dimensions of resilience. The ecosystem book focused on multi-stable states in large scale ecosystems (Gunderson and Pritchard 2002). The social science book was a lovely one on governance of and institutions for social-ecological systems (Berkes et al. 2003). The economic one concerned non-linear economics focused on renewable resource ecosystems (Dasgupta and Maler 2004).

Resilience and multi-stable states now seem to be pervading notable parts of ecosystem science and related social sciences, and even emerging in policy. Both features are affecting international policy of some nations. And I note in a bibliographic survey by Marco Janssen that the original 1973 resilience paper has been a central reference that links vulnerability and resilience research (Holling 1973). That is indeed pleasing since it took such a long time to happen. And it was delightful to have a major review paper on resilience appear in the same Annual Review series that my original paper did 31 years earlier (Folke et al. 2004).

The Resilience Alliance publication that was particularly novel was the synthesis volume where resilience and panarchy were offered as names to combine the adaptive cycle with hierarchical structures across scales in space and time (Gunderson and Holling 2002). I poured out all I had discovered in the first three chapters of Panarchy and in the two summary chapters. A complete, personal mind dump! It was densely written but of sustained value still, after 13 years. That was a lovely effort. A marvelous group of people became the heart of the panarchy component–Buz Brock, Steve Carpenter, Carl Folke, Lance Gunderson, Don Ludwig, Lin Ostrom, Garry Peterson, Martin Scheffer, Brian Walker and Frances Westley. This is a mix that is strongly ecosystemic but also has extensive economic, social and mathematical science expertise.

The development of Panarchy led to a number of other books of real consequence–first is the Foundations of Ecological Resilience, which identified key papers that started the process (Gunderson et al. 2010). Frances Westley's Getting to Maybe (Westley et al. 2006), which provided interpretations for social behavior, and Thomas Homer Dixon's The Upside of Down (Homer-Dixon 2006), which provided insights into interpretations of international politics and turbulence. Marten Scheffer's Critical Transitions in Nature and Society (2009) is a deeply revealing explication of theoretical foundations, and Terry Chapin's Principles of Ecosystem Stewardship (Chapin et al. 2009) provides the first perceptive textbook of resilience and transformation in regional resource systems.

In addition, Craig Allen has tested and significantly extended the discovery that ecosystems are structured in a clumped manner across scales (Peterson et al. 1998, Allen et al. 1999). After critiques from some macroecologists, a workshop of supporters, skeptics and complexity theorists explored the data and examples (Allen and Holling 2008), leading to new projects from across the spectrum of reactions. The key was to use your own data and ask the question. The result provided still further evidence of the lumpiness of ecosystems (Allen et al. 2006) and of regional human populations, institutions and economic systems as well (Garmestani et al. 2006, 2008).

To my mind one of the many good papers published during the early years of the Resilience Alliance was the paper by Steve Carpenter, Buz Brock and P. Hanson (Carpenter et al. 1999). It used a meta-model of a watershed, a lake, individualist farmers and intensive farmers, a market manager, a land manager and a governing board. It was a synthetic, appropriately simple meta-model built on the basis of the deep knowledge that Steve Carpenter has of water and phosphate dynamics and

Buz Brock has of microeconomics and decision theory. It showed multi-stable states, and generated the adaptive cycle in its three dimensions of net economic yield (potential), population of intensive farms (connectedness) and attractor width (resilience). It showed the inevitability of periodic collapses in the face of fixed policies and showed the results of a strategy where one individual achieved a persistent system without collapses in the only way possible—through the probing, monitoring and learning policy of adaptive management. The structure of the foundational model (game) is provided as downloadable software and it thereby becomes important for the training of any resource manager (available at http://www.ecologyand-society.org/vol3/iss2/art4/append9.html). As Si Levin said in a commentary, "The potential for this powerful combination of ordinary and extraordinary fare is just the sort of advance that sets Conservation Ecology (the name later changed to Ecology and Society) apart from standard journals" (personal communication).

Prediction, Uncertainty and What is Unknown

In response to a question in a press briefing on the Iraq war in December 2003, Donald Rumsfeld, U.S. Secretary of Defense said:

> Reports that say that something hasn't happened are always interesting to me, because, as we know, there are known knowns; there are things we know we know. We also know there are known unknowns; that is to say, we know there are some things we do not know. But there are also unknown unknowns- the ones we don't know we don't know.

Now that is a realistic statement for planning. It is all about behavior you can predict, behavior you can possibly expect, and behavior that is a complete surprise: prediction, uncertainty and surprise. That is what motivates the panarchy work and adaptive management. The surprises come from the way evolved systems are structured, the non-linearity that creates alternative stable conditions, and the influences that cross scales, from the fast and small to the big and slow.

I cannot prove it, but all our experience suggests that the key parts of regional systems can be captured with five to six sets of variables, similar to ecosystems. That makes understanding for policy actions difficult. It is easy to imagine the effects of single variables or of two, but three or more are a challenge to our minds and to those who live and endure those systems. That level of minimum-expected complexity helps create different lobby groups, each of which grabs a different small piece of the whole to explain the whole.

We deal with that in workshops involving people drawn from different organizations and lobbies. We initially mask the discussion and arguments of different goals by disaggregating the system into parts that can then be discussed, modeled and tested separately, with less emotional argument and more substantive ones. Combining all the modules makes an integrated model which opens the ability to discover what causes are important, what causes are unimportant, and what alternatives need further examination.

Organizations and Institutions Evolve

Adaptive management's significance, its failures and successes, depends on social organizations and their flexibility, transparency and responsibility. Sets of organizations express their roles at different ranges of scale (Ostrom 1990), from the neighborhood, local community, municipality, province or region, to nation and world. I have been fortunate to have worked with a particular set of local, community, national and international organizations of science during their phases of early innovation, growth, and consolidation to stages of collapse, tepid persistence or revival.

Those experiences made it clear that the large influences of wonderful, integrative scientific or policy organizations can come and go. I have been fortunate to be a working part of the initial innovative phase of a number of organizations, for example, IIASA, the Institute of Resource Ecology at the University of British Columbia, and the Beijer Institute of Ecological Economics, and also to have been on the boards of others, including the Santa Fe Institute (SFI) and the Canadian Center of Advanced Research. They often become burdened by their initial success and are rarely able to maintain the same liveliness and novelty over time. Instead, the novelty develops in one place and then typically shifts elsewhere, expanding, extending, testing and deepening the work as it moves. The intellectual area or topic becomes the evolving entity, but often not the founding organization itself.

What Might be Next?

"Panarchy"; an odd name, but one that is meant to capture the way living systems both persist and yet innovate (Gunderson and Holling 2002). It shows how fast and slow, small and big events and processes can transform ecosystems and organisms through evolution, or can transform humans and their societies through learning, or the chance for learning. The central question is what allows rare transformation, not simply change.

The aspect of Panarchy that is most novel and significant concerns the phase when social systems, ecosystems or economic systems start to break down or transform, releasing the chance for a renewed system to emerge. At that moment, novelty that had been simmering in the background can emerge and be debated, and new associations begin to develop among previously separate innovations. The big influence comes from discoveries that, at that moment in time, emerge from people's local experiments at small scales, discoveries that can emerge at times of big change, to trigger bigger changes at large scales. That process highlights the keys for the future.

One key is to recognize that the small, that is the individual human, can at times transform the big—that is the politics and institutions of governance. But there are traps, and their potential to occur needs some discussion. We identified two types of traps, a Poverty Trap where accumulated capital has been lost, and a Hierarchical Trap where isolation and control structures limit experiments and learning. Helen Allison identified a third trap, the Lock-in Trap, where sunk-costs are so high that resource degradation continues until all capital is gone (Allison and Hobbs 2004). Adaptive management can help us recognize and identify those traps, and the thresholds that could lead us into them.

Five Decades of Learning

More than five decades of observation, immersion and research into complex systems of people and nature have allowed me some insight. Below, in summary, I list those observations.

On Starting the Process of Adaptive Management

1. In each project, make the overall goal large and approachable, but ultimately unattainable, such as goals related to justice, equity and opportunity. Make the first step tough, but simple, doable and open. Design the second step from the success and failures of the first and retain the overall, impossible goal. Continue for each succeeding step.
2. In the best of projects, the preliminary analysis and communication phase can be designed to take a year or two, with three workshops involving a diverse community. Design several visualizations of the past and of alternative futures that can be quickly perceived. Dynamic models are important as ways to open awareness of these ever changing alternative futures. Then add one or a few open discussions with people representing combinations of every possible interest. Encourage them to discover stereotypes they have and, from that, begin to communicate across interests.

On Theory

3. On the side, continue questioning theory, conducting tests and inventing expansions to theory. Keep the theoretical underpinnings and assumptions of management in the forefront. Your actions change the world, but your knowledge is limited, and will always be so.
4. Much of established theory is severely limited at the time of conception—too simplistic, too static, too uniform in scale and perceived by the originators as too certain.
5. We live in a slice of time on a spot in space. Therefore we see myopically, but we adapt if we look across scales, recognize ignorance, monitor and innovate/invent. Lobbies fight that.

On Practice

6. Rarely do organizations experiment, monitor, abandon, and modify. They need to.

7. Design large experiments, with small parts and just-sufficient complexity. They will always fail (partly). Learn from the surprises when they fail.

On Implementation

8. The key failure is of implementation, not of evaluation, or understanding, or policy. That is because the politics of the lobbies freeze abilities to act.
9. Solutions for parts of the problem need to be developed and widely communicated, but kept in the "back pocket" until doors of opportunity open with the politics and the people.
10. Resilience comes when individuals can access diversity of opportunity at times of crisis or transformation.
11. Search for a persistent champion, a leader in one of the institutions or in the society who is broadly respected in the region. Often the failure of adaptive management comes at the very end of analysis and synthesis, when there are no such effective leaders to carry forward the lessons learned in a political arena.
12. Persistent monitoring of policies and management at several key scales is critical, expensive and often avoided. Innovative methods are now needed involving agencies, NGOs and communities in imaginative collaboration across scales in space and time. This is one of the most promising new initiatives needed.

Who Opposes

13. So often most environmentalists become narrow "lobbyists", pushing simplistic explanations, avoiding shared discovery, ignoring uncertainty and hostile to or fearful of adaptive experiments.
14. Transformation to test and implement new discoveries is opposed by narrow lobbies in science, in business, in industry, in public concerns. At the same time, however, other examples of the same lobbies can, if broadly perceived, open novel directions.

On Organizations

15. Organizations can begin with brilliance, pause with rigidity and then die or persist with irrelevance. They rarely transform, but some do. Therefore, find or invent new organizations and persist in introducing periodic phases of novelty and adventures. The Resilience Alliance is perhaps one such organization.

And Finally

16. Make your journey fun. Write limericks, make jokes, invent celebrations, sing and make music, paint and sculpt while learning about, and improving, the complex world we live in.

References

Allen, C. R., & Holling, C. S. (Eds.). (2008). *Discontinuities in ecosystems and other complex systems*. New York: Columbia University Press.

Allen, C. R., Forys, E. A., & Holling, C. S. (1999). Body mass patterns predict invasions and extinctions in transforming landscapes. *Ecosystems, 2,* 114–121.

Allen, C. R., Garmestani, A. S., Havlicek, T. D., Marquet, P. A., Peterson, G. D., Restrepo, C., Stow, C., & Weeks, B. E. (2006). Patterns in body mass distributions: Sifting among alternative hypotheses. *Ecology Letters, 9,* 630–643.

Allison, H. E., & Hobbs, R. J. (2004). Resilience, adaptive capacity, and the "Lock-in Trap" of the Western Australian agricultural region. *Ecology and Society, 9.*

Bell, D. E. (1972). A utility function for time streams having interperiod dependencies. *Operations Research, 25,* 448–458.

Bell, D. E. (1977). A decision analysis of objectives for a forest insect problem. In D. E. Bell, R. L. Keeney, & H. Raiffa (Eds.), *Conflicting objectives in decisions* (pp. 389–421). London: Wiley.

Berkes, F., Colding, J., & Folke, C. (Eds.). (2003). *Navigating social-ecological systems: Building resilience for complexity and change*. Cambridge: Cambridge University Press.

Carpenter, S. R., Brock, W. A., & Hanson, P. (1999). Ecological and social dynamics in simple models of ecosystem management. *Conservation Ecology, 3.*

Chapin, F. S., Kofinas, G. P., & Folke, C. (Eds.). (2009). Principles of ecosystem stewardship: Resilience-based natural resource management in a changing world. New York: Springer.

Clark, W. C., Jones, D. D., & Holling, C. S. (1979). Lessons for ecological policy design: A case study of ecosystems management. *Ecological Modelling, 7,* 1–53.

Dasgupta, P., & Maler, K.-G. (Eds.). (2004). *The economics of non-convex ecosystems*. Dordrecht: Kluwer Academic Publishers.

Folke, C. S., Carpenter, S. R., Walker, B. H., Scheffer, M., Elmqvist, T., Gunderson, L. H., & Holling, C. S. (2004). Regime shifts, resilience, and biodiversity in ecosystem management. *Annual Review of Ecology, Evolution, and Systematics, 35,* 557–581.

Garmestani, A. S., Allen, C. R., Mittelstaedt, J. D., Stow, C. A., & Ward, W. A. (2006). Firm size diversity, functional richness, and resilience. *Environment and Development Economics, 11,* 533–551.

Garmestani, A. S., Allen, C. R., & Gallagher, C. M. (2008). Power laws, discontinuities and regional city size distributions. *Journal of Economic Behavior and Organization, 68,* 209–216.

Gunderson, L. H., & Holling, C. S. (Eds.). (2002). *Panarchy: Understanding transformations in human and natural systems*. Washington, D.C.: Island Press.

Gunderson, L. H., & Pritchard, L. J. (Eds.). (2002). *Resilience and the behavior of large-scale systems*. Washington, D.C.: Island Press.

Gunderson, L. H., Holling, C. S., & Light, S. S. (Eds.). (1995). *Barriers and bridges to the renewal of regional ecosystems*. New York: Columbia University Press.

Gunderson, L. H., Allen, C. R., & Holling, C. S. (Eds.). (2010). *Foundations of Ecological Resilience*. Washington, D.C.: Island Press.

Holling, C. S. (1973). Resilience and stability of ecological systems. *Annual Review of Ecological Systems, 4,* 1–23.

Holling, C. S. (1978). *Adaptive environmental assessment and management*. Chichester: Wiley.

Holling, C. S. (1988). Temperate forest insect outbreaks, tropical deforestation and migratory birds. *Memoirs of the Entomological Society of Canada,* 21–32.

Holling, C. S., & Chambers, A. D. (1973). Resource science: The nurture of an infant. *BioScience, 23,* 13–20.

Holling, C. S., Jones, D. D., & Clark, W. C. (1977). Ecological policy design: A case study of forest and pest management. In G. A. Norton & C. S. Holling, (Eds.), *Proceedings of a conference on pest management. International Institute for Applied Systems Analysis* (pp. 12–90). Laxenburg.

Homer-Dixon, T. (2006). *The upside of down: Catastrophe, creativity, and the renewal of civilization*. Washington, D.C.: Island Press.

Lee, K. N. (1994). *Compass and gyroscope: Integrating science and politics for the environment*. Washington, D.C.: Island Press.

Lovelock, J. (1988). *The ages of Gaia*. Oxford: Oxford University Press.

Ludwig, D., Jones, D. D., & Holling, C. S. (1978). Qualitative analysis of insect outbreak systems: The spruce budworm and forest. *The Journal of Animal Ecology, 47,* 315–332.

Morris, R. F. (1963). The dynamics of epidemic spruce budworm populations. *Memoirs of the Entomological Society of Canada, 331,* 1–332.

Olsson, P., Folke, C., & Hahn, T. (2004). Social-ecological transformation for ecosystem management: The development of adaptive co-management of a wetland landscape in southern Sweden. *Ecology and Society, 9.*

Ostrom, E. (1990). *Governing the Commons: The evolution of institutions for collective action* (1–298). Cambridge: Cambridge University Press.

Peterson, G. D., Allen, C. R., & Holling, C. S. (1998). Ecological resilience, biodiversity, and scale. *Ecosystems, 1,* 6–18.

Scheffer, M. (2009). *Critical transitions in nature and society*. Princeton: Princeton University Press.

Walters, C. J. (1986). Adaptive management of renewable resources. New York: Macmillan.

Walters, C. J. (1997). Challenges in adaptive management of riparian and coastal ecosystems. *Conservation Ecology,* 1, 1.

Walters, C. J., & Martell, S. J. (2004). *Fisheries ecology and management*. Princeton: Princeton University Press.

Westley, F., Zimmerman, B., & Patton, M. Q. (2006). *Getting to maybe: How the world has changed*. Canada: Random House.

Chapter 3
Lessons from Adaptive Management: Obstacles and Outcomes

Lance Gunderson

Keywords Adaptive management · Ecology · Uncertainty · Policy · Obstacles

Introduction

The approach to resource management called adaptive management was proposed as a way for resource managers to integrate scientific understanding with the management of natural resources. Adaptive management has been applied in large-scale resource systems in the United States for at least three decades. Indeed, it has now become codified as the way in which major management agencies, such as those in the Department of Interior, operate (Williams et al. 2009). Even so, the application of adaptive management has not been without major impediments that arise from technical, scientific, institutional and governance barriers. Even with the impediments and failures of adaptive management, a major benefit of this approach has been a process that produces new and novel understandings of complex system dynamics as well as innovative policies and practices. Adaptive management provides a crucible for innovation because it maintains a focus on integrating scientific understanding, resolving and managing inherent uncertainties, and questioning assumptions, boundaries and policies.

Adaptive management is based on an assumption that managed resource systems are complex and dynamic with large uncertainties and high levels of unpredictability (Walters 1986). As such, management cannot understand and anticipate the effects of their actions, but rather must learn and adapt. Adaptive management therefore seeks to foster learning and develop new understanding, which confront inherent uncertainties and complexities of resource systems over time. Adaptive management is not trial and error management. That is, it doesn't blindly probe

L. Gunderson (✉)
Department of Environmental Sciences, Emory University,
524 Mathematics and Science Center, 400 Dowman Drive, Atlanta, GA 30322, USA
e-mail: lgunder@emory.edu

C. R. Allen, A. S. Garmestani (eds.), *Adaptive Management of Social-Ecological Systems*,
DOI 10.1007/978-94-017-9682-8_3

uncertainty, but uses scientific tools and methods to foster learning and understanding. Nor is adaptive management a form of management that changes actions and course when new information becomes available. While it has these elements, it is an approach that is structured to integrate understanding and foster learning (Walters 1986, Walters 2007).

Adaptive management was proposed to fill three perceived gaps in natural resource management (Holling 1978). The first is to bridge diverging assumptions (mental models or paradigms) of resource dynamics. That is, scientists have different and often competing models about how ecosystems operate, such as assumptions of equilibrium or steady states of population dynamics. The second gap adaptive management seeks to resolve is the differing perspectives that exist amongst scientific disciplines. Managed resource systems are studied and understood by a wide range of disciplinary scientists (i.e., earth scientists, hydrologists, biologists and ecologists to name a few), all of whom share different ideas about key variables and influences. The third gap was identified as the gap between knowledge and action. In such complex systems, with many confounding variables, both knowledge and practical experience of resource managers is often limited.

Adaptive management originally described the separate but linked processes of integrated assessment and active management (Holling 1978, Walters 1986). The main process during the integrative assessment aims to articulate assumptions of resource dynamics and integrate disciplinary perspectives and assess what is known and not known about resource issues. The assessment phase begins with developing hypotheses (or explanations) around specific resource issues that include: (1) how specific ecological dynamics operate and (2) how human interventions will affect the ecosystem. The development of those hypotheses is done using models to integrate understanding and alternative perspectives of ecosystem dynamics and evaluate a set of possible policy outcomes (Walters 1986).

Since its inception in the late 1970's, adaptive management approaches have been applied to hundreds of resource systems around the world. Yet, there are very few cases where the adaptive assessment led to adaptive management, defined as the design and execution of explicit experiments to resolve key resource uncertainties (Johnson 1999, Gunderson et al. 2006). Walters (1997) reviewed twenty cases of riparian ecosystems where adaptive assessments were done, and found only seven where experimental management ensued. Walters (1997) outlined obstacles to this transition to include: (a) the belief that further modeling and monitoring would resolve uncertainties; (b) that experimentation was costly and risky; (c) experimental approaches were opposed by special interests; and (d) value conflicts among scientists and/or stakeholders could not be resolved. Gunderson (1999) suggested that adaptive management would not occur unless sufficient ecological resilience exists in key resources (such as a population of endangered species), or unless sufficient trust exists among agencies and stakeholders. In sum, the history of applying an adaptive management approach indicates it has been effective in designing management experiments, but less so as a process for resolution of complex social and political uncertainty. In another review, Johnson (1999) found that adaptive management has proven to be 'technically capable and socially challenged'. The technically capable part of adaptive management reflects on the ability to develop integrative models that help scientists and managers to: (1) decide upon an agreed set of hypotheses to test; (2) design experiments to test both the understanding of system dynamics and the effects of subsequent interventions and policy implementation (Walters 1986, 1997). The socially challenged part of Johnson's (1999) statement refers to the non-scientific parts of adaptive management, in which complicated social and political uncertainties arise. Allen and Gunderson (2011) describe pathologies of adaptive management, to include technical, social and political obstacles to implementation. The paradox of implementation success highlights three areas for discussion, as described in the following paragraph.

Three key outcomes from examining the theory and practice of adaptive management help describe both obstacles and opportunities, and structure the remainder of this chapter. The first key outcome is that adaptive assessments have led to creative syntheses of scientific understanding while catalyzing and transforming management in a few settings. The second outcome is that experimentation, including active, explicit management treatments and passive monitoring of change has resulted in organizational learning in a few resource systems. The third outcome is that major obstacles to adaptive management are a result of institutional and governance resistance. Therefore, adaptive governance should complement adaptive management in order to overcome the political barriers and institutional obstacles observed in cases where adaptive management has been invoked. Each of these will be addressed in turn, following a short section on the processes and procedures that have been used to define frameworks for adaptive management.

Frameworks for Adaptive Management

Adaptive management is an ongoing process that combines assessment with management actions in order to learn about the complexities of system dynamics as well as to achieve intended social objectives (Holling 1978, Walters 1986). Assessing a system requires synthesizing available data to generate a set of competing alternative explanations about particular sets of resource problems and social objectives.

Management actions are designed by considering what actions are robust to uncertainties among alternative explanations and what actions will help test and reduce those uncertainties (Walters 1986). Management actions are evaluated by monitoring system indicators in a process that uses that information to promote learning. While these activities are described linearly, adaptive management is typically an iterative process that develops an ongoing dialogue about the understanding, evaluating, and functioning of the system and the goals of management.

At least two frameworks have been used to describe the processes and procedures that guide the implementation of adaptive management in resource systems. Both frameworks begin with an assessment of environmental or resource problems facing managers. The assessment phase helps define the resource issue and design a set of possible solutions or actions to resolve the resource issue. Both frameworks

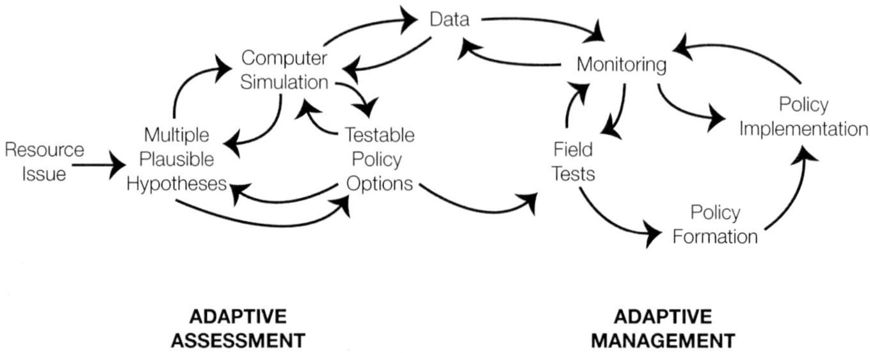

Fig. 3.1 Conceptual model of adaptive environmental assessment and management, indicating the integration of processes that assess, propose, test and evaluate hypotheses of ecosystem dynamics and policy implementation (Holling 1981)

then call for a set of management actions to be implemented. These actions can be either intentional interventions ("active adaptive management" *sensu* Walters 1986) or not. The non-intervention option is described as passive adaptive management (Walters 1986) and uses variation in natural ecosystem processes to create effects on managed resource variables. In both of these frameworks, the management actions (and non-management variations) are evaluated through ecosystem monitoring. These periodic evaluations are then used as diagnostic tools to determine if the initially defined resource issues have been resolved. Both frameworks involve sequential processes of assessment, actions and evaluation in an iterative or circular format, as described in the following paragraphs.

Holling (1981) developed a flow diagram (Fig. 3.1) into which he described at least three interacting spheres of activity. One is a resource problem assessment, primarily the left hand side of the figure, and management actions. The assessment phase of adaptive management uses models, simulation, deduction, etc., to help winnow among sets of hypotheses about resource issues. That information is used to develop a range of policy options that are tested and evaluated prior to policy formation and implementation. A second model was developed, and as in the Holling (1981) approach, problem assessment is the beginning of the adaptive management process, followed by design and implementation of particular policy options (Fig. 3.2). The model in Fig. 3.2 is more linear, and proposes a cyclical, iterative approach. Both diagrams emphasize the circular and iterative nature of adaptive management. Figure 3.1 focuses on the key role of alternative scientific hypotheses in developing and testing policies prior to selection for implementation. Figure 3.2 shows separation of model assessments of a policy to implementation as a management action as design features of the model. Both diagrams depict distinct steps in the process, i.e., that the framework involves shifting the focus of activities from one of assessment to implementation and then evaluation.

Interestingly, there is no distinct separate step in either model for the extensive traditional science inquiry process prior to implementation of an action. That is,

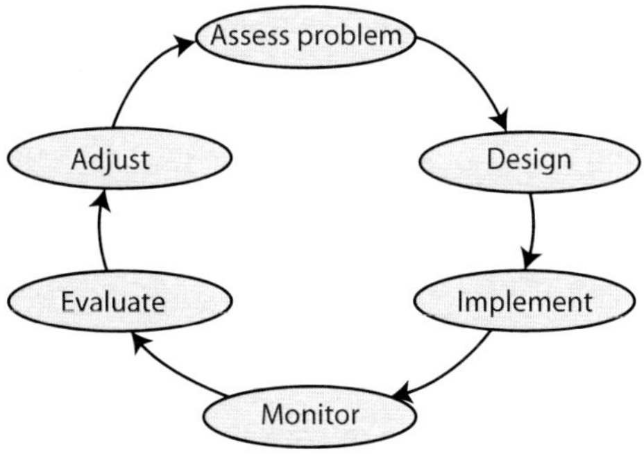

Fig. 3.2 U.S. Department of Interior diagram showing steps in the adaptive management process (Williams et al. 2009)

neither of the conceptual models depicts science as a process that is distinct (as an identifiable item or bubble) and unto itself. Elements of scientific processes are ever present, i.e., hypothesis formulation, simulation experiments, monitoring, modeling, etc. However, they are cast into the process of adaptive management policy experimentation. This is why many authors state that adaptive management blurs the distinction between science and management (Holling 1978, Walters 1986, Lee 1993, Gunderson and Pritchard 2002, Williams 2009). This relates to the principle in an adaptive management framework that there is no clear separation of science and management activities, and indeed they are both part of a more holistic model of management. Both diagrams emphasize the role of monitoring as a critical step in adaptive management. In the adaptive assessment phase of the process, one of the key outcomes is identification of critical ecosystem variables to monitor. Monitoring should evaluate the outcomes of management interventions, and as such is a critical part of adaptive management (Walters 1997). While monitoring is done for many reasons, it is in the context of adaptive management that monitoring helps to build understanding and provides the basis for learning.

Adaptive Assessment and Creative Syntheses

A critical, but often overlooked part of adaptive management is the environmental assessment process. This is indicated in Fig. 3.1 as the oval in which hypotheses are generated, and in Fig. 3.2 as the problem assessment and design modules. One of the important differences between adaptive assessments and other assessment approaches is how ecosystem understanding is integrated (or not). Scientific or ecosystem based assessments, are often based on piecemeal or disciplinary analysis of resource dynamics. Holling (1998) describes this problem as two different modes of science. He argues that one mode of science focuses on parts of the system and deals with analyses and experiments that narrow uncertainty to the point of acceptance by peers; it is conservative and unambiguous by being incomplete and fragmentary. The other view is integrative and holistic, searching for simple structures and relationships that explain much of ecological complexity (Holling 1994). This view provides the underpinnings for the adaptive approach, because surprises are inevitable and knowledge will always be incomplete.

One of the novel innovations in adaptive management was the use of computer models to structure the discourse in a series of workshops (Holling and Chambers 1973). There have been hundreds of environmental assessments in settings around the world (Walters 1997) which have used computer models to articulate what is known and not known, highlight competing claims about understanding ecosystem dynamics, and evaluate these alternatives. The construction of a computer model in a series of workshops has been a hallmark of adaptive management. First described by Holling and Chambers (1973) the workshops were structured to create an atmosphere where interdisciplinary gaps (among various 'ologies' or sciences) could be bridged. One design element of the workshops was decidedly "open", that is, a style in which the participants and the rules are allowed to co-evolve. Another design element of the workshops involved acknowledgment of failure. Since the territory was so new, the likelihood for failure was high, so the approach had to be robust or safe to fail. Part of this safe to fail design was that actions of the workshops were called a game and had three components; people, rules and tools. The use of computers as a communication device remains a staple of adaptive management today- four decades later. The computer models range from the simple to the sophisticated, but the key precept is that the models are developed as a translator among various perspectives and disciplines, and less for a predictive, deductive engine for forecasting impacts of proposed management actions. The computer displays information visually, which allows for people to instantly react and consume large amounts of information. The computers are used as "gaming devices", a safe environment in which the complexity of resource issues can be explored, and ideas tried, with no consequences other than learning (Holling and Chambers 1973).

The search for simplification is manifest in both the theory and practice of adaptive management. As mentioned above, computer models are built to help integrate and organize collective understanding of complex issues. The approach in constructing these models is to be parsimonious in the selection of variables and interactions. That is, only include enough complexity in the model to capture essential dynamics of the ecosystems. Otherwise, the model becomes as complicated as the 'real' world that is being assessed or managed, and as intractable (Clark et al. 1979, Walters, 1986).

A key step in the assessment process is to determine the credibility of models. The computer models are viewed as hypotheses, and as such cannot be validated, only invalidated in the Popperian view of science. The models are caricatures of reality, only including what is essential. Therefore what is important is model credibility, not validity. It is only after resisting attempts at invalidation that a model becomes credible. One way of attempting invalidation is to compare the model output with historical data (verified data, not interpreted). Another is that correlation between the model and historical data does not imply causation. Other means of invalidation include trial and error approaches that compare model predictions with what happens in the real world, natural trials where model output can be compared to natural experiments, and comparing the behavior of alternative models. Once the models (or sets of models) have resisted invalidation, they can be used to evaluate alternative polices.

Many cases around the world have undergone the assessment phase of adaptive management, but only a subset has moved through this phase to the management phase (Walters 1997, Gunderson et al. 2008). Among the reasons for this include the inability to discern among competing hypotheses. That is, rather than develop policies or management tests based on a single hypothesis, multiple hypotheses can lead to dramatically different actions. One such case arose in the assessment of the Florida Everglades.

A major unresolved environmental issue of the Everglades has been the decline of wading bird nesting (Davis and Ogden 1994). Among the explanations include a loss of early season habitat due to land use changes; a loss of food production due to wetland conversion to agriculture and development; changes in seasonal hydrology which affects food supply through nesting, increased predators; decreased water flow to estuarine habitats; changes in behavior due to heavy metals and other pollution; and an increase in feeding opportunities outside the Everglades ecosystem (distant magnet) among others (Davis and Ogden 1994). In the assessment process, it was clear that many of these hypotheses centered on hydrologic modifications of the ecosystem (Walters et al. 1992). Indeed, the ongoing Everglades restoration plan is based on an assumption that hydrologic changes are at the heart of wading bird nesting decline. If the alternative hypotheses of behavioral change due to pollution or distant magnets are valid, then the multi-billion dollar recovery plan that calls for hydrologic manipulation will not meet wading bird recovery goals.

While the Everglades adaptive assessment process was key to developing current restoration plans, the system has yet to move into active adaptive management (Gunderson and Light 2006). Indeed, it seems stuck in ongoing modeling and analysis to attempt to determine policy outcomes prior to any management action. In other words, the assessment process is key to designing policy actions that can be tested over time. Models are useful in the policy design phases, but should not be used to predict outcomes (Walters 1986). It is only through testing actions in an adaptive framework can system understanding be gained, not through extended modeling and monitoring (Williams 2009).

Adaptive Management: Learning Through Doing

The essence of adaptive management is the development of actions that are designed as much for learning as to meet other social objectives. The design of adaptive experiments or treatments is one of the outcomes of adaptive assessments (Walters 1986). The implementation of those experimental or treatment designs has been problematic, and can be stymied because of a number of reasons. Among the reasons include inability to control key variables at appropriate scales, unwillingness to risk the results of outcomes, costs of experiments and inability to monitor key resource responses, and lack of leadership (Walters 1997, Gunderson 1999).

One critical obstacle to adaptive management is the unwillingness to risk or tolerate failed experiments. Volkman and McConnaha (1993) were the first to indicate how application of the Endangered Species Act limited experimental actions on the Columbia River. Gunderson (1999) described this as a lack of ecological resilience in target variables, such as endangered populations. That is, once populations are defined as endangered, there is a tendency to minimize risks to the population from known and unknown environmental impacts. Other writers (Zellmer and Gunderson 2009, Williams 2009) suggest that experimental treatments are often limited by regulatory constraints, to the point where little or no treatment effect on target resources is achieved. While the inability to achieve learning through adaptive management has been documented for cases like the Everglades (Gunderson and Light 2006, Zellmer and Gunderson 2009), there are other cases such as the Adaptive Waterfowl harvest (Williams et al. 2009), and the Grand Canyon Adaptive Management Program (Hughes et al. 2007), where management experiments have led to institutional and organizational learning.

Gunderson and Holling (2002) proposed that at least three types of organizational and institutional learning can occur in these resource systems. The first is incremental learning, in which plans and policies are implemented and evaluated. Incremental learning is the type of learning proposed by many agencies, in which plans are updated and modified based on information gleaned from monitoring in evaluation. The U.S. Department of Interior framework for adaptive management is based upon an incremental learning model (Fig. 3.2). That is, policies are evaluated and updated based on new information. In this mode of learning plans, models and policies are assumed to be correct, and learning is characterized by collecting data or information to update and reinforce these models, and not overturn the policies should they prove to be wrong. In many resource systems the activity of learning is carried out by self-referential professionals or technocrats, who primarily view learning as problem solution (Westley 2002).

The second type of learning has been called episodic learning because it is sporadic and often surprising. Episodic learning has also been described as double-loop learning, where the underlying model or schema is questioned and rejected (Argyris and Schon 1978). This is also characterized as problem reformation (Westley 2002). In bureaucratic resource systems, this type of learning is facilitated by outside groups or charismatic integrators (Blann et al. 2003). This type of learning occurs after environmental crisis, where policy failure is undeniable (Gunderson et al. 1995). A key characteristic of episodic learning is when an underlying model of resource dynamics has been shown to change. In the Grand Canyon adaptive management program, periodic water releases or flood experiments allowed for episodic learning and led to reframing of resource issues around sediment dynamics and endangered species management (Hughes et al. 2007). In the case of sediment dynamics, the much publicized flood release experiments have led managers to conclude that short term floods can redistribute sand from the river to the banks (and rebuild beaches) over the short term, but these releases also move sediment downstream and out of the system. Mental models of sediment storage and movement were changed as a result of the flow experiments. These realizations occurred because of recurring knowledge assessment workshops that facilitated episodic learning (Melis et al. 2006).

The third type of learning is transformational. Chapin et al. (2009) describe this type of learning as triple loop learning. In some cases, transformational learning involves solving problems by identifying problem domains among sets of confounding and complex variables (Westley 2002). Parson and Clark (1995) call this evolutionary learning where not just new models or schema are developed, but also new paradigmatic structures.

Adaptive Governance: Experiments in Adaptive Management

Adaptive governance is a phrase that has been applied by many authors in the context of natural resource management. Chapin et al. (2009) use it to link the social context (policies, institutions and governance) with the natural resources context for adaptive management. Folke et al. (2005) argue that adaptive governance is a social-political framework that defines management of nonlinear, discontinuous or turbulent dynamics of ecosystem dynamics. Brunner et al. (2006) defines adaptive governance from a political science perspective and uses the phrase to explain bottom up, and other informal structures that have emerged in response to failures of top-down bureaucratic institutions. Brunner et al. (2006) describe adaptive governance as operating in a situation where science is contextual, knowledge is incomplete, multiple ways of knowing and understanding are present, policy is implemented to deal with modest steps and unintended consequences and decision making are both top-down (although fragmented) and bottom-up. Adaptive governance in this context deals with the complex social and political institutions that facilitate the implementation of adaptive management (Walters 1997, Gunderson 1999), as in the case of the Grand Canyon Dam adaptive management program.

When then Secretary of the Interior Bruce Babbitt established the Grand Canyon Dam adaptive management program, he started an experiment. Hence, the adaptive management program was proposed to not only resolve a set of complicated technical or resource issues, but also a process for managing political issues as well. The technical issues had not been resolved through normal or traditional scientific inquiry; therefore a new approach was needed. Great uncertainties remained because of the dynamic nature of the variables and interactions among those variables (e.g., how to best maintain sediments for beach renewal, cultural resources, sport fisheries and hydropower; how to resolve habitat and population declines of threatened and endangered species). The Secretary acknowledged that understanding of a complex, altered and poorly accessible ecosystem would be constantly changing.

He also determined that the best way to resolve difficult tradeoffs in management actions would be to require collaboration among some 25 different stakeholders, each representing divergent viewpoints and values. Management must be treated as an adaptive process by evaluating management policies through experiments to improve resources and learning. Hence, Grand Canyon Dam adaptive management program actually incorporates two experimental approaches: a new administrative management and science procedure for advising the Secretary on effects of dam operations, and evaluating significantly changed operating criteria and conservation measures for managing resources of concern to stakeholders.

Recent reviews of the Grand Canyon Dam adaptive management program (Zellmer and Gunderson 2009, Susskind et al. 2010), have pointed out issues and problems with the program. Acknowledging those criticisms, the Grand Canyon Dam adaptive management program has been a successful example of adaptive governance because it has facilitated both adaptive assessments and adaptive management experiments. The development of the flow treatments have informed managers about key resource dynamics and allowed them to learn that further releases are unlikely to help resolve endangered species management, as there is a long term sediment issue in the Grand Canyon. Experiments have allowed for learning that has reframed management policies. Other critiques are more concerned with imposing a regulatory or collaboration framework on the system at the expense of an adaptive management framework. In essence, critics argue that adaptive management has not solved the critical resource problems, so it should be abandoned in favor of other management alternatives. I contend that this is equivalent to 'throwing the baby out with the bathwater', and thus not a sound policy choice.

Conclusion

Winston Churchill is quoted as saying, "It has been said that democracy is the worst form of government except all the others that have been tried." I think that there are parallels with adaptive management, as it has been applied and implemented over the past four decades. The three phases or components of adaptive management: assessment, management and governance are all problematic. They face uphill struggles because of scientific and technical communities who are not trained in these approaches and see them as threats to their research and scholarly traditions. In the Everglades and Glen Canyon, adaptive assessments have provided imaginative syntheses that led to transformation of the programs. Adaptive management continues to face many obstacles, as the costs of experiments rise along with a continued fear of failure. I agree with Walters (2007) who argued that the presence of key people in leadership roles is the only factor that seems to separate success from failure in the implementation of adaptive management. It is these leaders that have overcome obstacles and created opportunities in implementing the worst form of natural resource management- except for all the others.

References

Allen, C. R., & Gunderson, L. H. (2011). Pathology and failure in the design and implementation of adaptive management. *Journal of Environmental Management, 92,* 1379–1384.

Argyris, C., & Schon, D. A. (1978). *Organizational learning: A theory of action perspective*. Reading: Addison-Wesley.

Blann, K., Light, S., & Musumeci, J. (2003). Facing the adaptive challenge. In F. Berkes & C. Folke (Eds.), *Linking social and ecological systems: Institutional learning for resilience* (pp. 210–240). Cambridge: Cambridge University Press.

Brunner, R. D., Steelman, T. D., Coe-Juell, L., Cromley, C. M., Edwards, C. M., & Tucker, D. W. (2006). *Adaptive governance: integrating science policy and decision making*. New York: Columbia University Press.

Chapin, F. S., Kofinas, G., & Folke, C. (2009). *Principles of ecosystem stewardship*. New York: Springer.

Clark, W. C., Jones, D. D., & Holling, C. S. (1979). Lessons for ecological policy design: A case study of ecosystem management. *Ecological Modelling, 7,* 1–53.

Davis, S. M., & Ogden, J. (Eds.). (1994). *The Everglades: The ecosystem and its restoration.* Boca Raton: St. Lucie Press.

Folke, C., Hahn, T., Olsson, P., & Norberg, J. (2005). Adaptive governance of social-ecological systems. *Annual Review of Environment and Resources, 30,* 441–473.

Gunderson, L. (1999). Resilience, flexibility and adaptive management—antidotes for spurious certitude? *Conservation Ecology, 3*(1), 7. http://www.consecol.org/vol3/iss1/art7/.

Gunderson, L. H. (2001). Managing surprising ecosystems in southern Florida. *Ecological Economics, 37,* 371–378.

Gunderson, L. H., & Holling, C. S. (Eds.). (2002). *Panarchy: Understanding transformations in human and natural systems*. Washington, D.C.: Island Press.

Gunderson, L. H., & Light, S. S. (2006). Adaptive management and adaptive governance in the Everglades. *Policy Sciences, 39,* 323–334.

Gunderson, L. H., & Pritchard, L. (Eds.). (2002). Resilience and the behavior of large scale ecosystems. Washington, D.C.: Island Press.

Gunderson, L. H., Carpenter, S. R., Folke, C., Olsson, P., & Peterson, G. D. (2006). Water RATs (resilience, adaptability, and transformability) in lake and wetland social-ecological systems. *Ecology and Society, 11*(1), 16. http://www.ecologyandsociety.org/vol11/iss1/art16/.

Gunderson, L. H., Holling, C. S., & Light, S. S. (Eds.). (1995). *Barriers and bridges to renewal of ecosystems and institutions*. New York: Columbia University Press.

Gunderson, L., Peterson, G., & Holling, C. S. (2008). Practicing adaptive management in social-ecological systems. In J. Norberg & G. Cumming (Eds.), *Complexity theory for sustainable futures* (pp. 223–245). New York: Columbia University Press.

Holling, C. S. (1978). *Adaptive environmental assessment and management*. Chichester: Wiley.

Holling, C. S. (1981). Highlights of adaptive environmental assessment and management. Report-23, Institute of Resource Ecology, University of British Columbia, Vancouver, Canada.

Holling, C. S. (1994). Simplifying the complex: The paradigms of ecological function and structure. *Futures, 26,* 598–609.

Holling, C.S. (1998). Two cultures of ecology. *Conservation Ecology, 2*(2), 4. http://www.consecol.org/vol2/iss2/art4/.

Holling, C. S., & Chambers, A. D. (1973). The nurture of an infant. *Bioscience, 23,* 13–20.

Holling, C. S., & Meffe, G. K. (1996). Command and control and the pathology of natural resource management. *Conservation Biology, 10,* 328–337.

Hughes, T. P., Gunderson, L. H., Folke, C., Baird, A. H., Bellwood, D., & Berkes, F. (2007) Adaptive management of the Great Barrier Reef and the Grand Canyon World Heritage areas. *Ambio, 36,* 586–592.

Johnson, B. L. (1999). The role of adaptive management as an operational approach for resource management agencies. *Conservation Ecology, 3*(2), 8. http://www.consecol.org/vol3/iss2/art8/.

Lee, K. N. (1993). *Compass and gyroscope*. Washington, D.C.: Island Press.

Melis, T. S., Wright, S. A., Ralston, B. E., Fairley, H. C., Kennedy, T. A., Andersen, M. E., Coggins, L. G. Jr., & Korman, J. (2006). Knowledge assessment of the effects of Glen Canyon Dam on the Colorado river ecosystem: An experimental planning support document. Report, Grand Canyon Monitoring and Research Center, US Geological Survey. Flagstaff, USA.

Parson, E. W., & Clark, W. C. (1995). Evolutionary learning. In L. H. Gunderson, C. S. Holling, & S. S. Light (Eds.), *Barriers and bridges to the renewal of ecosystems and institutions* (pp. 428–460). New York: Columbia University Press.

Susskind, L., Camacho, A. E., & Schenk, T. (2010). Collaborative planning and adaptive management in Glen Canyon: A cautionary tale. *Columbia Journal of Environmental Law, 35,* 1–54.

Volkman, J. M., & McConnaha, W. E. (1993). Through a glass, darkly: Columbia River salmon, the Endangered Species Act, and adaptive management. *Environmental Law, 23,* 1249–1272.

Walters, C. J. (1986). *Adaptive management of renewable resources*. New York: McGraw Hill.

Walters, C. J. (1997). Challenges in adaptive management of riparian and coastal ecosystems. *Conservation Ecology, 1*(2), 1. http://www.consecol.org/vol1/iss2/art1.

Walters, C. J. (2007). Is adaptive management helping to solve fisheries problems? *Ambio, 36,* 304–307.

Walters, C. J., & Holling, C. S. (1990). Large-scale management experiments and learning by doing. *Ecology, 71,* 2060–2068.

Walters, C., Gunderson, L., & Holling, C. S. (1992). Experimental policies for water management in the Everglades. *Ecological Applications, 2,* 189–202.

Westley, F. (2002). The devil in the dynamics: Adaptive management on the front lines. In L. H. Gunderson & C. S. Holling (Eds.), *Panarchy: Understanding transformations in human and natural systems* (pp. 333–360). Washington, D.C.: Island Press.

Williams, B. K., Szaro, R. C., & Shapiro, C. D. (2009). Adaptive management: The U.S. Department of the Interior technical guide. Adaptive Management Working Group, U.S. Department of the Interior, Washington, D.C.

Zellmer, S., & Gunderson, L. (2009). Why resilience may not always be a good thing: Lessons in ecosystem restoration from Glen Canyon and the Everglades. *Nebraska Law Journal, 87,* 893–949.

Chapter 4
Adaptive Management and Law

Melinda Harm Benson and Courtney Schultz

Keywords Adaptive management · Law · Uncertainty ·Enforceability ·Agencies

Adaptive Management: From Theory to Practice

Adaptive management is increasingly recognized as a valuable approach to natural resource and environmental management challenges that involve high degrees of uncertainty. The legal rules and requirements that drive environmental protection efforts in the United States, however, are often considered barriers to successful implementation of adaptive management (Allen et al. 2011). A recent survey of adaptive management practitioners found that over seventy percent (70%) feel hampered by legal and institutional constraints (Benson and Stone 2013). While adaptive management has been widely discussed in the fields of ecology and conservation biology for decades (Holling 1978, Walters and Holling 1990), its incorporation into natural resource management in the United States is relatively recent. Examples include the U.S. Department of Interior's development of a technical guide for adaptive management implementation (Williams et al. 2009), landowner based habitat conservation planning under the Endangered Species Act (Ruhl 2005, 65 Fed. Reg. 25242 [2000]), and the compensatory wetlands mitigation protection program under the Clean Water Act (U.S. Army Corps 2002). In the context of federal lands management a standard definition, as adapted from the National Research Council, is as follows:

M. H. Benson (✉)
Department of Geography & Environmental Studies, University of New Mexico,
Albuquerque, NM 87131-0001, USA
e-mail: mhbenson@unm.edu

C. Schultz
Department of Forest and Rangeland Stewardship, Colorado State University,
Fort Collins, CO 80523-1472, USA
e-mail: courtney.schultz@colostate.edu

C. R. Allen, A. S. Garmestani (eds.), *Adaptive Management of Social-Ecological Systems*,
DOI 10.1007/978-94-017-9682-8_4

> Adaptive management [is a decision process that] promotes flexible decision-making that can be adjusted in the face of uncertainties as outcomes from management actions and other events become better understood. Careful monitoring of these outcomes both advances scientific understanding and helps adjust policies or operations as part of an iterative learning process. Adaptive management also recognizes the importance of natural variability in contributing to ecological resilience and productivity. It is not a 'trial and error' process, but rather emphasizes learning while doing. Adaptive management does not represent an end in itself, but rather a means to more effective decisions and enhanced benefits. Its true measure is in how well it helps meet environmental, social, and economic goals, increases scientific knowledge, and reduces tensions among stakeholders. (Williams et al. 2009)

As a conceptual approach, adaptive management is a vehicle for operationalizing a systems-based understanding of social-ecological dynamics (Walters 2002). It is based on a recognition of non-equilibrium in social-ecological systems and the corresponding complexity, uncertainty and instability associated with both social and ecological systems and processes (Gunderson and Holling 2002, Folke et al. 2005). Adaptive management is therefore seen as a key strategy for fostering resilience of social-ecological systems (Lee 1999, Salafsky et al. 2001, McCarthy and Possingham 2007), and there is a growing area of scholarship examining practical applications of adaptive management (e.g., Berkes and Seixas 2005, King and Brown 2006, Allan et al. 2008, Brugnach et al. 2008, Schultz and Nie 2012).

Within this literature, scholars do not often directly consider the institutional constraints on adaptive management. Scholarship that does address institutional issues tends to emphasize other factors necessary for adaptive management to be successful, including polycentric governance, collaboration, social learning, and issues of scale (Bodin et al. 2006, Stringer et al. 2006, Pahl-Wostl 2007, Pahl-Wostl et al. 2007, Brugnach et al. 2008, Raadgever et al 2008, Folke et al. 2007). Jacobson et al. (2006) looked at barriers to adaptive management implementation by surveying the staff of the Florida Fish and Wildlife Conservation Commission. This study based its questionnaire on a literature review that identified 47 potential barriers to use of adaptive management. Legal and regulatory requirements were not listed explicitly, though related issues regarding management flexibility and availability of agency resources were listed among the categories of logistical and institutional barriers. This survey revealed that logistical issues were the most problematic of all barriers for respondents, who specifically cited lack of agency resources and the time consuming nature of adaptive management protocols. Similarly, Butler and Koontz (2005) surveyed 345 U.S. Forest Service managers regarding their experiences implementing the agency's ecosystem management objectives, of which adaptive management is one component (Grumbine 1994). Their results established that managers view adaptive management as the most difficult element of ecosystem management to implement. Among the reasons for this, according to managers, were the significant institutional changes required, the immense costs of monitoring and the lack of public and political support. One interviewee was quoted as stating: "Adaptive management happens, but is a reach for the agency. We don't have all the mechanisms in place to do it well, and there are legal, logistical, contractual and social constraints" (Butler and Koontz 2005).

While adaptive management is widely acknowledged as a valuable approach in theory, scholars are often critical of adaptive management in practice (Doremus 2002,

Zellmer and Gunderson 2009, Ruhl and Fischman 2010, Susskind et al. 2010). Within the legal scholarship on adaptive management, there have been two major areas of emphasis. First, there is an acknowledgement that virtually all of the efforts to integrate adaptive management strategies to date reflect attempts to fit adaptive management within existing legal mandates and protocols. While existing management mandates are usually sufficiently broad to encompass adaptive management approaches, "the disconnect between adaptive management in practice and adaptive management in law is quite palpable.... No other principle of natural resources law has so deeply permeated the practice on the basis of so little mention in law" (Ruhl 2008). As a result, adaptive management is being thrown like a blanket on top of existing authorizations and requirements, with little attention to how practitioners balance this new mandate in relation to other legal and institutional requirements. Critics of adaptive management have argued that without more specific legal grounding, adaptive management provides agencies with an undesirable amount of discretion (Doremus 2002, Houck 2009). In the same way, even adaptive management proponents have cautioned against lax standards for adaptive management that would, in essence, create a situation in which agencies use it as "rhetorical cover for requests for blanket preauthorization to reverse or revise policies should the agency later decide to change its mind" (Karkkainen 2005). In other words, unless adaptive management is given some legal definition and its application is enforceable in some way, the approach can be used as a smokescreen for open-ended and discretionary decision-making that fails to meet legal standards, lacks accountability, and fails to incorporate some of the most important aspects of the paradigm, including rigorous monitoring and feedback loops that inform an adaptive planning cycle (Schultz and Nie 2012).

The second shared observation is that current legal and regulatory requirements do not generally support the iterative processes required by adaptive management (Thrower 2006, Ruhl 2008, Craig 2010). For example, the National Environmental Policy Act (NEPA) is the major federal law that requires agencies to take a "hard look" at the environmental impacts of proposed agency action (Schultz 2008, see Box 1). NEPA is built upon a model of predictive and rational planning and makes a number of implicit assumptions that are at odds with adaptive management, including that there is a single, final "agency action," rather than a series of iterative processes and that resource managers already have knowledge of natural systems needed to assess environmental impacts (Benson and Garmestani 2011). Several scholars have highlighted the challenges associated with engaging in adaptive management while also navigating the NEPA process and other legal constraints (Angelo 2009, Benson 2009, 2010, Zellmer and Gunderson 2009, Susskind et al. 2010).

Summary of National Environmental Policy Act

Passed into law in 1970, the National Environmental Policy Act (NEPA) is one of the most influential environmental laws in the United States. It requires all federal agencies that propose a "major federal action significantly affecting the quality of the human environment" to first assess the potential impacts of the proposed action (42 USC § 4332).

The resulting document is an "environmental impact statement" (EIS) that informs both the agency and the public regarding possible environmental consequences. An EIS is generally comprised of several elements, including:

1. The environmental impact of the proposed action,
2. Any adverse environmental effects which cannot be avoided should the proposal be implemented,
3. Alternatives to the proposed action,
4. The relationship between local short-term uses [the] environment and the maintenance and enhancement of long-term productivity, and
5. Any irreversible and irretrievable commitments of resources which would be involved in the proposed action should it be implemented.

42 USC § 4332(1)(C)(a). There are several stages to the NEPA process, including determining when an EIS is necessary and opportunities for public comment. While essentially "procedural" in the sense that it does not specify the agency reach any particular outcome (i.e., it does not require the agency to avoid environmental impacts), the information gathered through the NEPA process is generally considered a valuable part of the decision-making process.

The following are some important concepts and terms associated with NEPA implementation:

Environmental Impact Statement (EIS). A formal NEPA document that conducts the required "hard look" at the environmental consequences of the agencies proposed action. Must include an analysis for the environmental impacts of the proposed action, a reasonable range of alternatives to the proposed action and identification of any irreversible and irretrievable commitments of resources which would be involved in the proposed action should it be implemented.

Environmental Assessment (EA). An analysis, provided in the form of a public document, often used by agencies to determine whether to prepare an EIS. Also used to "tier" a project-specific agency action to larger, programmatic EIS that has already conducted the required NEPA analysis.

Finding of No Significant Impact (FONSI). A determination that an EIS is not required. Often accompanies an EA as the final conclusion of NEPA compliance. Mitigation measures taken by an agency to reach a FONSI are legally enforceable

Supplemental Environmental Impact Statement (SEIS). Additional NEPA analysis required when significant new circumstances or information relevant to environmental concerns or substantial changes in the proposed action that are relevant to environmental concerns may necessitate preparation of a supplemental EIS.

Given the legal challenges and the propensity of agencies to pursue their own administrative discretion in the form of flexible decisions, adaptive management in practice often manifests as something less than adaptive management in theory. Ruhl and Fischman (2010) explain: "From theory to policy to practice, at each step forward in the emergence of adaptive management something has been lost in the translation. The end product is something we call 'a/m-lite,' a watered down version of the theory that resembles ad hoc contingency planning more than it does planned 'learning while doing.'"

In sum, there is a recognition that more collaboratively-based, iterative processes are needed to promote flexibility and facilitate adaptive management (Gunderson and Light 2006). At the same time, given the political context in which adaptive management is applied, some enforceable standards for adaptive management are preferable to open-ended guidance, so that adaptive management in practice incorporates some measure of accountability to legal standards and to the public (Schultz and Nie 2012). A brief examination of the structure of the federal government—and particularly the role of federal agencies within the realm of administrative law—provides insight into this inherent tension between flexibility and enforceability.

Use of Adaptive Management by Federal Agencies

The federal government in the United States is comprised of three branches: the legislative, executive and judicial. Often described as a "separation of powers," each branch has a role to play in governing the nation: the legislative branch (*i.e.*, Congress) creates laws; the executive branch implements and enforces them; and the judiciary assures that the other two branches are conducting themselves in accordance with both statutory and constitutional provisions. What can be considered "law" is actually a compilation of a number of types of legislative, judicial and executive enactments that can be seen as a hierarchical structure (see Fig. 4.1). At the top, there are constitutional provisions; these laws cannot be changed without the

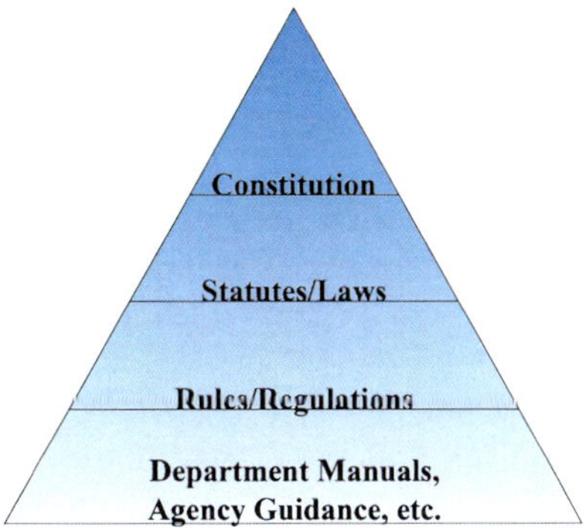

Fig. 4.1 Hierarchy of various laws and policies

rather onerous process of a constitutional amendment. Because there are currently no constitutional provisions for environmental protection *per se*, the second tier of law—statutes passed by Congress—are generally the highest level of legal authorization for environmental and natural resource management. Examples include NEPA, the Endangered Species Act, the Clean Water Act, etc. As statutes, these laws are generally enforceable in court, and litigation brought by concerned citizens (often referred to as "citizen suits") are in fact a primary means of environmental law enforcement.

Next in the hierarchy arc administrative rules and regulations. Rules and regulations are developed by the executive branch's numerous federal agencies responsible for the implementation and enforcement of various statutes. For example, the Environmental Protection Agency is primarily responsible for the development of regulations for the Clean Water Act; similarly, the U.S. Fish and Wildlife Service is responsible for developing the regulations that give additional specificity to the provisions of the Endangered Species Act. As an overarching statute, the Administrative Procedures Act guides the development and enforcement of rules and regulations by federal agencies by allowing for public involvement and judicial oversight of the executive branch's interpretation and implementation of laws from Congress.

When federal agencies go through formal rulemaking procedures, there is generally public notice—published in the Federal Register—and an opportunity for comment. The resulting rules and regulations provide the details needed to further define the interpretation and means of enforcing the overarching, but often vague, statutory language. For example, the Endangered Species Act prohibits the "take" of an endangered species and provides a definition of "take" of a species as actions that "harass, harm, pursue, hunt, shoot, wound, kill, trap, capture, or collect" such species (16 U.S.C. 1533[19]). Regulations promulgated by the U.S. Fish and Wildlife Service provide a more refined definition of "take," specifically expanding on the statutory definition of "harm" to include "any act which actually kills or injures fish or wildlife, and emphasizes that such acts may include significant habitat modification or degradation that significantly impairs essential behavioral patterns of fish or wildlife" (50 CFR 17.3). In this way, the regulatory definition further explains the meaning of the statute. When rules are finalized, the rulemaking process and the content of the rule are reviewable by the judiciary. For example, the Supreme Court upheld the U.S. Fish and Wildlife Service's regulatory definition of "harm" in the case *Babbitt v. Sweet Home Chapter of Communities for a Great Oregon* (1995). Once finalized, rules and regulations are generally legally enforceable in a court and represent the agencies official interpretation of the relevant statute.

Finally, there are a number of more informal agency policies, including departmental manuals, internal memoranda and guidance documents, etc., that are developed without formal rulemaking procedures under the Administrative Procedures Act. For example, the U.S. Forest Service has both a "manual" and "handbook" providing guidance to agency officials. The manual contains legal authorities, objectives, policies, responsibilities, instructions, and guidance, and the "handbook" provides more specialized guidance and instructions for carrying out the direction issued in the manual. Court decisions regarding the legal enforceability of these types of management tools are mixed. Determinations are made on a case-by-case basis, and the outcome often depends on a number of factors, including the procedures taken, whether the policies prescribe substantive or interpretive rules, the agency's intent, and the Congressional mandate involved (Fischman 2007). Generally speaking, however, guidance documents are not legally enforceable in a court of law (Ruhl and Fischman 2010). As Fischman (2007) explains: "The majority of courts that examine the question closely find agency manuals to be non-binding, internal guidance unless some special circumstance raises the legal status of the policy. The few manual provisions promulgated under notice-and-comment procedures, though, are regarded by courts as binding on agencies."

The relative enforceability of various types of law becomes of particular importance with regard to adaptive management. At present, no statute explicitly defines adaptive management and agency regulations that do are generally silent about how to implement the approach. For example, the U.S. Forest Service's regulations define adaptive management as, "A system of management practices based on clearly identified intended outcomes and monitoring to determine if management actions are meeting those outcomes; and, if not, to facilitate management changes that will best ensure that those outcomes are met or re-evaluated. Adaptive management stems from the recognition that knowledge about natural resource systems is sometimes uncertain" (30 CFR § 220.3). The specifics of integrating adaptive management into federal agency management and planning, however, takes place informally.

The U.S. Department of Interior's approach for implementing adaptive management provides another example. The formal regulatory provision related to adaptive management is in the agency's regulations for its environmental impact assessment procedures under NEPA, where it states that the agency "should use adaptive management, as appropriate, particularly in circumstances where long-term impacts may be uncertain and future monitoring will be needed to make adjustments in subsequent implementation decisions" (43 C.F.R. § 46.14). In 2007, the Secretary for Interior issued an order requiring agency officials to use adaptive management whenever possible. That same year, the agency released a technical guide (revised in 2009) in order to "aid U.S. Department of the Interior managers and practitioners in determining when and how to apply adaptive management" (Williams et al. 2009). The Department of Interior's technical guide has already undergone one revision, and the agency recently released a companion applications guide (Williams and Brown 2012). This approach leaves the agency with a great deal of discretion regarding both when adaptive management is "appropriate" and how to conduct adaptive management.

The current integration of adaptive management in federal agency decision-making highlights the tension between flexibility and enforceability. As Fig. 4.2 illustrates, there is generally an inverse correlation between a management approach's enforceability by those outside the agency and the flexibility with which the agency can interpret and implement the approach. The major advantage of using informal guidance is the flexibility it affords. At the same time, relegation of adaptive management to agency manuals and guides leaves much of the agency's approach unenforceable. This

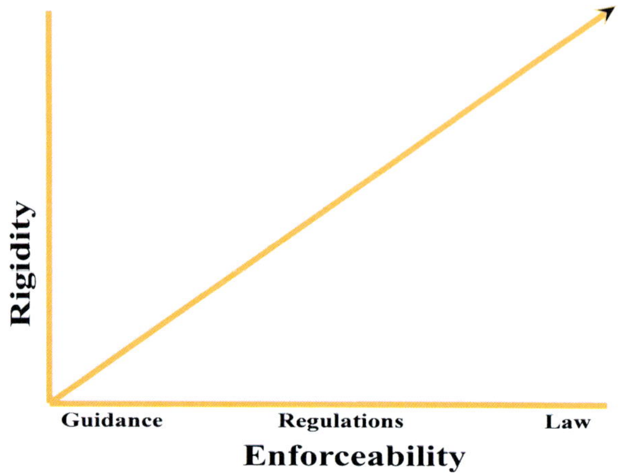

Fig. 4.2 The tradeoff between flexibility and enforceability

tension is worthy of further investigation, because, as demonstrated through an examination of several court cases below, successful implementation of adaptive management requires some measure of both.

Adaptive Management and the Courts

As we have explained, a key long-standing question is whether and how adaptive management can be incorporated in the U.S. legal framework, which relies heavily on *a priori* planning and includes a number of substantive legal standards. Some scholars have made the case that adaptive management is, to a large extent, incompatible with the framework of U.S. administrative law (Allen et al. 2011). However, in the United States, courts are beginning to outline the legal parameters of how adaptive management can be applied within the context of U.S. environmental and administrative law. Ruhl and Fischman (2010) recently published an overview of adaptive management case law. They analyzed thirty-one federal court decisions—which they refer to as "the first generation" of case law—in which the judiciary speaks directly to the legality of adaptive management. They found that federal agencies lost more than half of the cases in which they used adaptive management. Several key findings emerge from their analysis (Table 4.1).

One key theme to emerge out of the adaptive management jurisprudence is that the courts demand assurances that adaptive management plans meet substantive management criteria required by law. Ruhl and Fischman (2010) explain: "When agencies lose challenges to their adaptive management plans, it is often because their preference for management latitude runs afoul of the need to show they can meet substantive and procedural standards in statutes, regulations, or even their own earlier plans." For example, an important substantive legal standard is the re-

Table 4.1 Key court decisions regarding adaptive management in U.S. courts as of 2011 (Schultz and Nie 2012)

Case	Summary of Key Issues	Relevancy of Triggers/Thresholds
Center for Biological Diversity v. Rumsfeld, 198 F. Supp. 2d. 1139 (D. Az. 2002)	This case considered whether the Department of the Army's plan, outlined in its operating plan and associated Biological Opinion, to collaboratively develop a mitigation program to maintain minimum water levels was sufficient to satisfy its obligation under the ESA to not jeopardize species	The court found the Army's plan insufficient. It made several points: (1) Mitigation measures must be within the agency's power to implement; (2) Agencies must show that they will meet substantive requirements; and (3) Potential mitigation measures must be detailed and enforceable. As the court puts it, they must be "reasonably specific, certain to occur, and capable of implementation; they must be subject to deadlines or otherwise-enforceable obligations; and most important, they must address the threats to the species in a way that satisfies the jeopardy and adverse modification standards."
Natural Resources Defense Council v. Kempthorne, 506 F. Supp. 2d. 322 (E.D. Cal. 2007)	This case reviewed the biological opinion for the delta smelt, as affected by operation of two major California water projects. A key issue was whether the adaptive management framework to monitor and mitigate take of the species satisfied "no jeopardy" requirements under the ESA	The monitoring framework was clear, but triggered a discretionary process where actions *could* be taken but were not required. What was triggered in this case was an unenforceable and discretionary process, devoid of clear requirements to take action. This was legally insufficient for meeting requirements under Sect. 7 of the ESA
Pacific Coast Federation of Fishermen's Associations v. Gutierrez, 606 F. Supp. 2d. 1122 (E. Dist. Calif. 2008)	The court reviewed the BiOp for salmonid species affected by operation of the same California water projects. The question was the same: whether the adaptive management framework, put in place to deal with uncertainty about future effects, was sufficient to meet Sect. 7 requirements	In this case, triggered actions were an enforceable process under the terms conditions of the incidental take permit. Specific triggers points, including water temperatures at specific locations, were included that, if exceeded, would lead to violation of the terms of the permit and reinitiation of consultation prior to the announcement of the following year's water deliveries

Table 4.1 (continued)

Case	Summary of Key Issues	Relevancy of Triggers/Thresholds
Greater Yellowstone Coalition v. Servheen, 672 F. Supp. 2d 1105 (D. Mont. 2009)	The court reviewed the delisting decision for the Greater Yellowstone DPS of grizzly bears. At issue was whether the National Forest plan amendments and state management plans sufficed as adequate regulatory mechanisms to ensure long-term conservation of the species	Despite the presence of population standards and a monitoring program, the court ruled the strategy was unenforceable and non-binding. The monitoring program promised nothing more than good intentions for future actions. This is not an adequate regulatory mechanism if it cannot be enforced and there is no way to ensure anything will happen. The judge, citing Norton v. SUWA (2004), also noted that monitoring is generally not enforceable under the APA
Greater Yellowstone Coalition v. Kempthorne, 557 F. Supp. 2d. 183 (D. D.C. 2008)	In its ROD for its new Winter Use Plan, the National Park Service, determined that maintaining a higher level of snowmobiles would not impair resources, despite the fact that previously set thresholds for environmental impacts had been exceeded. Plaintiffs asked why the exceeding of these thresholds did not constitute an unacceptable impact	Without some "quantitative standard or qualitative analysis to support its conclusion that the adverse impacts of the [Winter Use Plan] are 'acceptable,'" the court found the justification in the ROD to be arbitrary. The lesson here is that all thresholds do not necessarily have to correlate with significance in terms of impacts; however, if thresholds are crossed and an agency nonetheless finds impacts to be less than significant, there must be a clear rationale offered as to how this evaluation is made
Klamath Siskiyou Wildlands Center v. Boody, 468 F.3d 549, 553 (9th Cir. 2006)	Plaintiffs challenged changes to the status of the red tree vole under survey and manage requirements of the NWFP, asking whether the changes required plan amendment and supplemental NEPA analysis. The question involved how much leeway an agency has to make changes under an adaptive management plan in light of new information	The court held that the changes in the vole's status contradicted what was contemplated in the NWFP's most recent amendments and associated NEPA analysis. When agencies make substantial changes to requirements in adaptive management plans, courts will require new analysis, in the form of plan amendments and supplemental NEPA analysis. This is the case when the new information or the permitted actions are outside the bounds of what was originally discussed in the NEPA document. Just because a plan contemplates possible future actions, this alone does not obviate the need to amend a plan or supplement NEPA analysis

Table 4.1 (continued)

Case	Summary of Key Issues	Relevancy of Triggers/Thresholds
Oregon Natural Resources Council Action v. USFS, 59 F. Supp. 2d 1085 (W.D. Wash. 1999)	In the context of the NWFP, new information emerged regarding water quality, and species status was changed under the ESA. The court considered whether this new information required supplemental NEPA analysis	In this case, the court held that possible changes in conditions, and associated changes in management practices, had been adequately analyzed in the original NEPA document and were covered as part of the adaptive framework of the NWFP. Flexibility can be built into a NEPA assessment that anticipates changes in conditions and gives an agency the opportunity to adjust activities within certain limits
In re Operation of the Missouri River System Litigation, 516 F.3d 688 (8th Cir. 2008)	Plaintiffs challenged determinations made by the Army Corps of Engineers in an EA that changes in their management actions fell within the scope of a previous EIS	The court upheld the Corps' decision. It noted a supplemental EIS is only required when the change in management direction is one that was not within the spectrum of alternatives analyzed in the prior EIS. Even if an agency decides to implement aspects of an alternative not originally selected, as long as the impacts have been analyzed and no significant new information has arisen, supplemental NEPA analysis is not required

quirement under Sect. 7 of the Endangered Species Act that federal agencies not cause jeopardy to listed species (16 USC § 1536[a][2]). When enforcing Sect. 7, the regulatory agencies responsible for implementing the Endangered Species Act, as part of a process called "consultation," issue a Biological Opinion to the agency planning the action; this document guides and constrains the action agency's activities so that it will not cause jeopardy to the protected species.

Substantive standards such as this play a critical role in legal challenges to adaptive management plans. For example, *Center for Biological Diversity v. Rumsfeld* (2002) revolved around the adequacy of monitoring and mitigation strategies in an adaptive management framework that involved aquatic species listed under the Endangered Species Act. At issue were provisions in the U.S. Army's Fort Huachuca 10-year operating plan and the associated Biological Opinion from the U.S. Fish and Wildlife Service for water savings and monitoring of species status. The court found the plan for future management actions ambiguous and unsatisfactory in light of requirements under Sect. 7; it explained, "Mitigation measures must be reasonably specific, certain to occur, and capable of implementation; they must be subject to deadlines or otherwise-enforceable obligations; and most important, they must address the threats to the species in a way that satisfies the jeopardy and adverse modification standards" (*Center for Biological Diversity v. Rumsfeld* 2002). These requirements, as articulated in this case, are now repeatedly cited in adaptive management case law.

A pair of cases reviewing adaptive management frameworks for operation of water projects on the Sacramento and San Joaquin Rivers is also instructive. At issue in *Natural Resources Defense Council v. Kempthorne* (2007) was the Biological Opinion issued for the Delta smelt *(Hypomesus transpacificus)*, a listed species under the Endangered Species Act. This adaptive management framework was designed to trigger management changes based on factors such as estimates of number of fish killed in water facilities, and spawning rates, and if thresholds were crossed, a working group could meet and submit recommendations that could potentially be undertaken by a separate management team. The court agreed with plaintiffs that this was too uncertain and unenforceable of a framework to support a "no jeopardy" conclusion. On the other hand, the same judge upheld the Biological Opinion for the anadromous fish species affected by the same water projects (*Pacific Coast Federation of Fisherman's Associations v. Guitierrez* 2008). In that case, the court determined that mitigation measures were adequately specific, requiring action if a certain water temperature was exceeded, and were included under the "Terms and Conditions" of the Incidental Take Statement, which, the court noted, is enforceable by law and therefore binding. The court was satisfied because mitigation measures were based on an enforceable standard, which triggered a non-discretionary mandate to reinitiate consultation with the regulatory agency before proceeding.

Another set of important lessons from the case law revolves around compliance with the procedural requirements of NEPA. Ruhl and Fischman (2010) note that larger-scale plans are often more suited to adaptive management then smaller projects or plans, due to the array of mitigation options available across large scales and the potential to "tier" analyses. Tiering of NEPA documents, where one NEPA document refers to analysis in another, often more broad and overarching NEPA document, appears to work well in the context of adaptive management. Adaptive management frameworks can be established at larger scales that consider cumulative impacts or programmatic standards, and more site-specific documents can tier to that analysis (Benson and Garmestani 2011). The challenge is striking the balance between adaptability in these large-scale plans with a satisfactory level of commitment to monitor results and take action if thresholds or trigger points are reached.

For example, adaptive management plans have survived legal review in cases involving the Northwest Forest Plan and the Sierra Forest Framework, both large-scale land management plans completed by the U.S Forest Service and done in accordance with NEPA. Each of these management plans acknowledges uncertainty, includes monitoring and adaptation, and employs tiering and supplementation, whereby additional environmental impact analysis is conducted in accordance with NEPA in light of new information or a change of circumstances, to balance the need for a broad planning framework with site-specific analysis (Ruhl and Fischman

2010). An instructive case is *Klamath Siskiyou Wildlands Center v. Boody* (2006), which revolved around the issue of when, under an adaptive management plan, supplemental NEPA analysis is required. The Northwest Forest Plan of 1994 amended all National Forest plans and resource management plans for Bureau of Land Management districts in the Pacific Northwest; it also established "Survey and Manage" requirements for individual species that would not be adequately protected as a result of the land management allocations. In 2000, the Bureau of Land Management (BLM) and the U.S. Forest Service (USFS) issued a Final Environmental Impact Statement (2000 FEIS) for amendments to the Northwest Forest Plan. The 2000 FEIS discussed the status of the red tree vole (*Arborimus longicaudus)* and stated that approximately five years of data collection would likely be necessary prior to contemplating any changes to its status under Survey and Manage requirements. In the summer of 2002, after doing the first annual review for red tree voles, the BLM downgraded the species' status, and in December 2003 the BLM removed the vole from the Survey and Manage designation completely. Neither of these decisions was accompanied by any NEPA document, and plaintiffs brought challenges under NEPA and the Federal Land Policy and Management Act. With regard to the NEPA claim, the BLM argued that the 2000 FEIS contemplated changes in Survey and Manage designations as part of an adaptive management framework. The court disagreed, holding that simply because an adaptive management plan *contemplates* potential changes, this does not obviate the need to comply with NEPA. The court explained:

> BLM is partly correct: the 2001 [decision] contemplated that moving a species from one survey strategy to another or dropping Survey and Manage protection for any species whose status is determined to be more secure than originally projected could occur under the plan. However, merely because the 2001 [decision] contemplated this type of change, it does not necessarily follow that all contemplated changes fall under the narrow definition of plan maintenance in § 1610.5-4 [BLM planning regulations]. If that were the law, BLM could circumvent the mandates of § 1610.5-5 (i.e., requiring environmental assessments and impact statements, public disclosure, etc.) by merely designing a management plan that "contemplates" a wide swath of future change. (Klamath Siskiyou Wildlands Center v. Boody 2006)

The court held that if an agency takes action contrary to what they found in a previous NEPA document, it must explain the rationale for the action and complete a supplemental NEPA analysis. In this case, the original FEIS did not provide any basis for the BLM's decisions; therefore, the judge explained, the decisions were plainly inconsistent with the prior plan and FEIS. NEPA also requires supplementation when there is significant new information, as there was in this case.

On the other hand, in cases such as *Oregon Natural Resources Council Action v. USFS* (1999), courts have indicated that an agency does not always need to prepare supplemental analyses if the adaptive management actions and collection of additional information were covered in a prior, programmatic environmental impact statement (EIS). In this case, where new information emerged regarding water quality and the status of some species under the Endangered Species Act, the court explained, "The plan's adaptive management approach is adequate to deal with any new information plaintiffs have identified. If circumstances warrant, the [decision] gives the Forest Service and the BLM the flexibility to reduce or halt logging in order to comply with their statutory mandates" (*Oregon Natural Resources Council Action v. USFS* 1999). In other words, flexibility can be built into a NEPA assessment that anticipates changes in conditions and gives an agency the opportunity to adjust activities within certain limits. New information does not always require the preparation of a supplemental EIS, unless it fundamentally alters the predictions in the original EIS or if the response to the new information is plainly contrary to what was planned or predicted in the original EIS.

Finally, a suite of adaptive management cases involving the Army Corps' management of Missouri River dams (Ruhl and Fischman 2010) provide several lessons related to tiering and supplemental NEPA analysis. In a 2008 hearing, the court ruled that it was appropriate for the Corps to utilize an environmental assessment (EA), a less detailed type of environmental impact assessment, to determine whether impacts resulting from changes in its springtime water release strategies were consistent with management strategies that had been analyzed in a 2004 Final EIS (*In re* Operation of the Missouri River System Litigation 2008). The Corps determined that the impacts resulting from the new bimodal springtime release strategy were within the range of impacts considered in the 2004 Final EIS and determined that no supplemental EIS was necessary. At the same time, they also determined that a finding of no significant impact (FONSI) was not appropriate, because significant impacts, which had already been analyzed in the 2004 Final EIS, were predicted. The court ruled that the Corps' method of complying with NEPA while navigating the incorporation of a change in management strategy was adequate. It noted a supplemental EIS is only required when the change in management direction is one that was not within the spectrum of alternatives analyzed in the prior EIS. Even if an agency decides to implement aspects of an alternative not originally selected, as long as the impacts have been analyzed and no significant new information has arisen, supplemental NEPA analysis is not required.

Although it may take artful navigation of legal requirements, the case law indicates that adaptive management is not entirely incompatible with the framework of administrative decision-making and environmental law. Even where clear substantive standards are relevant, adaptive management can survive judicial review, but only when mechanisms are built into a plan that require clear and meaningful actions that are triggered when specific conditions are met. Several large scale plans, including the Northwest Forest Plan, species management on the Sacramento and San Joaquin Rivers, and flood control on the Missouri River, have all seen several rounds of litigation regarding their approaches to adaptive management (Ruhl and Fischman 2010). These types of plans may be best suited to an adaptive management approach because they involve ongoing decisions with iterative monitoring and often require further NEPA analysis that is tiered to a programmatic plan.

In summary, to satisfy legal requirements: (1) agencies must show that they will meet substantive standards; (2) if agencies acknowledge uncertainty, they must show that they have a clear monitoring and mitigation strategy that is within their power to implement if unexpected or unacceptable effects are detected; (3) tiering can be an appropriate tool for pursuing adaptive management while complying with NEPA; and (4) courts do not always require additional NEPA analysis when new information comes to light, as long as any changes in action and predicted effects are within the range of what was analyzed in the original NEPAdocument.

Legal Enforceability of Commitments in Adaptive Management Plans

Although adaptive management can be written into a plan and survive legal review, a critical question is whether monitoring and adaptive planning commitments are enforceable under the parameters of administrative law once an adaptive management plan is underway. This is an important issue, given concerns that adaptive management in name can be used as a vehicle by agencies to pursue open-ended and discretionary decision-making. What if promises made to conduct monitoring or undertake mitigation are not kept? When are they enforceable?

In the NEPA context, mitigation measures, which may be promised as part of an adaptive management strategy, are not necessarily legally binding. Agencies are not required under NEPA to implement mitigation measures that are discussed in an EIS. Mitigation measures are scrutinized more closely when agencies make mitigation promises as a way to justify a FONSI in lieu of preparing an EIS. However, even in these cases, courts have not "required absolute certainty or any binding legal commitment to mitigation measures" (*Robertson v. Methow Valley Citizens Council* 1989). The general judicial trend is to require a "moderately high level of assurance" that mitigation measures will be performed, with the recognition that funding for monitoring and mitigation often must materialize after the decision point has passed (Owen 2009/2010).

A number of courts have held that NEPA does not give rise to a "private right of action" to enforce promises made in EISs (McGarity 1990). In some cases, courts have acknowledged that commitments in a decision are legally binding, but generally in cases where agencies issued a FONSI. It may be challenging, in either case, to bring a claim that an agency has not fulfilled commitments in a decision document. If there is no remaining federal action, courts may not intervene to require compliance with a record of decision for an action that has been completed (McGarity 1990). The Council on Environmental Quality, which interprets NEPA, explains, in cases where mitigation measures have not taken place, "if there is Federal action remaining, it is appropriate for agencies to consider preparing supplemental NEPA analysis and documentation and to pursue remaining opportunities to address the effects of that remaining action" (Council on Environmental Quality 2010). If there is federal action remaining, NEPA sometimes requires supplementation where the assumptions or commitments in an EA or EIS and the associated decision document are no longer valid; still, this is different than requiring that agencies do what they said they were going to do. Nonetheless, if NEPA supplementation is triggered when an agency fails to conduct promised mitigation, this could potentially stop further action until the agency has completed the supplemental analysis. Although this may lend adaptive management plans some accountability, some scholars point to NEPA's supplementation requirements as a real and potential obstacle to practicing adaptive management (Ruhl 2008, Benson 2009). Agencies practicing monitoring and information-intensive adaptive management could find that new information repeatedly triggers additional NEPA analysis, which is not cheap or quick. However, in theory supplementing NEPA analysis when significant new information arises could be an appropriate vehicle for meshing adaptive management and NEPA.

This issue of supplementing NEPA analysis and revising plans based on new information is particularly complex in the context of land use planning. The Supreme Court ruled in *Norton v. Southern Utah Wilderness Alliance* (2004) (*SUWA)* that a land use plan is not an "ongoing" major federal action requiring supplementation. In this case, the Bureau of Land Management did not have to write a supplemental EIS due to increased off-road vehicle use in the planning area. Several district courts have followed *SUWA* and ruled that there is no ongoing action requiring NEPA supplementation once an agency approves a land use plan or issues a license, even if the assumptions in the plan are no longer valid (Blumm and Bosse 2007). In these cases, new information came to light, such as an Endangered Species Act listing or evidence that protective wildlife measures were not working as predicted, but still the courts did not require a supplemental EIS to be prepared. One review of post-*SUWA* case law summarizes that "federal agencies have experienced considerable, if not universal, success in arguing that they have no obligation to supplement their NEPA analysis after SUWA," particularly when it comes to decisions in land-use plans (Blumm and Bosse 2007).

Also as a result of *SUWA*, agency commitments to monitor are especially suspect when they are made in a land use plan. The Supreme Court ruled that the Bureau of Land Management's commitment to monitor off-road vehicle use—"like other 'will do' projections of agency action set forth in land use plans—are not a legally binding commitment enforceable under [the Administrative Procedures Act]," because a broad commitment to monitor is not a discrete action reviewable under the Administrative Procedures Act. The result is that discretionary processes such as the implementation of monitoring and subsequent mitigation are not generally justiciable when they are written into land use plans. However, the Court acknowledged that monitoring commitments *could* be written in a way that is enforceable if the action was written as a clear and binding commitment. If commitments in plans are written in ways such that monitoring is required *before* an action can be taken, this is still actionable under the Administrative Procedures Act. For example, Survey and Manage requirements under the Northwest Forest Plan required some species to be surveyed prior to ground disturbing activities. A failure to comply with such guidelines would be reviewable in court. Likewise, environmental groups have successfully challenged the Bureau of Land Management in court when it approved grazing leases without monitoring resource conditions, when the land use plan explicitly stated that the monitoring would occur prior to the authorization of grazing (Blumm and Bosse 2007).

Even outside the context of land use planning, the courts are often reluctant to force agencies to conduct monitoring. Biber explains that there are three primary reasons for this: "an agency monitoring program is neither a 'final' nor specific agency 'action' that a court can review or mandate under the [Administrative Procedures Act]; the level of compliance by an agency with a mandatory duty is not for the court to review, as long as at least some compliance exists; or, the apparently mandatory language in the statute, regulation, or plan is in fact only hortatory" (Biber 2011). As was the case with *SUWA*, courts will make a distinction between the reviewability of discrete agency actions and ongoing agency operations or conduct, with which they are reluctant to interfere. Courts are also unlikely to review the quality and extent of monitoring taking place, as long as some monitoring is occurring. For these reasons, and because intermittent court decisions are unlikely to lead to an effective ongoing monitoring program: "courts are more willing to step in when a monitoring duty can be framed as a precondition to the agency being able to pursue some other activity that it seeks to accomplish (such as a timber sale or road construction)" (Biber 2011).

The lesson is that monitoring and mitigation commitments made as part of an adaptive management framework can be made enforceable, and in some cases, with the cases involving Biological Opinions for fish species on the Sacramento and San Joaquin river systems, they must be made enforceable for an adaptive management plan to survive a legal challenge. Generally speaking, in order to be enforceable, plans must include specific monitoring requirements and timelines tied, through the use of explicit trigger points, to clear mitigation requirements, along with specific implementation timelines. When such a monitoring/mitigation program is part of a legally binding agreement, such as in the case of a permit issued under the Endangered Species Act, enforcement is more possible, especially where monitoring serves as a precondition for renewal. If monitoring is written into a land-use plan or project level decision in a way that it serves as a precondition for future actions, this can also be legally enforceable. Furthermore, if such a program served as the basis for a FONSI and was not implemented, NEPA supplementation could be triggered. In other cases, there may be a requirement for supplementation under NEPA if commitments in a record of decision are not kept. Other statutes with clear legal standards may provide additional vehicles for challenges to a promised monitoring/mitigation program that is either not succeeding or not occurring at all. However, enforceability within the parameters of administrative law is a significant challenge and one that requires concerted attention to the details of the adaptive management strategy and the legal context within which commitments are made.

Conclusion

Adaptive management holds great promise as an approach to complex social-ecological challenges that involve high degrees of uncertainty. When placed within the context of already well developed legal systems and institutions, the challenge becomes how to best take advantage of the strategies and practices adaptive management has to offer while also complying with existing laws and requirements. There have been several suggestions for explicit congressional action to better facilitate adaptive management. One major recommendation is to provide an adequate and constant source of funding of adaptive management via annuities or some other method. In making this suggestion, Ruhl and Fischman (2010) observe that "[I]n the absence of congressional action, agencies should at least use NEPA to disclose funding needs for adaptive management and the environmental effects that would result from failure to find the means for implementation of monitoring, mitigation, or adjustment." This is in line with proposals that recommend NEPA as a "regulatory home" for adaptive management, which, among other advantages, would encourage more uniform implementation of adaptive management across federal agencies (Benson and Garmestani 2011).

In addition to funding resources, Ruhl and Fischman (2010) provide three other recommendations for Congress, in addition to reforming the appropriations process:

> Congress could substantially improve the practice of adaptive management in natural resource administration. It is possible to establish clearer standards to ensure that an agency purporting to employ adaptive management actually does an adequate job. Congress should explicitly require adaptive management plans to (1) clearly articulate measurable goals, (2) identify testable hypotheses (or some other method of structured learning from conceptual models), and (3) state exactly what criteria should apply in evaluating the management experiments. These requirements would address the vast majority of non-budgetary problems with a/m-lite.

These recommendations highlight the need to provide more explicit guidance for agencies, while also allowing adaptive management to be tailored to specific contexts. A recent Congressional Research Service report entitled *Adaptive Management for Ecosystem Restoration: Analysis and Issues for Congress* (Stern et al. 2011) also provided a number of recommendations for congressional action that echo those made above. These include: (1) designation of a federal representative or agency to be in charge of implementing an adaptive management program; (2) assignment of specific groups or numbers of stakeholders to committees to oversee and recommend changes to adaptive management efforts; and (3) congressionally specified procedures for carrying out adaptive management, "including how the results from adaptive management research and monitoring are to be tied to operational or project-based changes" (Stern et al. 2011). All of these suggestions are compatible with proposals from those who argue that Congress should enact a National Environmental Legacy Act that would be reflective of resilience principles and provide an overarching framework for the administration of adaptive management (Flournoy 2008).

Given the unlikelihood of congressional action, however, the tension in adaptive management implementation between the need for administrative flexibility and accountability will continue. In the absence of legislation, more explicit and enforceable regulatory provisions—that provide public notice and an opportunity to comment—would be a significant step towards establishing the legal context of adaptive management procedures and protocols. The challenge would be to design such regulations in a way that successfully avoids a "one size fits all" approach and allows agencies to adjust their management actions to the specific task at hand.

In the absence of some more specific regulatory grounding, continued use of informal agency guidance will leave adaptive management as more of an implementation tool than a management approach. Courts will continue to look to the underlying legislative mandates and assess whether adaptive management strategies are sufficiently rigorous and detailed to achieve legal compliance. This is perhaps the most likely outcome with respect to the relationship between adaptive management and law, and also the most disappointing. In order for adaptive management to reach its potential to transform environmental and natural resource management, a more substantial integration of adaptive management principles into legal and institutional requirements is necessary.

References

Allan, C., Curtis, A., Stankey, G., & Shindler, B. (2008). Adaptive management and watersheds: A social science perspective. *Journal of the American Water Resources Association, 44,* 166–174.

Allen, C., Fontaine, J., Pope, K., & Garmestani, A. (2011). Adaptive management for a turbulent future. *Journal of Environmental Management, 92,* 1339–1345.

Angelo, M. J. (2009). Stumbling toward success: A story of adaptive management law and ecological resilience. *Nebraska Law Review, 87,* 951–952.

Babbitt v. Sweet Home Chapter of Communities for a Great Oregon. (1995). 115 U.S. 2407.

Benson, M. H. (2009). Integrating adaptive management and oil and gas development: Existing obstacles and opportunities for reform. *Environmental Law Reporter, 39,* 10962–10978.

Benson, M. H. (2010). Adaptive management approaches by resource management agencies in the United States: Implications for energy development in the interior West. *Journal of Energy and Natural Resources Law, 28,* 87–118.

Benson, M. H., & Garmestani, A. S. (2011). Embracing panarchy, building resilience and integrating adaptive management through a rebirth of the National Environmental Policy Act. *Journal of Environmental Management, 92*, 1420–1427.

Benson, M. H., & Stone, A. B. (2013). Practitioner perceptions of adaptive management implementation in the United States. *Ecology and Society, 18*(3), 32. http://dx.doi.org/10.5751/ES-05613-180332.

Berkes, F., & Seixas, C. (2005). Building resilience in lagoon social-ecological systems: A local-level perspective. *Ecosystems, 8,* 967–974.

Biber, E. (2011). The problem of environmental monitoring. *University of Colorado Law Review, 83,* 66–79.

Blumm, M. C., & Bosse, S. L. (2007). Norton v. SUWA and the unraveling of federal public land planning. *Duke Environmental Law and Policy Forum, 18,* 105–160.

Bodin, Ã. R., Crona, B., & Ernstson, H. (2006). Social networks in natural resource management: What is there to learn from a structural perspective? *Ecology and Society, 11*(2), r2. http://www.ecologyandsociety.org/vol11/iss2/resp2/.

Brugnach, M., Dewulf, A., Pahl-Wostl, C., & Taillieu, T. (2008). Toward a relational concept of uncertainty: About knowing too little, knowing too differently, and accepting not to know. *Ecology and Society, 13*(2), 30. http://www.ecologyandsociety.org/issues/view.php?id=71#Research.

Butler, K., & Koontz, T. (2005). Theory into practice: Implementing ecosystem management objectives in the USDA Forest Service. *Environmental Management, 35,* 138–150.

Center for Biological Diversity v. Rumsfeld. (2002). 198 F. Supp. 2d. (D. Arizona).

Council on Environmental Quality (CEQ). (2010). Final guidance for federal departments and agencies on the appropriate use of mitigation and monitoring and clarifying the appropriate use of mitigated findings of no significant impacts. CEQ, Washington, D.C.

Craig, R. K. (2010). Stationarity is dead: Long live transformation: Five principles for climate change adaptation law. *Harvard Environmental Law Review, 31,* 9–75.

Doremus, H. A. (2002). Adaptive management, the Endangered Species Act, and the institutional challenges of "new age" environmental protection. *Washburn Law Journal, 41,* 50–89.

Fischman, R. L. (2007). From words to action: The impact and legal status of the 2006 National Refuge System Management Policies. *Stanford Environmental Law Journal, 26,* 77–135.

Flournoy, A. C. (2008). Protecting a natural resource legacy while promoting resilience: Can it be done? *Nebraska Law Review, 87,* 1009–1035.

Folke, C., Hahn, T., Olsson, P., & Norberg, J. (2005). Adaptive governance of social-ecological systems. *Annual Review of Environment and Resources, 30,* 441–473.

Folke, C., Pritchard, L., Berkes, F., Colding, J., & Svedin, U. (2007). The problem of fit between ecosystems and institutions: Ten years later. *Ecology and Society, 12*(1), 30. http://www.ecologyandsociety.org/issues/view.php?id=67#Insight.

Greater Yellowstone Coalition v. Kempthorne. (2008). 557 F. Supp. 2d. 183 (D. D.C.).

Greater Yellowstone Coalition v. Servheen. (2009). 672 F. Supp, 2d, 1105 (D. Montana).

Grumbine, E. (1994). What is ecosystem management? *Conservation Biology, 8,* 27–38.

Gunderson, L., & Light, S. S. (2006). Adaptive management and adaptive governance in the everglades ecosystem. *Policy Sciences, 39,* 323–334.

Gunderson, L., & Holling, C. S. (Eds.). (2002). *Panarchy: Understanding transformations in human and natural systems*. Washington, D.C.: Island Press.

Holling, C. S. (Ed.). (1978). *Adaptive environmental assessment and management.* Chichester: Wiley.

Houck, O. A. (2009). Nature or nurture: What's wrong and what's right with adaptive management. *Environmental Law Reporter, 39,* 10923–10924.

Jacobson, S. K., Morris, J. K., Sanders, J. S., Wiley, E. N., Brooks, M., Bennetts, R. E., Percival, H. F., & Marynowski, S. (2006). Understanding barriers to implementation of an adaptive land management program. *Conservation Biology, 20,* 1516–1527.

Karkkainen, B. C. (2005). Panarchy and adaptive change: Around the loop and back again. *Minnesota Journal of Law, Science and Technology, 7,* 59–77.

King, J., & Brown, C. 2006. Environmental flows: Striking the balance between development and resource protection. *Ecology and Society, 11*(2): 26 http://www.ecologyandsociety.org/issues/view.php?id=66#Research.

Klamath Siskiyou Wildlands Center v. Boody. (2006). 468 F.3d 549, 553 (9th Circuit).

Lee, K. N. (1999). Appraising adaptive management. *Ecology and Society, 3*(2), 3. http://www.ecologyandsociety.org/issues/view.php?id=6#Perspective.

McCarthy, M. A., & Possingham, H. P. (2007). Active adaptive management for conservation. *Conservation Biology, 21,* 956–963.

McGarity, T. O. (1990). Judicial enforcement of NEPA-inspired promises. *Environmental Law, 20,* 569–609.

Natural Resources Defense Council v. Kempthorne. (2007). 506 F. 2d. 322 (E.D. California).

Norton v. Southern Utah Wilderness Alliance. (2004). 542 U.S. 55.

In re: Operation of the Missouri River System Litigation. (2008). 516 F.3d 688 (8th Circuit).

Oregon Natural Resources Council Action v. USFS. (1999). 59 F. Supp. 2d 1085 (W.D. Washington).

Owen, D. (2009/2010). Probabilities, planning failures, and environmental law. *Tulane Law Review, 84,* 265–335.

Pacific Coast Federation of Fisherman's Associations v. Gutierrez. (2008). 606 F. 2d. 1122 (E.D. California).

Pahl-Wostl, C. (2007). The implications of complexity for integrated resources management. *Environmental Modelling and Software, 22,* 561–569.

Pahl-Wostl, C., Sendzimir, J., Jeffrey, P., Aerts, J., Berkamp, G., & Cross, K. (2007). Managing change toward adaptive water management through social learning. *Ecology and Society, 12*(2), 30. http://www.ecologyandsociety.org/issues/view.php?id=68#Research.

Raadgever, G. T., Mostert, E., Kranz, N., Interwies, E., & Timmerman, J. G. (2008). Assessing management regimes in transboundary river basins: Do they support adaptive management? *Ecology and Society, 13*(1), 14. http://www.ecologyandsociety.org/issues/view.php?id=69#Research.

Robertson v. Methow Valley Citizens Council. (1989). 490 U.S. 322.

Ruhl, J. B. (2005). Regulation by adaptive management - is it possible? *Minnesota Journal of Law, Science and Technology, 7,* 21–57.

Ruhl, J. B. (2008). Adaptive management for natural resources-inevitable, impossible, or both? *Rocky Mountain Law Institute, 54,* 1–6.

Ruhl, J. B., & Fischman, R. L. (2010). Adaptive management in the courts. *Minnesota Law Review, 95,* 424–484.

Salafsky, N., Margoluis, R., & Redford, K. (2001). Adaptive management: A tool for conservation practitioners. Biodiversity Support Program, Washington, D.C.

Schultz, C. (2008). Dealing with scientific uncertainty in forest policy and planning. *Environmental Science and Policy, 11,* 253–271.

Schultz, C., & Nie, M. (2012). Decision making triggers, adaptive management, and natural resources policy and planning. *Natural Resources Journal, 52,* 443–521.

Susskind, L. E., Camacho, A. E., & Schenk, T. (2010). Collaborative planning and adaptive management in Glen Canyon: A cautionary tale. *Columbia Journal of Environmental Law, 35,* 2–53.

Stern, C. V., Sheikh, P. A., & Brass, C. T. (2011). Adaptive management for ecosystem restoration: Analysis and issues for congress. Congressional Research Service, Washington, D.C.

Stringer, L. C., Dougill, A. J., Fraser, E., Hubacek, K., Prell, C., & Reed, M. S. (2006). Unpacking "participation" in the adaptive management of social-ecological systems: A critical review. *Ecology and Society, 11*(2), 39. http://www.ecologyandsociety.org/issues/view.php?id=66#Research.

Thrower, J. (2006). Adaptive management and NEPA: How a nonequilibrium view of ecosystems mandates flexible regulation. *Ecology Law Quarterly, 33,* 871–895.

U.S. Army Corps. (2002). Guidance on compensatory mitigation projects for aquatic resource impacts under the Corps Regulatory Program pursuant to Sect. 404 of the Clean Water Act and Sect. 10 of the Rivers and Harbors Act of 1899. http://www.usace.army.mil/CECW/Documents/cecwo/reg/rgls/RGL2-02.pdf.

Walters, C. J. (2002). *Adaptive management of renewable resources.* Caldwell: The Blackburn Press.

Walters, C. J., & Holling, C. S. (1990). Large-scale management experiments and learning by doing. *Ecology, 71,* 2060.

Williams, B. K., & Brown, E. D. (2012). Adaptive management: The U.S. Department of the Interior applications guide. Adaptive Management Working Group, U.S. Department of the Interior, Washington, D.C.

Williams, B. K., Szaro, R. C., & Shapiro, C. D. (2009). Adaptive management: The U.S. Department of the Interior technical guide adaptive management working group, U.S. Department of the Interior, Washington, D.C.

Zellmer, S., & Gunderson, L. (2009). Why resilience may not always be a good thing: Lessons in ecosystem restoration from Glen Canyon and the Everglades. *Nebraska Law Review, 87,* 894–947.

Chapter 5
A Decision-Analytic Approach to Adaptive Resource Management

Fred A. Johnson and Byron K. Williams

Keywords Adaptive management · Decision analysis · Uncertainty · Resilience thinking · Modeling

Introduction

Judging by the number of references in planning documents and professional journals over the last three decades, adaptive management has been an incredible success story. Its core principles of managing to learn and learning to manage are so self-evident that virtually all resource professionals today view adaptive management as a creed of sensible resource management. Yet there are legitimate concerns that adaptive management is becoming little more than a slogan, regarded by cynics as devoid of any real meaning. It probably hasn't helped that much of the recent literature can make adaptive management appear as a bewildering edifice encompassing ecology, sociology, economics, institutional theory, governance, and complexity science, rather than as an apprehensible approach to decision making under uncertainty. Consequently, there are far more adaptive management plans than there are cases of implementation (McFadden et al. 2011). Documented success stories are rare, and touted examples often fail to meet one or more basic requirements of adaptive management (Moir and Block 2001, Schreiber et al. 2004, Susskind et al. 2010). The barriers to implementation, learning, and adaptation have become legendary (McLain and Lee 1996, Walters 1997, Lee 1999, Allan and Curtis 2005, Gregory et al. 2006, Walters 2007, Allen and Gunderson 2011), and there are serious concerns about the applicability of adaptive management to "wicked problems" in resource conservation (Gunderson et al. 2008). On the whole, Kai Lee's (1999)

F. A. Johnson (✉)
U.S. Geological Survey, Southeast Ecological Science Center,
7920 NW 71 Street, Gainesville, FL 32653, USA
e-mail: fjohnson@usgs.gov

B. K. Williams
The Wildlife Society, 5410 Grosvenor Lane, Suite 200, Bethesda, MD 20814-2144, USA
e-mail: ken.williams@wildlife.org

C. R. Allen, A. S. Garmestani (eds.), *Adaptive Management of Social-Ecological Systems*,
DOI 10.1007/978-94-017-9682-8_5

observation that "adaptive management has been more influential, so far, as an idea than as a practical means of gaining insight into the behavior of ecosystems utilized and inhabited by humans" is as relevant today as it was over 10 years ago.

There are, however, hopeful signs that adaptive management may yet live up to its promise. We suggest that the basic concept of adaptive management is helping change the culture of resource management and, thus, is having an impact far broader than any improvements in resource conditions that may or may not have been achieved in particular applications. The popularity of the concept of adaptive management has made it acceptable (even fashionable) to acknowledge uncertainty and its consequences for resource management (Doremus 2007, Ruhl and Fischman 2010). The limits of reductionist science, and the inefficacy of the traditional relationship between research and management, have been broadly recognized (Costanza 1993, Hilborn and Ludwig 1993, Mooney and Sala 1993). The opportunities to learn through management are being explored if not yet routinely exploited, and a greater focus on uncertainty is producing a revolution in the application of decision analysis to inform conservation planning (Burgman 2005, Conroy and Carroll 2009, Moilanen et al. 2009). Having been directly involved in resource management throughout most of our careers, we are all too familiar with the way in which resource management decisions are often made, i.e., using intuition as a substitute for systematic analysis focused on problem formulation, management objectives, alternative actions, explicit predictions, and key sources of uncertainty. Obviously, smart decisions don't always produce good outcomes, but they should produce better outcomes on average than intuition, especially when problems are complex or there is conflict among decision-makers. If nothing else, the application of decision analysis to resource conservation is an effective mechanism for providing the transparency and accountability that is increasingly demanded by stakeholders and the courts (Ruhl 2006, Keene and Pullin 2011).

If the promise of adaptive management is yet to be fully realized, we nevertheless believe that the concept is improving the process by which resource managers make decisions. This improvement is manifest in an increasing number of examples of applied decision analysis (Howard 1968, Raiffa 1968) in conservation, and so we discuss here how recent advances in decision theory have been made more relevant to the management of renewable natural resources. The explicitness demanded by decision analysis, particularly as it concerns the specification of key uncertainties, has been a natural entrée into thinking about when an adaptive approach to management might be needed. We also believe that more bona fide examples of adaptive management would be forthcoming if decision makers relied on the principles of decision analysis to assess the value of reducing uncertainty and the capacity for learning. We suggest that impediments to implementation of adaptive management are more easily overcome when an assessment of the value of information is used to determine the potential effectiveness of an adaptive management program. Finally, we briefly explore the notion that "the era of management is over" (Ludwig 2001), whereby the application of decision analysis to complex natural-resource problems has been questioned on principle (Holling and Meffe 1996, Linkov et al. 2006, Walker and Salt 2006).

Application of Decision Analysis

Decision analysis has been widely used in business and government decision making (Keefer et al. 2004), but its application to problems in natural resource management has mostly been a phenomenon of the last two decades (Huang et al. 2011). Though decision-analytic approaches vary considerably, environmental decision making typically involves (1) properly formulating the decision problem; (2) specifying feasible alternative actions; and (3) selecting criteria for evaluating potential outcomes (Tonn et al. 2000). Traditional approaches to decision making, which tend to focus on alternatives and predicted outcomes, can be distinguished from modern methods by an emphasis on the fundamental values and the multiple-objective tradeoffs inherent in natural resource management (Arvai 2001). The emphasis on values rather than outcomes helps decision makers understand whether disagreements are over predicted outcomes or how those outcomes are valued (Lee 1993), and it clarifies the role for analysts and scientists in resource decision making as "honest brokers" rather than as advocates (Pielke 2007). Multi-criteria decision analysis that accounts for outcomes and values is now widely used in natural resource management, and is seen as contributing to better decisions through a formal structuring of decision problems that accommodates conflicts in fundamental values among stakeholders (Hajkowicz and Collins 2007, Huang et al. 2011).

A noteworthy aspect of the trend toward formal decision analysis in natural resource management has been the increasing application of dynamic optimization methods to analyze recurrent decisions (Walters and Hilborn 1978, Williams 1989, Possingham 1997) (Table 5.1). Recurrent decision problems are ubiquitous in conservation, ranging from obvious examples like harvesting or prescribed burning, to less obvious ones like development of a biological reserve system or the control of invasive plants and animals. The growing number of resource-management examples that rely on dynamic optimization methods is testament to the general applicability of these methods, and the rapid increase in computing power has made it feasible to analyze problems of at least moderate complexity.

Dynamic optimization methods, with their focus on recurrent decisions and the uncertainties attendant to future outcomes, are particularly well suited for formulating adaptive management strategies. They combine models of ecological system change with objective functions that value present and future consequences of alternative management actions. The general resource management problem involves a temporal sequence of decisions, where the optimal action at each decision point depends on time and/or system state (Possingham 1997). The goal of the manager is to develop a decision rule (or management policy) that prescribes management actions for each time and system state that are optimal with respect to the objective function. Under the assumption of Markovian system transitions, the optimal management policy satisfies the Principle of Optimality (Bellman 1957), which states that:

> An optimal policy has the property that, whatever the initial state and decision are, the remaining decisions must constitute an optimal policy with regard to the state resulting from the first decision.

Table 5.1 Examples of the problems in natural resource management addressed with dynamic optimization methods

Resource problem	Goal	Source
Harvesting	Sustainable use	Milner-Gulland (1997), Johnson et al. (1997), Kulmala et al. (2008)
Translocation	Endangered species persistence	Tenhumberg et al. (2004)
Pest management	Control	Sells (1995), Bogich and Shea (2008)
Management of human disturbance	Endangered species occupancy	Martin et al. (2011)
Fire management	Biodiversity conservation	Richards et al. (1999), McCarthy et al. (2001)
	Endangered species persistence	Johnson et al. (2011)
Forest management	Endangered species persistence	Moore and Conroy (2006)
Reservoir management	Water supply	Alaya et al. (2003), Eum et al. (2011)
Landscape reconstruction	Endangered species persistence	Westphal et al. (2003)
Allocation of conservation resources	Biodiversity conservation	Wilson et al. (2006)

A key advantage of dynamic optimization is its ability to produce a feedback policy specifying optimal decisions for possible future system states rather than expected future states (Walters and Hilborn 1978). In practice this makes optimization appropriate for systems that behave stochastically, absent any assumptions about the system remaining in a desired equilibrium or about the production of a constant stream of resource returns. The analysis of recurrent decision problems with dynamic optimization methods also allows for the specification of the relative value of current and future management returns through discount rates. By properly framing problems, dynamic optimization methods have been used successfully to address a broad array of important conservation issues. It seems clear from the wide applicability of these methods that it is not optimization per se that leads to unsustainable policies as some authors seem to suggest (e.g., Walker and Salt 2006), but rather the use of outdated methods that assume the existence of equilibrium in resource state or use, and the tendency to heavily discount the future.

A framework for dynamic optimization requires specification of (1) an objective function for evaluating alternative management policies; (2) predictive models of system dynamics that are formulated in quantities relevant to the stated management objectives; (3) a finite set of alternative management actions, including any constraints on their use; and (4) a monitoring program to follow the system's evolution and responses to management. More formally, let:

$$\underline{x}_{t+1} = \underline{x}_t + f(\underline{x}_t, a_t, \underline{z}_t) \tag{1}$$

characterize system dynamics, where $\underline{x}_t$ represents the system state at time t and a_t and $\underline{z}_t$ represent management actions and environmental variation respectively. Random demographic and environmental variation induces Markovian transition probabilities $p(\underline{x}_{t+1}|\underline{x}_t, a_t)$. Let policy A_t specify an action for every system state $\underline{x}_t$ at every time in the time frame $\{t, t+1, \ldots T\}$. Benefits and costs attend management actions, which are included in returns $R(a_t|\underline{x}_t)$ that in turn are accumulated in an objective or value function:

$$V(A_t|\underline{x}_t) = E\left\{\sum_{\tau=t}^{T} \alpha^{\tau-1} R(a_\tau|\underline{x}_\tau)|\underline{x}_t\right\}, \tag{2}$$

where the expectation is with respect to stochastic influences on the process and α discounts future returns. This function can be decomposed into current returns and future values by:

$$V(A_t|\underline{x}_t) = R(a_t|\underline{x}_t) + \alpha \sum_{\underline{x}_{t+1}} p(\underline{x}_{t+1}|\underline{x}_t, a_t) V(A_{t+1}|\underline{x}_{t+1}), \tag{3}$$

which makes clear that future values are conditioned on the effect of current actions on future states. A value $V(A_t|\underline{x}_t)$ can be obtained for every possible policy A_t over the time frame, and the optimal policy satisfies:

$$V^*(\underline{x}_t) = \max_{a_t} \left\{R(a_t|\underline{x}_t) + \alpha \sum_{\underline{x}_{t+1}} p\left(\underline{x}_{t+1}|\underline{x}_t, a_t\right) V^*\left(\underline{x}_{t+1}\right)\right\} \tag{4}$$

(Bellman 1957, Puterman 1994, Bertsekas 1995). The discount factor α in Eqs. 3 and 4 highlights the influence of myopia in decision making, i.e., the effect of discounting the future relative to the present.

A key consideration in dynamic optimization of natural resource problems is the uncertainty attendant to management outcomes, which adds to the demographic and environmental variation of stochastic resource changes. This uncertainty may stem from errors in measurement and sampling of ecological systems (partial system observability), incomplete control of management actions, and incomplete knowledge of system behavior (structural uncertainty) (Williams et al. 1996). A failure to recognize and account for these uncertainties can significantly depress management performance and in some cases can lead to severe environmental and economic losses (Ludwig et al. 1993). In recent years there has been an increasing emphasis on methods that can account for uncertainty about the dynamics of ecological systems and their responses to both controlled and uncontrolled factors (Walters 1986, Williams 2001).

Model uncertainty, an issue of special importance in adaptive management, can be characterized by continuous or discrete probability distributions of model parameters, or by discrete distributions of alternative model forms that are hypothesized or estimated from historic data (e.g., Walters and Hilborn 1978, Johnson et al. 1997). Important advances have followed from the recognition that these probability distributions are not static, but evolve over time as new observations of system behaviors are accumulated from the management process. Indeed, the defining characteristic of adaptive management is the attempt to account for the temporal dynamics of this uncertainty in making management decisions (Walters 1986, Williams 2001, Allen et al. 2011).

It thus appears adaptive management and decision analysis have been self-reinforcing concepts, with one driving advances in the other and both playing increasingly important roles in how resource managers approach decision making. However, we suggest here that adaptive management efforts can fail if an adaptive approach is advocated before there is a careful, systematic analysis of the decision problem. Not all problems are suitable for adaptive management (Gregory et al. 2006, Allen and Gunderson 2011), and the deliberate structuring of a problem based on the principles of decision analysis can help discern good from bad candidates. For example, if the primary impediment to decision making is a conflict of values, adaptive management may have little to offer. Its use under these circumstances can become little more than displacement behavior that avoids the difficult challenges of developing more effective institutional and governance structures to resolve disputes (Susskind et al. 2010). Nor is an adaptive approach needed if management choices are insensitive to structural sources of uncertainty (although even here dynamic optimization may be useful). Finally, a failure of management choices to discriminate among competing system models means that adaptive management will not result in learning, an essential element of adaptive decision making.

Decision analysis provides a systematic framework for exploring these issues, and it is difficult to imagine how adaptive management could be planned or implemented absent this structure. In this light, perhaps it is not surprising that the clarion call for adaptive resource management over 30 years ago was followed more by an expanding use of decision analysis than by bona fide examples of system probing and management experiments. Fortunately, that focus has taught us much about how best to proceed with management problems that are characterized by various sources and degrees of uncertainty. In turn, a focus on learning has spurred innovations in decision analysis that make it increasingly useful for real-world problems in natural resource management.

Advances in Decision Analysis

Walters (1986) recognized the potential of dynamic optimization to identify strategies that account for uncertainty in system dynamics. His insight provided a framework by which managers could effectively attack the "dual-control problem," in which management for short-term objective values must somehow be balanced with the learning necessary to improve future returns. At the time, the additional dimensionality introduced by the need to track structural uncertainty along with system state made almost all realistic problems computationally intractable. However, over the last three decades the development of efficient computing algorithms and increases in computer speed and memory have dramatically expanded the class of feasible resource problems.

Computing Algorithms

Many important advances have followed from acknowledgment that the process controlling state transitions is uncertain, and that uncertainty can be incorporated directly into decision making. Here we express uncertainty with alternative system models that are characterized by parameter β:

$$\underline{x}_{t+1} = \underline{x}_t + f_\beta\left(\underline{x}_t, a_t, \underline{z}_t\right), \tag{5}$$

with random demographic and environmental variation inducing model-specific Markovian transition probabilities: $p_\beta(\underline{x}_{t+1}|\underline{x}_t, a_t)$. Policy value is again given in terms of accumulated returns, except in this case the returns are averaged over alternative models:

$$\begin{aligned} V(A_t|\underline{x}_t, q_t) &= \sum_\beta q_t(\beta) E\left\{\sum_{\tau=t}^{T} \alpha^{\tau-1} R_\beta(a_\tau|\underline{x}_\tau)|\underline{x}_t\right\} \\ &= \sum_\beta q_t(\beta) V_\beta(A_t|\underline{x}_t). \end{aligned} \tag{6}$$

The distribution q_t represents model-specific probabilities that evolve through time according to Bayes theorem:

$$q_{t+1}(\beta) = \frac{q_t(\beta)\, p_\beta\left(\underline{x}_{t+1}|\underline{x}_t, a_t\right)}{\overline{p}\left(\underline{x}_{t+1}|\underline{x}_t, a_t, q_t\right)}, \tag{7}$$

with

$$\overline{p}(\underline{x}_{t+1}|\underline{x}_t, a_t, q_t) = \sum_\beta q_t(\beta) p_\beta(\underline{x}_{t+1}|\underline{x}_t, a_t). \tag{8}$$

The optimal policy satisfies:

$$V^*(\underline{x}_t, q_t) = \max_{a_t}\left\{R(a_t|\underline{x}_t, q_t) + \alpha \sum_{\underline{x}_{t+1}} \overline{p}\left(\underline{x}_{t+1}|\underline{x}_t, q_t\right) V^*\left(\underline{x}_{t+1}, q_{t+1}\right)\right\}, \tag{9}$$

with

$$R\left(a_t|\underline{x}_t, q_t\right) = \sum_\beta q_t(\beta) R_\beta(a_t|\underline{x}_t) \tag{10}$$

and

$$V^*\left(\underline{x}_{t+1}, q_{t+1}\right) = \max_{a_{t+1}} \sum_\beta q_{t+1}(\beta) V_\beta\left(\underline{x}_{t+1}|\underline{x}_t, a_{t+1}\right) \tag{11}$$

(Williams 2009). Note that current returns in Eq. 9 are averaged using the prior probabilities of the alternative models (Eq. 10), whereas future returns are weighted by the posterior probabilities (Eq. 11). A management action is chosen at each point in time depending on resource state and the parameter state q_t, and the action in turn influences future resource state as well as changes in the parameter state. Optimal management consists of actions that maximize objective returns, not learning per se, with model discrimination (i.e., learning) pursued only to the extent that it increases long-term returns. This approach can be described as active adaptive management, in that it explicitly accounts for the effect of management actions on learning and the effect of that learning on future returns. We characterize this form of adaptive management as "learning while doing," as opposed to an experimental approach to adaptive management (Walters and Holling 1990, Walters and Green 1997) that might be characterized as "learn, then do."

The calculation of optimal adaptive management policies can impose large demands on computing resources. One way to relieve those demands is to use passive adaptive optimization, an approach that accounts for structural uncertainty while eliminating the need to carry distribution q_t in the optimization algorithm. In this case, at each time q_t is treated as fixed and is used in both terms of Eq. 9. As with actively adaptive optimization, q_t is updated periodically using a comparison of observed and model-predicted system responses, but in passive adaptive management that updating is not anticipated in the optimization algorithm (Williams et al. 2002). This approach is quite different from the more prominent description of passive adaptive management in the literature, in which management actions are derived using the single best model and then management experience is used to revise or replace it (Walters and Hilborn 1978, Schreiber et al. 2004, Williams 2011a).

One of the most successful examples of large-scale adaptive management is the U.S. Fish and Wildlife Service's program to regulate mallard (*Anas platyrhynchos*) harvests, which relies on passive adaptive optimization (Johnson and Williams 1999, Nichols et al. 2007). The optimization algorithm explicitly accounts for environmental variation and partial controllability of harvests, and admits four alternative models of population dynamics, with process error attendant to each. In 1995, the prior probability for each of the four models was set to 0.25 by consensus of the management community. Probabilities associated with the alternative models have changed substantially since implementation of the program and, as a result, there has been considerable change in management policy (Fig. 5.1) (Johnson 2010). The change in harvest policy from 1995 to 2007 is a striking example of the efficacy of a passive adaptive management program that involves model averaging to calculate policies, and a monitoring program to periodically update the probabilities of alternative models.

Recent Advances

Thus far we have focused on resource systems in which system state $\underline{x}_t$ is completely observable and system dynamics can be characterized by stationary Markov chains. These simplifying assumptions greatly facilitate the calculation of optimal policies, but the potential for suboptimal management performance can be signifi-

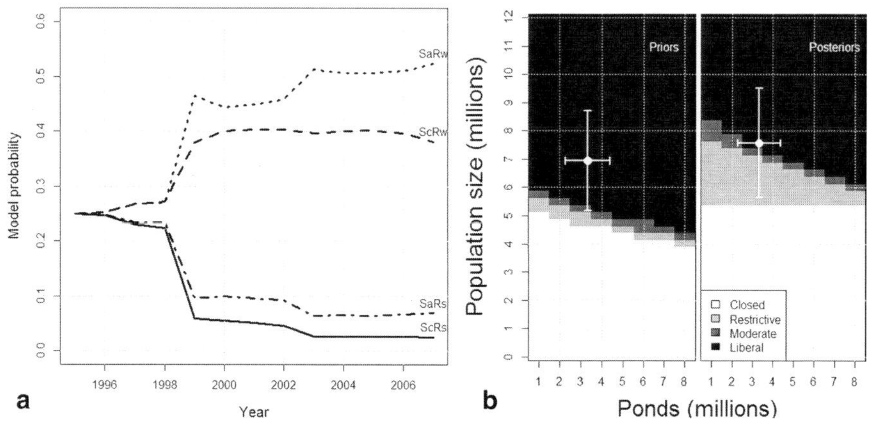

Fig. 5.1 (**a**) The sport harvest of mallards in the United States is a long-running, successful application of adaptive management. At right are the probabilities of four alternative models of mallard population dynamics, which have been updated each year based on Bayes' theorem and a comparison of model-specific predictions of population size with that observed via a monitoring program (from Johnson 2010) (**b**) At left are the passively adaptive, regulatory harvest policies for mallards based on four alternative models of population dynamics, conditioned on prior (*1995*) and posterior (*2007*) model probabilities. The *cross-hairs* on each plot indicate the expected mean (*plus and minus one standard deviation*) of population size and pond numbers. The optimal policies seek to balance accumulated harvest with a desire to keep the population from falling below a goal of 8.8 million (from Johnson 2010)

cant in some natural resource problems. There is active interest in ways to relax the assumption of a completely observable system while maintaining computational tractability. For a problem involving a space of discrete system states, partial observability transforms the problem into a partially observable Markov decision process (POMDP) with a continuous space of state probability distributions at each time (Williams et al. 2011). POMDP approaches appear to be particularly applicable to questions of monitoring design (White 2005, Chades et al. 2008). The development and subsequent modification of monitoring programs has been an under-scrutinized element of adaptive management efforts (Nichols and Williams 2006, Lyons et al. 2008). More generally, similarities in the way partially observed and structurally uncertain systems are modeled allow computing algorithms for POMDPs to be applicable for resolving structural uncertainties in systems with large dimensionality in the distribution q_t (Williams 2009, 2011b). Finding efficient computing algorithms for POMDPs is an active area of research (e.g., Pineau et al. 2006).

To date, most of the work on both partially observable and structurally uncertain systems has assumed Markov state transitions that are stationary over time. Non-Markovian systems, which contain time-lags or other forms of history-dependent transitions, can greatly increase computational demands, but there are no theoretical obstacles to including non-Markovian transitions in state dynamics (Williams 2007). Similarly, non-stationary dynamics can in principle be handled in computing optimal management policies, assuming analysts are clever enough to capture plausible ideas about the degree and rate of change in a set of alternative models (Conroy et al. 2011). A shift in system dynamics is the signature feature of climate change, and dynamic optimization offers a way to plan and adapt conservation strategies in the presence of a changing but uncertain climate (McDonald-Madden et al. 2011, Nichols et al. 2011).

Finally, we recognize that decision analysis focusing on maximizing the expected return from a management process is not possible if the probabilities attendant to stochastic outcomes cannot be specified or the values of those outcomes cannot easily be assigned (or agreed upon). Information-gap decision theory (Ben-Haim 2001) replaces the focus on management alternatives that produce the highest return with one that seeks alternatives that are robust to uncertainty in outcomes or values. In a context of static decision making with model uncertainty, the necessary elements for info-gap are: (1) a "guesstimate" $\tilde{q}_\beta$ of a probability distribution for the models (i.e., some notion about model credibilities); (2) a value function $V(a \mid \underline{x}, q) = \Sigma_\beta \, q(\beta) V_\beta\,(a \mid \underline{x})$ that averages model-specific values; (3) a minimum performance criterion V_c for the value function; and (4) an uncertainty horizon α that defines a range of model states about $\tilde{q}$ by:

$$U(\alpha, \tilde{q}) = \left\{ q : \sum_\beta q(\beta) = 1 \text{ and} \max\left[0, (1-\alpha)\tilde{q}(\beta)\right] \le q(\beta) \le \min\left[1, (1-\alpha)\tilde{q}(\beta)\right] \right\}. \tag{12}$$

For a given action a and system state x, the idea is to find the largest uncertainty horizon α for which $V(a \mid x, q) \ge V_c$ for every model state q in the associated range $U(\alpha, \tilde{q})$, and then to choose the action that maximizes that range. The decision rule is choose a to maximize: $\tilde{\alpha}(a|\tilde{q}, \underline{x}) = \max_a \left[\min_{q \in U(\alpha, \tilde{q})} V(a \mid \underline{x}, q) \ge V_c \right]$. Analogous decision rules can be identified for uncertain observation states, and for uncertainty in both model and observation states. The decision rule also can be extended to dynamic decision making (Williams and Johnson *this volume*).

Conceptually, the intent with an info-gap approach is to choose the action for which the minimum expected performance is greater than the critical criterion over the greatest range of model or observation states. Thus, info-gap theory is closely related to the idea of minimax in classic decision theory, where the idea is to maximize the likelihood of some minimal return. The key difference is that minimax relies on known values and probabilities, whereas info-gap theory focuses on the sensitivity of a decision to unbounded uncertainties and the selection of an action that is robust to those uncertainties. Info-gap theory has been used to explore alternatives for reserve design (Nicholson and Possingham 2007), conservation strategies for species at risk (Regan et al. 2005, van der Burg and Tyre 2011), water resource management (Hipel and Ben-Haim 1999) and watershed management (McCarthy and Lindenmayer 2007).

Prospects for the Future

Advances in decision analysis over the last three decades have been impressive, but one could legitimately question whether they have contributed materially to improved conservation of natural resources. Indeed, some authors argue that that gap between environmental problems and our ability to solve them is growing (Holling et al. 2002, Berkes et al. 2003). Though that may be true, we suggest that the expanding use of decision-theoretic approaches is contributing to an increased capacity for rational decision making and adaptation in resource use and conservation. The structuring of a decision-making process—i.e., bounding and focusing the debate over choices, outcomes, and values—is a key element that has been lacking in many complex and contentious natural-resource issues. Absent this structure, environmental scientists and policy makers can assert that they are independent and objective experts acting unilaterally in the best interests of the public. In such cases it would not be unusual for such expert advice to be ignored (Ludwig 2001).

Our view is that decision analysis (in its broadest sense) is almost always needed for addressing natural resource issues. We recognize, however, that it is not sufficient. Difficult questions remain about how to design processes to formulate, evaluate, and modify environmental policies, in which the engagement of stakeholders, scientists, and decision makers can be nurtured and sustained, and in which governing bodies and institutions can promote discourse, transparency, accountability, learning, and a shared stewardship of the environment (Berkes 2010). Indeed, much of the recent adaptive management literature focuses on the need for so-called "double-loop" and "triple-loop" learning (Pahl-Wostl 2009), in which extant problem formulations, laws and regulations, institutional norms, and power relationships are called into question (e.g., Armitage et al. 2009, Allen et al. 2011, Benson and Garmestani 2011). In contrast, "single-loop" learning is sometimes used to characterize adaptive management, in which incremental improvements in policy occur as uncertainty about system response to management is reduced. Whether these issues should be subsumed within the rubric of adaptive management is unclear; however, what does seem clear is that procedural issues are the principal obstacles to successful implementation of many, perhaps most, adaptive management applications (Allen et al. 2011). While these obstacles are formidable and merit considerable attention, there still remains the question of how an adaptive approach to management, with its additional costs, is to be motivated in the first place.

Valuing Information

There has been a great deal written about why adaptive management programs are not commonplace, but perhaps too little attention has been paid to whether adaptive management is the appropriate tool for a specific resource issue (Gregory et al. 2006). Doremus (2011) made an effective case that adaptive management is an information problem, in that the key question to be addressed is whether the lack of information about ecological processes and system responses to human intervention are the principal impediments to decision making and effective management. Adaptive management can be expensive, and decision makers need some assurance that those costs can be offset by improvements in management performance resulting from a reduction in uncertainty. Uncertainty in resource conservation is ubiquitous, but not all uncertainties matter in terms of choosing the best management actions, and not all uncertainties that matter can be reduced through the management experience. Decision makers require some way to identify pertinent and reducible uncertainties, so as to determine whether a particular resource conservation issue is a good candidate for adaptive management, whether learning through management is possible, and whether an effective adaptive management program can be designed. As with management decisions themselves, these questions can be addressed through a systematic framing of the decision problem in terms of choices, values, and predictions.

A useful tool for addressing questions about the nature and implications of uncertainty is the expected value of information (Clemen 1996). In particular, the expected value of perfect information (EVPI) expresses the gain in the value expected from optimal management if uncertainty were to be eliminated. Obviously, uncertainty can never be eliminated in resource management problems, but EVPI nonetheless provides a useful heuristic for determining the extent to which a specified source of uncertainty is relevant to management decisions. EVPI is simply the difference between the objective return expected if there were no uncertainty and the best that could be expected with values that are averaged over uncertain outcomes. EVPI is often expressed in dollars, but any relevant performance metric will suffice. Expressing EVPI in dollars is useful, however, for determining what managers should be willing to spend on monitoring and other data-collection programs designed to reduce the uncertainty.

Using notation introduced previously, EVPI for a single decision can be defined as:

$$\begin{aligned}\text{EVPI}(\underline{x},q) &= E_\beta\left[\max_{(a|\underline{x},\beta)} R_\beta(a\mid\underline{x})\right] - \max_{a|\underline{x}}\left[E_\beta\left[R_\beta(a\mid\underline{x})\right]\right]\\ &= \sum\nolimits_\beta q(\beta)\left[\max_{(a|\underline{x},\beta)} R_\beta(a\mid\underline{x})\right] - \max_{a|x}\left[\sum\nolimits_\beta q(\beta)\left[R_\beta(a\mid\underline{x})\right]\right].\end{aligned} \tag{13}$$

Thus, EVPI compares two terms that use expectation and optimization of returns, but in reverse order (Williams et al. 2002). The first term optimizes model-specific returns over actions and then averages these over the parameter state β. The best that could be achieved if the most appropriate model were known is an expectation because the best model is not known a priori. The second term averages action-specific returns over the uncertain parameter state, and then optimizes the action. Figure 5.2 provides an example of EVPI based on information provided by Runge et al. (2011) concerning management of a population of endangered whooping cranes (*Grus americana*).

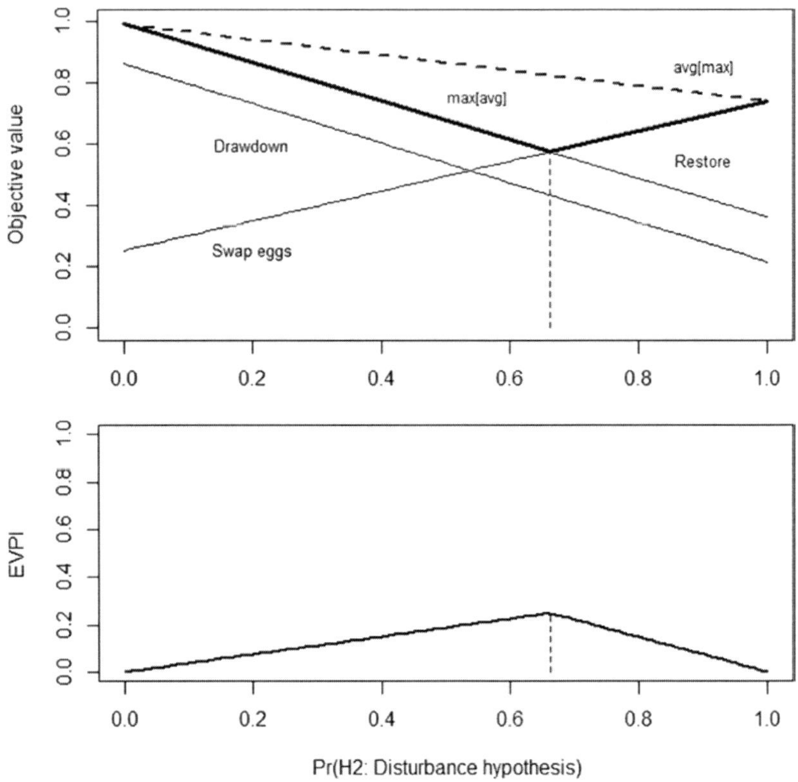

Objective values (normalized and weighted) for each action and each hypothesis are found in the following table.

Action	H1: nutrients	H2: disturbance
Swap eggs	0.254	0.740
Restore meadows	0.992	0.363
Draw-downs	0.863	0.216

Fig. 5.2 Runge et al. (2011) explored the implications of uncertainty for restoring an eastern population of whooping cranes. Several hypotheses were considered to explain reproductive failures at the breeding site in Wisconsin, of which we consider two: (1) nutrient limitation; and (2) human disturbance. Several management actions were designed to increase demographic rates of cranes, of which we consider three: (1) swap eggs in nests for those further along in incubation (from a captive flock); (2) restore meadows; and (3) conduct spring draw-downs in impounded wetlands. To see how EVPI varies with the probabilities associated with each hypotheses, we can plot the values for each action for varying levels of Pr(H2) (*with Pr(H1)* = *1-Pr(H2)*) (*top figure*). Notice that the restoration action is optimal until the Pr(H2) = 0.66, after which the swap-eggs action is optimal. The draw-downs action is never optimal for any hypothesis probability. The *dashed line* represents the value that could be expected if uncertainty were eliminated, given an a priori Pr(H2). EVPI is the difference between this value and the best that could be achieved under uncertainty (*bottom figure*). Notice that EVPI goes to zero as Pr(H2) approaches zero or one. Also notice that the maximum EVPI = 0.25 is at precisely the Pr(H2) which would cause a switch in the optimal action. Thus, the value of information is highest where uncertainty leads to ambiguity as to the best action

It is noteworthy that EVPI depends on available actions, returns that are action- and model-specific, and probabilities for the alternative models. From Eq. 13 it also is seen to depend on system state. A change in any one of these factors induces a corresponding change in EVPI. For example, Johnson et al. (2002) demonstrated how EVPI in mallard harvest management differed substantially depending on the set of available actions and how associated outcomes were valued, irrespective of model uncertainty. Of special relevance to adaptive management, EVPI is also applicable in a context of dynamic decision making. In this situation EVPI is given by:

$$\mathrm{EVPI}\left(\underline{x}_t, q_t\right) = \sum_{\beta} q_t(\beta) V_{\beta}^{*}\left(\underline{x}_t\right) - V^{*}\left(\underline{x}_t, q_t\right), \tag{14}$$

which is simply the difference in expected performance between the average of optimal values for individual models and the optimum value for an actively adaptive policy, with the optima in each term derived using dynamic optimization (Williams et al. 2011).

Understanding how EVPI changes over time in response to system state and actions taken can be useful in designing an experimental adaptive management strategy. For example, Walters and Holling (1990) demonstrated how to calculate the optimum number of experimental replicates using dynamic optimization, which combined an information state consisting of model probabilities, transition probabilities for the information state as a function of the number of replicates (via Bayes theorem), and model- and action-specific management returns. Also, Walters and Green (1997) demonstrated how information valuation could be used to plan both the size and duration of adaptive management experiments. Notably, their results suggested that adaptive management should be pursued only to the extent that it is expected to improve long-term management performance; absent in their results is any reference to statistical hypothesis testing.

Also of potential use in the design of adaptive management programs is the notion of the expected value of partial information, in which the value of eliminating one of multiple sources of uncertainty is assessed. Letting β and $\acute{\beta}$ represent two different sources of uncertainty (e.g., in the intrinsic growth rate and carrying capacity of a wildlife population), with $q_t(\beta, \acute{\beta}) = q_t(\acute{\beta} | \beta) q_t(\beta)$ the joint information state, the marginal value associated with eliminating uncertainty concerning β is:

$$\mathrm{EVPXI}\left(\underline{x}_t, q_{\beta,t}\right) = \sum_{\beta} q_t(\beta) V_{\acute{\beta}|\beta}^{*}\left(\underline{x}_t\right) - V^{*}\left(\underline{x}_t, q_t\right), \tag{15}$$

with $V_{\acute{\beta}|\beta}^{*}\left(\underline{x}_t\right)$ the optimal value for state $\underline{x}_t$ based on the parameter distribution $q_t(\acute{\beta} | \beta)$ (Williams et al. 2011). This form of the value of information recognizes two sources of uncertainty in the parameterization, but focuses on the value of reducing only one of the sources while accounting for the other. Runge et al. (2011) used the expected value of partial information to help focus an adaptive management program by prioritizing eight competing hypotheses concerning reproductive failure in a population of endangered whooping cranes.

Some authors (e.g., Walters 1986, Moore and McCarthy 2010) have observed that EVPI is often low in practice. EVPI will be low if uncertainty is low or if optimal management actions are insensitive to model choice. In some cases, management may be constrained (e.g., by laws or cultural norms) in such a way that it is not possible to capitalize on what is learned. Clearly, EVPI will be low where time horizons are short (Hauser and Possingham 2008), or where the future is heavily discounted (Moore et al. 2008). Interestingly, the work of Moore et al. (2008) suggests that EVPI may be higher in those cases where variability in objective returns are considered (e.g., some minimal performance is desired), because learning may have more influence on the variance of a parameter estimate than on its expected value.

EVPI can be particularly useful for the design and implementation of effective monitoring programs to support adaptive management (Moir and Block 2001). Even if a rigorous assessment of information value is not possible, the expected-value heuristic can be helpful for bringing clarity of thought and purpose to questions concerning monitoring design (Wintle et al. 2010). For example, because of the direct and opportunity costs of monitoring, some authors have begun to explore the optimal frequency of resource monitoring. Here the notion of optimality concerns the ability of a monitoring program to provide information that will improve management performance in a demonstrable and cost-effective way (Hauser et al. 2006, McDonald-Madden 2010).

In summary, we believe that assessing the value of information (or the cost of continued uncertainty) is an important antecedent for an adaptive management program. Questions concerning key uncertainties, their relative importance, and the costs attendant to their reduction can be effectively addressed within a decision-theoretic framework, which in turn requires critical thinking about the resource problem and the nature of the management decision(s). Calls for adaptive management absent a demonstration of the tangible benefits of such a program are unlikely to be heeded. Obviously, there are many other factors affecting the viability of an adaptive management program, but they may be of little concern if the program can't be justified in the first place.

Is the Era of Management Over?

Decision analysis is not a panacea for all the difficult natural resource problems we face. In fact, for some resource problems it is not values, consequences, and decision-making authority that are ambiguous, but a different set of impediments that prevent progress. On the other hand, we believe that for many important problems a useful place to begin is by providing a decision-analytic framework for the exploration of values and value tradeoffs. Such a framework can focus the compilation and application of existing information, and lend transparency and repeatability to the process. Importantly in the context of this book, it can provide a means to identify key uncertainties and how they might be reduced through adaptive management.

Whatever the methodological approach, the complementary processes of predicting and valuing outcomes seem fundamental for an effective investigation of resource use and conservation.

It is noteworthy that not all those concerned with the prudent use of natural resources necessarily agree, at least about the linkage between more effective decision making and sustainability. Natural resource problems increasingly have a cross-scale nature (both in space and time), in which global changes have important local implications and vice-versa (Holling 2001), and the concept of a single decision-maker optimizing a finite set of utilities no longer applies (Berkes 2010). The notion that top-down control, with its focus on maximization at the expense of natural variability, can deliver sustainable ecosystem goods and services is seen as highly unlikely (Holling and Meffe 1996). The unrealistic assumption of equilibrium in natural systems, the challenging problem of specifying possible system states and associated probabilities, and the possibility of thresholds and alternative stable states, have led to serious questions about the utility of classic decision-analytic approaches (Holling 1973, Ludwig et al. 1997, Gunderson 1999, Peterson et al. 2003, Polasky et al. 2011a).

We agree with Ludwig (2001) that "the application of any method may do more harm than good if carried out beyond its limits." On the other hand, we take exception to the opinion that the "wicked" problems we collectively face suggest "the era of management is over" (Ludwig 2001), or that optimization "is a large part of the problem, not the solution" (Walker and Salt 2006). We prefer a more balanced view that appreciates the limits of decision analysis, that restricts its use to appropriate settings, that seeks advances in its generality, and explores alternative approaches when necessary (Polasky et al. 2011a). Decision analysis has come a long way from its dubious debut in natural resource management via the concept of maximum sustained yield (Larkin 1977). It now can account for systems not in equilibrium and for unpredicted changes in system state (Williams 2001), for the presence of lagged effects and thresholds (Hauser et al. 2007, Williams 2007, Martin et al. 2009), and it certainly can handle objectives other than maximizing the expected return (Ben-Haim 2001, Williams and Johnson *this volume*). Perhaps one of the biggest challenges facing formal decision analysis is the specification of models to describe potential ecosystem behaviors that have not been observed previously. The dangers of focusing on parameter uncertainty at the expense of model (or functional) uncertainty have been demonstrated (e.g., Runge and Johnson 2002, Peterson et al. 2003), but there is still too little attention paid to the formulation of plausible models outside the range of experience (Carpenter 2002). Of growing interest are models that incorporate thresholds and alternative stable states, and their implications for optimal decision making and adaptive management.

The idea of unforeseen changes in ecological systems that may be resistant to reversal has been at the core of "resilience thinking" (Holling 1973, Carpenter et al. 2001, Folke et al. 2004, Walker and Salt 2006). Resilience is defined as the magnitude of disturbance a system can absorb while still retaining essentially the same function, structure, identity, and feedbacks (Walker et al. 2004) or as the disturbance that can be absorbed without shifting the system to an alternative stability regime (or "domain of attraction") (Holling 1973). Important concerns for ecosystem management are: (1) the loss of resilience as the system state approaches a (perhaps unknown) threshold, and the attendant increase in probability that some disturbance will shift the system to a less desirable stability regime; and (2) changes in the parameters governing the size and shape of the domains of attraction that make system shifts more or less likely (Beisner et al. 2003). Systems with alternative stable states can exhibit hysteresis, in which a loss of resilience is followed by a system change and thereafter with an increase in resilience so that reversing the change is difficult (Ludwig et al. 1997, Scheffer et al. 2001). Although a number of researchers have begun to formulate simple models that can be used to explore these properties (e.g., Scheffer et al. 2001, Carpenter 2002, Scheffer and Carpenter 2003), more needs to be done to develop models that can be used to provide practical advice for those concerned with ecosystem management. In particular, modelers need to become more adept in describing cross-scale dynamics, particularly where feedbacks between processes operating at different scales are important to system organization and function (Kerkhoff and Enquist 2007, Cumming and Norberg 2008). And we still have much to learn about how to take a more holistic view of ecosystems and humans, such that the social-ecological system becomes the analytical unit (e.g., Janssen and Carpenter 1999).

It must be noted, however, that not all regime shifts in ecological systems are catastrophic, and not all systems exhibit pronounced hysteresis (Scheffer et al. 2001, Beisner et al. 2003). In systems where gradual change is the rule, classic decision analysis and its variants remain valuable tools for those concerned with resource use and conservation. Of course, the trick is in knowing whether the system of interest has the potential for rapid regime shifts. Where plausible models can be asserted, decision analysis can still be useful. For example, Polasky et al. (2011b) used simple models to demonstrate how optimal management differs under various assumptions about the nature of regime changes. Interestingly, optimal management was seen to be precautionary if a potential regime shift causes changes in system dynamics, and if management affects the probability of a regime shift. With an exogenous probability of a regime shift, the optimal management policy is unaffected except that it will change in response to the regime shift. These results provide valuable insights, and we suggest that more of these investigations are warranted.

The implications of potential regime shifts for adaptive management are less clear. Intuition suggests that when resilience is low and the costs of undesirable states are high, system probing or experimentation to facilitate learning is unlikely to be prudent (Gunderson 1999, Allen and Gunderson 2011). A productive line of inquiry involves understanding how various sources and degrees of uncertainty in the mechanics of regime shifts influence optimal prescriptions for adaptive management. Methods of decision analysis that focus on variability in objective returns and on robust decision making are more likely to be relevant in these cases than classic methods that focus on maximizing expected values.

Resilience thinking, with its focus on interacting ecological and social systems, on non-linear dynamics, thresholds, hysteresis, and on ecosystem services as well as goods, has much to offer those concerned with the sustainability of natural resources. But resilience-based management is a work in progress, with much more work needed to move it from theoretical concept to practical advice for decision makers. It is perhaps time for resilience thinkers and decision analysts to ask how the strengths of both philosophies can be brought to bear on pressing environmental issues (Fischer et al. 2009). We agree with Berkes (2010) that rather than declaring the era of management over, we should seek to redefine "resources" and "management" in ways that set them apart from their overly simplistic and reductionist traditions. In the end, however, resource managers will still have to make decisions, and decision analysis can provide an effective antidote for our inherent limitations in making rational decisions in complex situations (Lehrer 2009, Gilbert 2011).

Conclusion

We suggest that adaptive management and decision analysis have been self-reinforcing concepts, with one driving advances in the other and with both playing increasingly important roles in how resource managers approach decision making. The structuring of a decision making process—i.e., bounding and focusing the debate over choices, outcomes, and values—has clearly been lacking in many complex and contentious natural-resource issues, though it is hard to imagine how effective science-based management could be planned or implemented in its absence. Of particular relevance are recent advances in the analysis of dynamic decision problems, in which the effects of today's decisions on system states and actions in the future are considered explicitly. Under uncertainty about the most appropriate model describing system behaviors, dynamic optimization provides a way to address the problem of dual control, in which short-term objective returns are balanced with the learning necessary to increase long-term returns. Decision analysis has advanced considerably since its original uses in resource management, which tended to rely on unrealistic assumptions like equilibrium in system states or management returns, and certainty in system dynamics and the valuation of outcomes. Moreover, we believe that more bona fide examples of adaptive management might be forthcoming if analysts estimated the value of reducing uncertainty in tangible terms, and then used these estimates to motivate and design adaptive management programs. Adaptive management is expensive, and decision makers deserve to know whether expected improvements in management returns are worth the additional cost. Of course, classic decision analysis has its limitations, but novel approaches are applicable to ecological systems that can exhibit sudden and unanticipated shifts to alternative regimes. Given that the way ecological systems are managed can reduce resilience and increase the probability of an undesirable regime change, a fruitful line of investigation concerns the appropriate use of adaptive management for systems that exhibit these properties. We suggest that decision analysts to date have not done much "resilience thinking," and resilience thinkers have not done much to provide practical advice to decision makers. That needs to change if we are to narrow the gap between environmental problems and our ability to solve them.

Acknowledgements We would like to acknowledge the numerous friends and colleagues who have influenced our thinking about decision making in resource management. In particular, we thank G. Scott Boomer (U.S. Fish and Wildlife Service), Michael J. Conroy, Clinton T. Moore, James D. Nichols, Michael C. Runge, (all U.S. Geological Survey), Hugh P. Possingham (University of Queensland), and Carl J. Walters (University of British Columbia). Funding for this research was provided by the U.S. Fish and Wildlife Service and the U.S. Geological Survey. Any use of trade, product, or firm names in this article is for descriptive purposes only and does not imply endorsement by the U.S. Government.

References

Alaya, A. B., Souissi, A., Tarhouni, J., & Ncib, K. (2003). Optimization of Nebgana Reservoir water allocation by stochastic dynamic programming. *Water Resources Management, 17,* 259–272.

Allan, C., & Curtis, A. (2005). Nipped in the bud: Why regional scale adaptive management is not blooming. *Environmental Management, 36,* 414–425.

Allen, C. R., & Gunderson, L. H. (2011). Pathology and failure in the design and implementation of adaptive management. *Journal of Environmental Management, 92,* 1379–1384.

Allen, C. R., Fontaine, J. J., Pope, K. L., & Garmestani, A. S. (2011). Adaptive management for a turbulent future. *Journal of Environmental Management, 92,* 1339–1345.

Armitage, D. R., Plummer, R., Berkes, F., Arthur, R. I., Charles, A. T., Davidson-Hunt, I. J., Diduck, A. P., Doubleday, N. C., Johnson, D. S., Marschke, M., McConney, P., Pinkerton, E. W., & Wollenburg, E. K. (2009). Adaptive co-management for social-ecological complexity. *Frontiers in Ecology and the Environment, 7,* 95–102.

Arvai, J. L., Gregory, R., & McDaniels, T. L. (2001). Testing a structured decision approach: Value-focused thinking for deliberative risk communication. *Risk Analysis, 21,* 1065–1076.

Beisner, B. E., Haydon, D. T., & Cuddington, K. (2003). Alternative stable states in ecology. *Frontiers in Ecology and the Environment, 1,* 376–382.

Bellman, R. (1957). *Dynamic programming.* Princeton: Princeton University Press.

Ben-Haim, Y. (2001). *Information gap decision theory: Decisions under severe uncertainty.* San Diego: Academic.

Benson, M. H., & Garmestani, A. S. (2011). Embracing panarchy, building resilience and integrating adaptive management through a rebirth of the National Environmental Policy Act. *Journal of Environmental Management, 92,* 1420–1427.

Berkes, F. (2010). Shifting perspectives on resource management: resilience and reconceptualization of 'natural resources' and 'management'. *Marine Studies (MAST), 9,*13–40.

Berkes, F., Colding, J., & Folke, C. (2003). Introduction. In F. Berkes, J. Colding & C. Folke (Eds), *Navigating social-ecological systems: Building resilience for complexity and change* (pp. 1–32). Cambridge: Cambridge University Press.

Bertsekas, D. P. (1995). *Dynamic programming and optimal control* (Vol. 1). Belmont: Athena Scientific.

Bogich, T., & Shea, K. (2008). A state-dependent model for the optimal management of an invasive metapopulation. *Ecological Applications, 18,* 748–761.

Burgman, M. (2005). *Risks and decisions for conservation and environmental management.* Cambridge: Cambridge University Press.

Carpenter, S. R. (2002). Ecological futures: Building an ecology of the long now. *Ecology, 83,* 2069–2083.

Carpenter, S., Walker, B., Andreries, J. M., & Abel, N. (2001). From metaphor to measurement: Resilience of what to what? *Ecosystems, 4,* 765–781.

Chades, I., McDonald-Madden, E., McCarthy, M. A., Wintle, B., Linkie, M., & Possingham, H. (2008). When to stop managing or surveying cryptic threatened species. *Proceedings of the National Academy of Science, 105,* 13936–13940.

Clemen, R. T. (1996). *Making hard decisions: An introduction to decision analysis* (2nd ed.). Pacific Grove: Duxbury.

Conroy, M. J., & Carroll, J. P. (2009). *Quantitative conservation of vertebrates.* Oxford: Wiley-Blackwell.

Conroy, M. J., Runge, M. C., Nichols, J. D., Stodola, K. W., & Cooper, R. J. (2011). Conservation in the face of climate change: The roles of alternative models, monitoring, and adaptation in confronting and reducing uncertainty. *Biological Conservation, 144,* 1204–1213.

Costanza, R. (1993). Developing ecological research that is relevant for achieving sustainability. *Ecological Applications, 3,* 579–581.

Cumming, G. S., & Norberg, J. (2008). Scale and complex systems. In J. Norberg & G. S. Cumming (Eds), *Complexity theory for a sustainable future* (pp. 246–276). New York: Columbia University Press.

Doremus, H. (2007). Precaution, science, and learning while doing natural resource management. *Washington Law Review, 82,* 547–579.

Doremus, H. (2011). Adaptive management as an information problem. *North Carolina Law Review, 89,* 1455–1495.

Eum, H., Kim, Y., & Palmer, R. N. (2011). Optimal drought management using sampling stochastic dynamic programming with a hedging rule. *Journal of Water Resources Planning and Management, 137,* 113–122.

Fischer, J., Peterson, G. D., Gardner, T. A., Gordon, L. J., Fazey, I., Elmqvist, T., Felton, A., Folke, C., & Dovers, S. (2009). Integrating resilience thinking and optimisation for conservation. *Trends in Ecology & Evolution, 24,* 549–554.

Folke, C., Carpenter, S., Walker, B., Scheffer, M., Elmqvist, T., Gunderson, L., & Holling, C. S. (2004). Regime shifts, resilience, and biodiversity in ecosystem management. *Annual Review of Ecology and Systematics, 35,* 557–581.

Gilbert, D. (2011). Buried by bad decisions. *Nature, 474,* 275–277.

Gregory, R., Ohlson, D., & Arvai, J. (2006). Deconstructing adaptive management: Criteria for applications to environmental management. *Ecological Applications, 16,* 2411–2425.

Gunderson, L. H. 1999. Resilience, flexibility and adaptive management—antidotes for spurious certitude? *Conservation Ecology, 3*(1), 7. http://www.consecol.org/vol3/iss1/art7/.

Gunderson, L., Peterson, G., & Holling, C. S. (2008). Practicing adaptive management in complex social-ecological systems.In J. Norberg & G. S. Cumming (Eds), *Complexity theory for a sustainable future* (pp. 223–240). New York: Columbia University Press.

Hajkowicz, S., & Collins, K. (2007). A review of multiple criteria analysis for water resource planning and management. *Water Resources Management, 21,* 1553–1566.

Hauser, C. E., & Possingham, H. P. (2008). Experimental or precautionary? Adaptive management over a range of time horizons. *Journal of Applied Ecology, 45,* 72–81.

Hauser, C. E., Pople, A. R., & Possingham, H. P. (2006). Should managed populations be monitored every year? *Ecological Applications, 16,* 807–819.

Hauser, C. E., Runge, M. C., Cooch, E. G., Johnson, F. A., & Harvey, W. F. IV. (2007). Optimal control of Atlantic population Canada geese. *Ecological Modelling, 201,* 27–36.

Hilborn, R., & Ludwig, D. (1993). The limits of applied ecological research. *Ecological Applications, 3,* 550–552.

Hipel, K. W., & Ben-Haim, Y. (1999). Decision making in an uncertain world: Information-gap modeling in water resources management. *IEEE Transactions on Systems, Man, and Cybernetics —Part C: Applications and Reviews, 29,* 506–517.

Holling, C. S. (1973). Resilience and stability of ecological systems. *Annual Review of Ecology and Systematics, 4,* 1–23.

Holling, C. S. (2001). Understanding the complexity of economic, ecological, and social systems. *Ecosystems, 4,* 390–405.

Holling, C. S., & Meffe, G. K. (1996). Command and control and the pathology of natural resource management. *Conservation Biology, 10,* 328–337.

Holling, C. S., Gunderson, L. H., & Ludwig, D. (2002). In quest of a theory of adaptive change. In L. H. Gunderson & C. S. Holling (Eds), *Panarchy: Understanding transformations in human and natural systems* (pp. 3–22). Washington: Island Press.

Howard, R. A. (1968). The foundations of decision analysis. *IEEE Transactions on Systems Science and Cybernetics SSC, 4,* 211–219.

Huang, I. B., Keisler, J., & Linkov, I. (2011). Multi-criteria decision analysis in environmental sciences: Ten years of applications and trends. *Science of the Total Environment, 409,* 3578–3594.

Janssen, M. A., & Carpenter, S. R. (1999). Managing the resilience of lakes: a multi-agent modeling approach. *Ecology and Society, 3*(2), 15. http://www.consecol.org/vol3/iss2/art15/.

Johnson, F. A. (2010). Analysis and management of migratory bird populations in North America. Ph. D. Dissertation, University of Florida, Gainesville, USA.

Johnson, F. A., & Williams, B. K. (1999). Protocol and practice in the adaptive management of waterfowl harvests. *Conservation Ecology, 3*(1), 8. http://www.ecologyandsociety.org/vol3/iss1/art8/

Johnson, F. A., Moore, C. T., Kendall, W. L., Dubovsky, J. A., Caithamer, D. F., Kelley, J. R. Jr., & Williams, B. K. (1997). Uncertainty and the management of mallard harvests. *Journal of Wildlife Management, 61,* 202–216.

Johnson, F. A., Kendall, W. E., & Dubovsky, J. A. (2002). Conditions and limitations on learning in the adaptive management of mallard harvests. *Wildlife Society Bulletin, 30,* 176–185.

Johnson, F. A., Breininger, D. R., Duncan, B. W., Nichols, J. D., Runge, M. C., & Williams, B. K. (2011). A Markov decision process for managing habitat for Florida scrub-jays. *Journal of Fish and Wildlife Management, 2,* 234–246.

Keefer, D. L., Kirkwood, C. W., & Corner, J. L. (2004). Perspective on decision analysis applications. *Decision Analysis, 1,* 4–22.

Keene, M., & Pullin, A. S. (2011). Realizing an effectiveness revolution in environmental management. *Journal of Environmental Management, 92,* 2130–2135.

Kerkhoff, A. J., & Enquist, B. J. (2007). The implications of scaling approaches for understanding resilience and reorganization in ecosystems. *BioScience, 57,* 489–499.

Kulmala, S., Laukkanen, M., & Michielsens, C. (2008). Reconciling economic and biological modeling of migratory fish stocks: Optimal management of the Atlantic salmon fishery in the Baltic Sea. *Ecological Economics, 64,* 716–728.

Larkin, P. A. (1977). An epitaph for the concept of maximum sustained yield. *Transactions of the American Fisheries Society, 106,* 1–11.

Lee, K. N. (1993). *Compass and gyroscope: Integrating science and politics for the environment.* Washington, D.C: Island.

Lee, K. N. (1999). Appraising adaptive management. *Conservation Ecology, 3*(2), 3. http://www.ecologyandsociety.org/vol3/iss2/art3/

Lehrer, J. (2009). *How we decide.* Boston: Houghton Mifflin Harcourt.

Linkov, I., Satterstrom, F. K., Kiker, G. A., Bridges, T. S., Bejamin, S. L., & Belluck, D. A. (2006). From optimization to adaptation: Shifting paradigms in environmental management and their application to remedial decisions. *Integrated Environmental Assessment and Management, 2,* 92–98.

Ludwig, D. (2001). The era of management is over. *Ecosystems, 4,* 758–764.

Ludwig, D., Hilborn, R., & Walters, C. (1993). Uncertainty, resource exploitation, and conservation: Lessons from history. *Science, 260*(17), 36.

Ludwig, D., Walker, B., & Holling, C. S. (1997). Sustainability, stability, and resilience. *Conservation Ecology, 1*(1), 7. http://www.consecol.org/vol1/iss1/art7/.

Lyons, J. E., Runge, M. C., Laskowski, H. P., & Kendall, W. E. (2008). Monitoring in the context of structured decision-making and adaptive management. *Journal of Wildlife Management, 72,* 1683–1692.

Martin, J., Runge, M. C., Nichols, J. D., Lubow, B. C., & Kendall, W. L. (2009). Structured decision making as a conceptual framework to identify thresholds for conservation and management. *Ecological Applications, 19,* 1079–1090.

Martin, J., Fackler, P. L., Nichols, J. D., Runge, M. C., McIntyre, C. L., Lubow, B. C., McCluskie, M. C., & Schmutz, J. A. (2011). An adaptive-management framework for optimal control of hiking near golden eagle nests in Denali National Park. *Conservation Biology, 25,* 316–323.

McCarthy, M. A., & Lindenmayer, D. B. (2007). Info-gap decision theory for assessing the management of catchments for timber production and urban water supply. *Environmental Management, 39,* 553–562.

McCarthy, M. A., Possingham, H. P., & Gill, A. M. (2001). Using stochastic dynamic programming to determine optimal fire management for Banksia ornata. *Journal of Applied Ecology, 38,* 585–592.

McDonald-Madden, E., Baxter, P. W. J., Fuller, R. A., Martin, T. G., Game, E. T., Montambault, J., & Possingham, H. P. (2010). Monitoring does not always count. *Trends in Ecology & Evolution, 25,* 547–550.

McDonald-Madden, E., Runge, M. C., Possingham, H. P., & Martin, T. G. (2011). Optimal timing for managed relocation of species faced with climate change. *Nature Climate Change, 1,* 261–265.

McFadden, J. E., Hiller, T. L., & Tyre, A. J. (2011). Evaluating the efficacy of adaptive management approaches: Is there a formula for success? *Journal of Environmental Management, 92,* 1354–1359.

McLain, R. J., & Lee, R. G. (1996). Adaptive management: Promises and pitfalls. *Environmental Management, 20,* 437–448.

Milner-Gulland, E. J. (1997). A stochastic dynamic programming model for the management of the saiga antelope. *Ecological Applications, 7,* 130–142.

Moilanen, A., Wilson, K. A., & Possingham, H. P. (Eds.). (2009). *Spatial conservation prioritization: Quantitative methods & computational tools*. Oxford: Oxford University Press.

Moir, W. H., & Block, W. M. (2001). Adaptive management on public lands in the United States: Commitment or rhetoric? *Environmental Management, 28,* 141–148.

Mooney, H. A., & Sala, O. E. (1993). Science and sustainable use. *Ecological Applications, 3,* 564–566.

Moore, C. T., & Conroy, M. J. (2006). Optimal regeneration planning for old-growth forest: Addressing scientific uncertainty in endangered species recovery through adaptive management. *Forest Science, 52,* 155–172.

Moore, A. L., & McCarthy, M. A. (2010). On valuing information in adaptive-management models. *Conservation Biology, 24,* 984–993.

Moore, A. L., Hauser, C. E., & McCarthy, M. A. (2008). How we value the future affects our desire to learn. *Ecological Applications, 18,* 1061–1069.

Nichols, J. D., & Williams, B. K. (2006). Monitoring for conservation. *Trends in Ecology & Evolution, 21,* 668–673.

Nichols, J. D., Runge, M. C., Johnson, F. A., & Williams, B. K. (2007). Adaptive harvest management of North American waterfowl populations: A brief history and future prospects. *Journal of Ornithology, 148*(Suppl 2), S343–349.

Nichols, J. D., Koneff, M. D., Heglund, P. J., Knutson, M. G., Seamans, M. E., & Lyons, J. E., Morton, J. M., Jones, M. T., Boomer, G. S., & Williams, B. K. (2011). Climate change, uncertainty, and natural resource management. *Journal of Wildlife Management, 75,* 6–18.

Nicholson, E., & Possingham, H. P. (2007). Making conservation decisions under uncertainty for the persistence of multiple species. *Ecological Applications, 17,* 251–265.

Pahl-Wostl, C. (2009). A conceptual framework for analyzing adaptive capacity and multi-level learning processes in resource governance regimes. *Global Environmental Change, 19,* 354–365.

Peterson, G. D., Carpenter, S. R., & Brock, W. A. (2003). Uncertainty and the management of multistate systems: An apparently rational route to collapse. *Ecology, 84,* 1403–1411.

Pielke, R. A. Jr. (2007). *The honest broker: Making sense of science in policy and politics*. Cambridge: Cambridge University Press.

Pineau, J., Gordon, G., & Thrun, S. (2006). Anytime point-based approximations for large POMDPs. *Journal of Artificial Intelligence Research, 27,* 335–380.

Polasky, S., de Zeeuw, A., & Wagener, F. (2011a). Optimal management with potential regime shifts. *Journal of Environmental Economics and Management, 62,* 229–240.

Polasky, S., Carpenter, S. R., Folke, C., & Keeler, B. (2011b). Decision-making under great uncertainty: Environmental management in an era of global change. *Trends in Ecology and Evolution, 26,* 398–404.

Possingham, H. (1997). State-dependent decision analysis for conservation biology. In S. T. A. Pickett, R. S. Ostfeld, M. Shachak, & G. E. Likens (Eds), *The ecological basis of conservation: Heterogeneity, ecosystems, and biodiversity* (pp 298–304). New York: Chapman & Hall.

Puterman, M. L. (1994). *Markov decision processes: Discrete stochastic dynamic programming.* New York: John Wiley and Sons.

Raiffa, H. (1968). *Decision analysis: Introductory lectures on choices under uncertainty.* Reading: Addison-Wesley.

Regan, H. M., Ben-Haim, Y., Langford, B., Wilson, W. G., Lundberg, P., Andelman, S. J., & Burgman, M. A. (2005). Robust decision-making under severe uncertainty for conservation management. *Ecological Applications, 15,* 1471–1477.

Ruhl, J. B. (2006). Regulation by adaptive management - Is it possible? *Minnesota Journal of Law, Science & Technology, 7,* 21–57.

Ruhl, J. B., & Fischman, R. L. (2010). Adaptive management in the courts. *Minnesota Law Review, 95,* 424–484.

Richards, S. A., Possingham, H. P., & Tizard, J. (1999). Optimal fire management for maintaining community diversity. *Ecological Applications, 9,* 880–892.

Runge, M. C., & Johnson, F. A. (2002). The importance of functional form in optimal control solutions of problems in population dynamics. *Ecology, 83,* 1357–1371.

Runge, M. C., Converse, S. J., & Lyons, J. E. (2011). Which uncertainty? Using expert elicitation and expected value of information to design an adaptive program. *Biological Conservation, 144,* 1214–1223.

Scheffer, M., & Carpenter, S. R. (2003). Catastrophic regime shifts in ecosystems: Linking theory to observation. *Trends in Ecology & Evolution, 18,* 648–656.

Scheffer, M., Carpenter, S., Foley, J. A., Folke, C., & Walker, B. (2001). Catastrophic shifts in ecosystems. *Nature, 413,* 591–596.

Schreiber, E. S. G., Bearlin, A. R., Nicol, S. J., & Todd, C. R. (2004). Adaptive management: A synthesis of current understanding and effective application. *Ecological Management & Restoration, 5,* 177–182.

Sells, J. E. (1995). Optimizing weed management using stochastic dynamic programming to take account of uncertain herbicide performance. *Agricultural Systems, 48,* 271–296.

Susskind, L., Camacho, A. E., & Schenk, T. (2010). Collaborative planning and adaptive management in Glen Canyon: A cautionary tale. *Columbia Journal of Environmental Law, 35,* 1–54.

Tenhumberg, B., Tyre, A. J., Shea, K., & Possingham, H. (2004). Linking wild and captive populations to maximize species persistence: Optimal translocation strategies. *Conservation Biology, 18,* 1304–1314.

Tonn, B., English, M., & Travis, C. (2000). A framework for understanding and improving environmental decision making. *Journal of Environmental Planning and Management, 43,* 163–183.

van der Burg, M. P., & Tyre, A. J. (2011). Integrating info-gap decision theory with robust population management: A case study using the Mountain Plover. *Ecological Applications, 21,* 303–312.

Walker, B., & Salt, D. (2006). *Resilience thinking: Sustaining ecosystems and people in a changing world.* Washington, D.C.: Island Press.

Walker, B., Holling, C. S., Carpenter, S. R., & Kinzig, A. (2004). Resilience, adaptability and transformability in social-ecological systems. *Ecology and Society, 9*(2), 5. http://www.ecologyandsociety.org/vol9/iss2/art5.

Walters, C. J. (1986). *Adaptive management of renewable resources.* New York: MacMillan.

Walters, C. J. (1997). Challenges in adaptive management of riparian and coastal ecosystems. *Conservation Ecology, 1*(2), 1. http://www.ecologyandsociety.org/vol1/iss2/art1/.
Walters, C. J. (2007). Is adaptive management helping to solve fisheries problems? *Ambio, 36,* 304–307.
Walters, C. J., & Green, R. E. (1997). Valuation of experimental management options for ecological systems. *Journal of Wildlife Management, 61,* 987–1006.
Walters, C. J., & Hilborn, R. (1978). Ecological optimization and adaptive management. *Annual Review of Ecology and Systematics, 9,* 157–188.
Walters, C. J., & Holling, C. S. (1990). Large-scale experimentation and learning by doing. *Ecology, 71,* 2060–2068.
Westphal, M. I., Pickett, M., Getz, W. M., & Possingham, H. P. (2003). The use of stochastic dynamic programming in optimal landscape reconstruction for metapopulations. *Ecological Applications, 13,* 543–555.
White, B. (2005). An economic analysis of ecological monitoring. *Ecological Modelling, 189,* 241–250.
Williams, B. K. (1989). Review of dynamic optimization methods in renewable natural resource management. *Natural Resource Modeling, 3,* 137–216.
Williams, B. K. (2001). Uncertainty, learning, and the optimal management of wildlife. *Environmental and Ecological Statistics, 8,* 269–288.
Williams, B. K. (2007). Optimal management of non-Markovian biological populations. *Ecological Modelling, 200,* 234–242.
Williams, B. K. (2009). Markov decision process in natural resources management: Observability and uncertainty. *Ecological Modelling, 220,* 830–840.
Williams, B. K. (2011a). Passive and active adaptive management: Approaches and an example. *Journal of Environmental Management, 92,* 1371–1378.
Williams, B. K. (2011b). Resolving structural uncertainty in natural resources management using POMDP approaches. *Ecological Modelling, 222,* 1092–1102.
Williams, B. K., Johnson, F. A., & Wilkins, K. (1996). Uncertainty and the adaptive management of waterfowl harvests. *Journal of Wildlife Management, 60,* 223–232.
Williams, B. K., Nichols, J. D., & Conroy, M. J. (2002). *Analysis and management of animal populations: Modeling, estimation, and decision making*. San Diego: Academic.
Williams, B. K., Eaton, M. J., & Breininger, D. R. (2011). Adaptive resource management and the value of information. *Ecological Modelling, 222,* 3429–3436.
Wilson, K. A., McBride, M. F., Bode, M., & Possingham, H. P. (2006). Prioritizing global conservation efforts. *Nature, 440,* 337–340.
Wintle, B. A., Runge, M. C., & Bekessy, S. A. (2010). Allocating monitoring effort in the face of unknown unknowns. *Ecology Letters, 13,* 1325–1337.

Chapter 6
Measuring Success of Adaptive Management Projects

Brian C. Chaffin and Hannah Gosnell

Keywords Adaptive management · Uncertainty · Natural resource management · Conservation · Success

Introduction

Measuring success is a critical waypoint in the growth of any natural resource management (NRM) program. In a climate of increasingly scarce funding for conservation and environmental management, measures of success aid funding agencies and managers in determining whether investments will yield desired results. In protected area conservation for example, the cost of selecting, designating, and managing protected areas is only feasible if these areas yield desired ecological and social benefits in the future (Hockings 2003). Resources for conservation are often inadequate relative to the scope and scale of environmental degradation, and thus managers are tasked with using resources as efficiently as possible. Identifying successful approaches to NRM is a critical step in this process (Kapos et al. 2008). In addition, granting organizations and agencies have increased demands for reporting and accountability towards clearly demonstrated benefits of funding (Parrish et al. 2003, Stojanovic et al. 2004). Examples linking NRM with improved ecological health are few compared to the dominant narrative of environmental degradation—more robust evaluation and reporting of these successes may increase the scope and scale of future NRM and conservation efforts (Burbridge 1997, Olsen et al. 1997). In many cases, critical research on how to evaluate a given management paradigm lags significantly behind the wave of experimentation and promotion of the approach (Kenney 2001) and this appears to be the case with adaptive management (AM), as well.

For these reasons, a critical evaluation of efforts to measure success in adaptive management seems timely. Its use as an NRM technique is widespread and

B. C. Chaffin (✉) · H. Gosnell
Oregon State University, 104 CEOAS Administration Building, Corvallis, OR 97331, USA
e-mail: chaffinb@geo.oregonstate.edu

H. Gosnell
e-mail: gosnellh@geo.oregonstate.edu

C. R. Allen, A. S. Garmestani (eds.), *Adaptive Management of Social-Ecological Systems*,
DOI 10.1007/978-94-017-9682-8_6

growing, with a demonstrated commitment to the approach in NRM and conservation literature, agency direction, and in statutory language, as exemplified by the numerous examples of adaptive management projects and processes described elsewhere in this book. Like many innovative techniques or tools, measuring the success of implementation is critical to adaptive management's continued application, performance, and refinement. Measuring success in adaptive management projects is imperative for managers to demonstrate fiscal accountability and the need for continued funding. As one example, federal, state, and local level grants are increasingly recognizing adaptive management as a desirable and thus fundable approach to NRM, and as such, government grantees need to demonstrate tangible benefits of an adaptive management approach (Fontaine 2011). Further, measuring success in adaptive management will help document good, bad, and ugly examples of adaptive management implementation, catalyzing a set of 'lessons learned' to help improve desired outcomes of adaptive management such as restoration, rehabilitation, and reintroduction of species (Weinstein et al. 1997, Downs and Kondolf 2002, Alexander and Allan 2007).

Although the concept of adaptive management appears widely in the NRM literature, there is relatively little peer-reviewed literature evaluating adaptive management success in a universal manner. We echo Williams and Brown's (2012) assertion that at this point in time "there are no broadly accepted standards by which to recognize and measure success" in adaptive management. Bearlin et al. (2002) observe that the inability to measure progress toward adaptive management objectives (lack of success) creates a weak argument for the continued use of adaptive management as a preferred NRM and policy directive.

In this chapter, we review scholarly and gray literature that deals with the enigmatic concept of adaptive management success, and ways to measure it. We look to related literature on measuring the success of other types of NRM such as integrated coastal management, protected area studies, and stream and wetland restoration research. Drawing on this synthesis, we articulate a new framework for measuring success in adaptive management programs and projects—including a means to assess whether adaptive management is an appropriate approach for a given context in the first place and, if it is, a series of metrics for measuring the efficacy of each phase in the adaptive management cycle. Finally, we demonstrate the utility of our framework and associated metrics by using it to evaluate the most thoroughly discussed adaptive management program in the United States, the Glen Canyon Dam Adaptive Management Program.

Measuring Success in Natural Resource Management

Measuring success has been a consistent and sometimes elusive theme in NRM research and corresponding literature. Many NRM paradigms including that of integrated coastal management (Burbridge 1997, Stojanovic et al. 2004), protected area conservation (Hockings 2003, Kapos et al. 2008), and collaborative watershed management have confronted the "thorny" issue of evaluating success (Kenney 2001). Here we take a broad look at approaches to evaluating NRM, then consider their applicability to measuring success in adaptive management more specifically.

Measuring success in NRM (as in other arenas) is often discussed in the context of inputs, outputs, and outcomes. In large-scale global conservation efforts undertaken by multinational NGOs, measuring success has always been a priority due in part to the need to satisfy the curiosity of big donors—and thus continue to cultivate revenues. Until recently measures of conservation program success consisted mostly of inputs such as money and time spent or resources utilized (Kapos et al. 2008). Over time this has shifted to a measurement of outputs such as projects completed, activities performed, or specific products of money and time spent including acres restored or number of species recovered (Kapos et al. 2008). In a discussion of measuring success in integrated coastal management (ICM), however, Stojanovic et al. (2004) point out the dangers in measuring outputs; depending on the nature of the project objectives, simply counting products without directly relating them back to an objective may not address whether or not those products actually contributed to project success. To clearly measure success, Day (2008) suggests that evaluations need to further shift toward evidence-based assessments of whether or not projects meet the goals, objectives, or expectations they were designed to achieve. However, if project objectives are framed as the achievement of certain measurable outputs, then output measurement may legitimately lead to a declaration of success. For example, if an explicit objective of a stream restoration project was to increase aquatic habitat by increasing woody debris in 20 miles of stream channel, the project could be determined to be successful if it met this criteria. However, if the objective was written more broadly—say, to increase functional habitat or to restore an aquatic species—then a simple output measurement of miles 'restored' would not legitimize a declaration of success.

Although evidence-based assessments can be accomplished by measuring outputs, many argue instead for looking to outcomes as measures of success. What changed as a result of action in relation to the project goals and objectives? Stojanovic et al. (2004) offer a useful discussion of outputs vs. outcomes in the context of measuring the success of ICM efforts. While outputs such as licenses issued or plans produced are often easier to measure, "[to] lose sight of the outcomes from ICM initiatives is to forget the very purpose of coastal management and its environmental context" (Stojanovic et al. 2004). Kapos et al. (2008) take evaluation a bit further and draw a distinction between outcomes and overall effect. For example, outcomes may demonstrate progress toward goals and objectives, but did the project have the intended effect of the overall vision? Did the project in question improve environmental health or sustainably produce goods or services desired by society? Kenney (2001) critiques attempts to measure success in the watershed management movement by citing the failure to relate watershed group accomplishments with the effect of improved environmental quality—an overarching goal which many watershed groups share, if only generally. The key to avoiding such a critique is to articulate and bound unambiguous project goals and objectives from the outset, particularly objectives that demonstrate a clear understanding of current and desired environmental, social, and economic conditions (Burbridge 1997).

What is clear from a brief review of measures of success in NRM is that no one category of metrics emerges as comprehensive. Outputs are relatively easy to measure, but might not provide valuable insight. Outcomes are much tougher to capture due to time lags and the lack of standardized metrics, but they hold the promise of demonstrating progress toward goals and objectives. Overall effect may be near impossible to define, but underlies the goals of all NRM efforts. What is needed to measure success in NRM projects is a method for combining the measurement of outputs, outcomes, and overall effect. Comparing outcomes against a set of clear objectives and a larger goal can provide insight into areas where modifications of a project are needed—in the implementation of the process, or in the guiding objectives themselves.

Measuring Success in Adaptive Management: Efforts to Date

While literature on adaptive management performance is somewhat limited, a number of authors have presented cases in which at least some 'success' in adaptive management projects was achieved (e.g., Stankey et al. 2003) while others have pointed directly at 'failures' (Gregory et al. 2006, Allen and Gunderson 2011). McFadden et al. (2011) present the most robust attempt at a universal review of measuring success in adaptive management to date. The authors performed a meta-analysis of peer-reviewed publications ($n=96$) containing mention of "adaptive management" between 2000 and 2009. Each publication reviewed was binned into one of six categories indicating the level of the engagement with adaptive management concepts: mention, theory, suggest, framework, implement, or against. As expected, the "implement" category contained the lowest number of published studies at 17 %, as compared to 42 % of studies categorized as "suggest". The authors posit there are a range of ways to declare success in adaptive management. To further qualify (and quantify) that range, they suggest asking the following four questions of adaptive management projects: "Was an explicit formal analysis of the decision conducted? Does the resulting management plan include an iterative cycle? Was a management action implemented? Did the implemented action achieve the desired outcome?" (McFadden et al. 2011). This rather general approach supports our observation that measuring success in adaptive management projects presents a unique set of challenges which helps explain why scholars have only just begun to comment on promising strategies (Gregory et al. 2006, Morghan et al. 2006, McFadden et al. 2011).

First, 'success and failure' are relative to how success in adaptive management projects is defined. Because adaptive management objectives and outcomes (both actual and desired) will vary significantly in virtually every adaptive management project according to context, it is extremely difficult to define success and identify a uniform set of indicators and criteria. As such, evaluators of adaptive management programs and projects are unsure of what to measure. Is success defined only as measurable progress toward project objectives, or can success also be characterized by valuable institutional changes or learning that takes place, despite progress (or lack thereof) toward on-the-ground objectives? Or, perhaps institutional change and learning should be explicitly articulated as a measurable objective and thus a potential outcome of an adaptive management project. What does success in adaptive management projects look like?

O'Donnell and Galat (2008) conducted a survey of resource managers and identified a 'degree of success' in 70 river enhancement projects (some, but not all of which used adaptive management as the approach to management) in the upper Mississippi River basin. Thirty-four percent of projects surveyed incorporated quantifiable measures of success. The authors argue that quantifiable success criteria may be the most important factor guiding decision making during later project phases including the monitoring and evaluation phases (O'Donnell and Galat 2008). According to this reasoning, measurable success criteria should be integrated into project goals and objectives at the outset of the project. The authors account for the roughly two-thirds of river enhancement projects without quantifiable success criteria by pointing to a lack of funding for long-term projects, lack of data (historic reference) for criteria, and lack of necessary technical expertise (O'Donnell and Galat 2008).

Much of the literature on measuring success in adaptive management comes from evaluations of restoration or ecosystem rehabilitation projects, the most common arena in which adaptive management is applied, as goals and objectives guiding restoration work are often more measureable and thus more attainable than in other management scenarios. As is the case with many NRM projects, however, there is difficulty in connecting ecological outcomes to specific management practices, procedures, and processes in the context of complex ecosystems that are functions of multiple variables that change over time and space. While the idea of being able to quantify successful outcomes is appealing, it is interesting to note that most success criteria in the river enhancement projects profiled by O'Donnell and Galat (2008) defaulted to measuring outputs of the adaptive management process such as area restored, habitat created, or number of a fish species caught by anglers (O'Donnell and Galat 2008). With a focus on improving the performance of adaptive management as a management technique, Weinstein et al. (1997) created quantitative and context specific measures of success based on a "bound of expectation" for successful restoration of a large-scale wetland restoration project. The authors combined expectations for specific parameters (including timeframe) into indices to guide adaptive management in a restoration project, comparing actual monitoring results against thresholds identified in the constructed indices that would trigger further management actions (Weinstein et al. 1997).

Second, it is difficult to effectively compare the performance of adaptive management programs or projects due to the highly contextualized nature of their individual goals and objectives—the context of one project's success will not necessarily mirror another's, even in similar ecological or social management scenarios.

McFadden et al. (2011) recognize that success, defined simply as a favorable or desired outcome, is relative to the ecological, social, and management context of each adaptive management project and can even vary within projects or between phases of the adaptive management cycle. Agencies such as the U.S. Department of the Interior (DOI) have addressed this challenge by publishing qualitative criteria for evaluating the success of outcomes related to adaptive management projects (Williams et al. 2009).

Another approach is to assess the process of implementing adaptive management. In the DOI *Technical Guide* for adaptive management, Williams et al. (2009) state, "the implementation of adaptive management is defined as successful if progress is made toward achieving management goals through a learning-based (adaptive) decision process." While somewhat vague, this is the most clearly articulated definition of successful adaptive management found in the literature to date. Focusing on process is promising because it promotes more successful implementation of adaptive management, and offers a potentially universal set of metrics since, regardless of context, all adaptive management efforts should be attempting to implement the adaptive management cycle. This approach is complicated, however, by the fact that there are very few examples of full iterations of the adaptive management cycle to evaluate. Indeed, most peer-reviewed articles on adaptive management projects discuss their application hypothetically (e.g., McCarthy and Possingham 2007, Broderick 2008). Also, the details of adaptive management processes may not be easily accessible to researchers or auditors interested in evaluating success. For example, dialogue and the micro-situation politics of the adaptive management goal setting or objective drafting process may be lost without adequate qualitative records or investigative research. However, some of the most well-known examples of longstanding adaptive management in action have successfully completed full iterations of the adaptive management cycle and may offer critical insight into the process. One of these examples is the Glen Canyon Dam Adaptive Management Program (GCDAMP) on the Colorado River in the southwest U.S.

In their review of the GCDAMP, Susskind et al. (2012) focus on process by identifying four strategies for successfully implementing adaptive management: "(1) clear overarching goals as well as concrete and measurable objectives to guide the management process; (2) well-defined fact-finding protocols to promote shared learning and manage scientific uncertainty; (3) tools and incentives that facilitate participation and foster collaboration; and (4) clear procedures for managing the programme [sic] adaptively and cultivating long-term capacity building." In theory, successful adaptive management follows a predetermined path (adaptive management cycle) with recognizable features (criteria) to work towards a shared management goal in the face of uncertainty.

In the DOI *Technical Guide* for implementing adaptive management, Williams et al. (2009) propose four indicators of success that they suggest could be universally applied to adaptive management projects: (1) "Stakeholders are actively involved and committed to the process"; (2) "Progress is made toward achieving management objectives"; (3) "Results from monitoring and assessment are used to adjust and improve management decisions"; and (4) "Implementation is consistent with applicable laws." These indicators are suggested to "promote successful implementation" of adaptive management, not necessarily to evaluate whether adaptive management resulted in successful outcomes (Williams et al. 2009). The authors recognize the importance of process in promoting adaptive management success, suggesting a nine-step model used to instill "learning-based management" (Williams et al. 2009). They present a comprehensive list of key questions to ask of each step in their model—questions that are aimed at helping practitioners more successfully implement adaptive management, but which could also be used to evaluate phases of the adaptive management cycle for adequate completion. In this manner, Williams et al. (2009) have created a path forward toward measuring success using an explicit evaluation of phases of the adaptive management cycle. While the DOI guide recognizes the value of focusing on process, there is a lack of explicit attention to systematically evaluating the individual phases of the adaptive management cycle as a means of holistically assessing an adaptive management program. What is needed to further this approach? We suggest that the next step in the evolution of efforts to evaluate adaptive management success is the development of explicit criteria for each phase including metrics that can be measured, compared, and peer-reviewed.

In the remainder of this chapter, we build on efforts to shift thinking about 'success' in adaptive management projects from a focus on measurable progress toward objectives, to a measure of programmatic success through an evaluation of the adaptive management cycle. Successful iterations of the adaptive management cycle will theoretically promote measurable progress toward overall objectives by institutionalizing the learning process inherent in an adaptive management approach. We suggest that a systematic evaluation of the individual phases of the adaptive management cycle will effectively uncover and explain potential miscues or oversights that may hinder adaptive management in subsequent phases. We emphasize the need for rigorous quantitative and qualitative research on the adaptive management process in a programmatic evaluation separate from the biophysical monitoring undertaken as a phase within the adaptive management process itself.

Measuring Success in the Adaptive Management Process

As stated above, adaptive management is confounded by an uncertainty in how success is defined, the contextual differences between adaptive management projects, and the difficulty in connecting adaptive management projects with outcomes and effects. However, if success is also defined in terms of rigorously completed phases of the adaptive management cycle, then the degree to which the process (adaptive management cycle) is completed effectively can be compared amongst adaptive management projects and across adaptive management programs, possibly yielding a more universal measure of success. We suggest that success can be measured by focusing on the process itself—outputs such as plans produced, project objectives adjusted, and iterations of the adaptive management cycle completed—rather than exclusively on outcomes relative to the substantive goals of the process.

But prior to any type of evaluation, it is helpful to review adaptive management programs to ensure appropriate situational application. Building upon a questionnaire presented by Williams et al. (2009), we present a decision table to assist in determining whether a NRM program has the potential to successfully execute adaptive management (Table 6.1). If a program does not meet the conditions presented, those conditions should be sought and attained where possible, since the likelihood of failure to produce meaningful results through adaptive management dramatically increases without them. Table 6.1 can also be used to evaluate existing adaptive management programs to determine potential weak components or to help diagnose problems.

Table 6.1 Evaluation criteria for the suitability of projects for adaptive management

	Condition for AM is met, success likely	Success will be fraught with challenges but is achievable	AM project not likely to be successful, potentially abandon AM as an approach to management
Was/Is there a mandate requiring a natural resource management decision?	Yes: mandate may be motivated by law, policy, agency directive, social pressure, competition for resources, or critical resource scarcity	Yes, but it is weak: mandate may be informal or critical mass to make a decision has not been achieved thus stalling the process in later stages	No
Is there uncertainty about the system in question, i.e. are consequences (both social and ecological) of management alternatives unknown?	Yes	No	
What is the scale and complexity of the AM project?	Small-scale or simple: for example, a single-species context in a protected area managed by one agency	Large-scale or complex: for example, a water allocation debate in a large, multi-jurisdictional basin with unsettled water rights	
Is there strong baseline data for the system in question for an AM project?	Yes	No, but it is easily attained through research. Peform the necessary research and proceed. If possible, include relevant stakeholders in baseline information gathering to promote learning processes in this early stage	No

Table 6.1 (continued)

	Condition for AM is met, success likely	Success will be fraught with challenges but is achievable	AM project not likely to be successful, potentially abandon AM as an approach to management
Can the system in question be manipulated easily, i.e. is there high controllability?	Yes	No	
Are all the relevant stakeholders engaged?	Yes	Some, but not all. If the AM process can engage stakeholders directly affected by AM decisions and those with ability to stall or block the process, AM can proceed	Not enough. If stakeholders with a direct stake in resource decisions to be made through the AM process cannot engage, the process is likely to fail
Are objectives of AM clear, measureable, and economically, politically, and ecologically feasible?	Yes	No. Reformulate goals and objectives until they are explicit, measureable, and feasible	No and there is no ability to modify them
Do stakeholders agree on AM objectives?	Yes	No. Stakeholders can seek conflict resolution avenues to reach agreement and proceed	No and conflict resolution is not an option or has been exhausted
Can the ecological system including current resource management regime and potential alternatives be modeled?	Yes	Maybe but capacity to do so is lacking. Attain necessary capacity (knowledge, skills, funding, etc.) and proceed	No
Can monitoring of resource management be designed, funded, and maintained?	Yes	Maybe but capacity to do so is lacking. Attain necessary capacity (knowledge, skills, funding, etc.) and proceed	No
Can management actions be modified (including implementation of necessary policy changes) from monitoring data?	Yes	Potentially, but certain modifications to law or policy are required. If those modifications can be made or adjusted for, proceed with AM	No

Assessing the Suitability of Adaptive Management in Different Contexts

Given a general definition of adaptive management success—measurable progress toward explicit objectives through strict attention to rigorous process—it is helpful to review factors that may or may not lead to success in adaptive management projects. Knowing when (and when not) to apply adaptive management has important implications prior to any attempt to measure success in adaptive management applications—including the conservation of precious (and often scarce) evaluation resources. If a situation does not merit the approach, there is a high likelihood that the approach will not be successful. Although this feels like common sense, adaptive management literature clearly points out that the most common reason for a lack of measureable success in adaptive management projects is the inappropriate application of the technique (Gregory et al. 2006, Allen and Gunderson 2011, McFadden et al. 2011). So what are the characteristics of a management scenario in which adaptive management would be an appropriate approach? Applying adaptive management as an approach to NRM management is appropriate when uncertainty and controllability are high (Allen and Gunderson 2011). If implemented correctly, adaptive management creates significant potential for learning through implementing experiments as management actions, thus enabling decision making in the face of uncertainty and allowing managers to adjust these decisions according to what is learned from rigorous monitoring of the experimentation. It is critical to have high levels of ecological and social controllability in a situation ripe for an adaptive management approach. Without a significant ability to manipulate the environment, adjusting management from monitoring of adaptive management experiments would not be possible and thus the adaptive management approach could not function. Knowing when to appropriately apply an adaptive management approach is the initial key to ensuring success in adaptive management.

Although not generally where the most complexity and uncertainty in management is found (e.g. Lee 1993), small and simple management contexts (for example, a one-species management focus in a small protected-area) have been shown to support more successful applications of AM (McConnaha and Paquet 1996, Gregory et al. 2006, Morghan et al. 2006). The more complete the data surrounding the social and ecological components of the system, the more likely adaptive management experimentation will produce meaningful results.

In addition, the approach to adaptive management, passive vs. active, may have some influence on the success rate of adaptive management projects. Morghan et al. (2006) suggest that the first attempt at adaptive management should be a "pilot study" for the *real* adaptive management implementation; a more passive approach with heavy modeling and introductory collaboration which can be used to generate the necessary conditions for active adaptive management. McCarthy and Possingham (2007) describe such an approach by using Bayesian statistics to passively model active adaptive management alternatives prior to experimental implementation in a small watershed context. In this way, passive adaptive management can be thought of as what Williams and Brown (2012) refer to as the "set-up phase"—part of the adaptive management cycle, but a precursor to the active experimentation of adaptive management. Adaptive management projects that engage in passive adaptive management prior to active adaptive management may be more likely to succeed because they have had an opportunity to 'work out the kinks' (Zellmer and Gunderson 2009).

Since adaptive management by definition necessitates stakeholder involvement, the type, level, and coordination of this involvement is a significant factor that either contributes to or detracts from adaptive management success. Building a community of shared understanding rather than opting for strict consensus in collaborative decision-making is more likely to foster adaptive management success (Zellmer and Gunderson 2009). Further, developing a shared understanding of the adaptive management *process* as well as the ecological context and management objectives is critical (Broderick 2008). Occasionally, institutional barriers to adaptive management will be presented by the lead agency or coordinating organization. One way to avoid this is to rely heavily on a stakeholder group, an appointed neutral third-party, or a series of elected/appointed advisory committees to carry out much of the adaptive management decision-making and implementation (Zellmer and Gunderson 2009, Smith 2011). Zellmer and Gunderson (2009) specifically suggest that a technically-based agency, such as U.S. Army Corps of Engineers (ACOE) is not the best choice for a coordinating or lead agency on an adaptive management project due to the narrow nature of its mission and expertise. If a management scenario is ripe for adaptive management according to an assessment based on Table 6.1, adaptive management should proceed with strict attention to process laid out in the adaptive management cycle (see Fig. 6.1 and Table 6.2). To support adaptive management success, we suggest a thorough review of each phase of the adaptive management cycle prior to moving on to subsequent phases. That review can be conducted as a self-assessment, but a robust peer-review will be more effective in determining potential areas of concern.

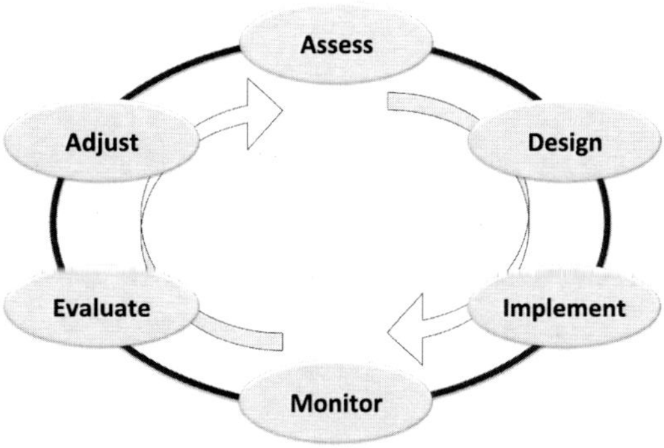

Fig. 6.1 The adaptive management cycle

Table 6.2 Criteria organized by phase of the adaptive management cycle

	Phase of AM Cycle					
	Assess	Design	Implement	Monitor	Evaluate	Adjust
Questions to ask of phase (qualitative inquiry)	Are all relevant stakeholders involved or engaged?	Are objectives explict, prioritized, shared, and measurable?	Has the program moved to action?	Who is in charge of monitoring? Is there benefit to or danger in joint-monitoring responsibility?	Was any new information learned?	Were adjustments made to experiments (management policies) in light of new information learned?
	Has clear and comprehensive baseline information been established?	Can management alternatives be posed as testable hypotheses?	Are experiments performed with the rigor of the scientific method?	Is there funding secured for the necessary and realistic monitoring timeframe?	Do the results match what was expected?	Do objectives still make sense in light of new knowledge or should adjustments be made?
Potential Metrics for measuring each phase (quantitative inquiry)	Stakeholder mapping: map the management scenario to determine individuals and groups who may be affected by a management decision; inform map through interviews and/or spatial analysis	1. Public and stakeholder surveys on objectives; 2. model management alternatives to determine potential outcomes; 3. design measurable indicators for measuring progress toward goals	Measure number and duration of completed experiments. What is the length and duration of committed funding for experimental policies?	Measure consistency of the monitoring program including overseeing agency, funding, and any interruptions during the duration of monitoring	Quantify new information learned. Integrate into models. Quantiatively compare observed vs. predicted data	Measure the number and breadth of adjustments made to management policies
External Peer-review	Review of baseline ecological data for management scenario	Review of models used	Review the procedures for implementing policies as experiments- was there a control, were they replicable?	Review of monitoring procedures- were the best possible practices used for an adequate duration of time?	Review collected data to ensure accurate interpretations were made	Review adjustments to management policies as accurate interpretations of monitoring as well as reasonable iterations of the AM cycle

Evaluating Phases in the Adaptive Management Cycle

The process of adaptive management is embodied in the adaptive management cycle (Fig. 6.1). Many variations of the adaptive management cycle have been proposed in the literature (see Murray and Marmorek 2003, Stankey et al. 2003, Duxbury and Dickinson 2007, Pahl-Wostl 2007, Williams et al. 2009, Fulton 2010). Although there are significant differences among some of the proposed models, there are consistencies as well, in that they all generally include six core phases: assess, design, implement, monitor, evaluate, and adjust. The value of this simplified six-phase conceptualization of the adaptive management process is its succinct portrayal of the feedback loop of learning that is created through iterations of the adaptive management cycle. Significant literature exists on the performance of each phase in the cycle and for the ease of use here, we have organized this information into a single table (see Table 6.2). We suggest that progress in adaptive management can be measured in terms of completion of the adaptive management cycle and the presence or absence of feedback (i.e. learning) in the next cycle evidenced in new information being applied to objectives, experimental design, and ultimately NRM policies (Schreiber et al. 2004). Any complications or failures in individual phases of the adaptive management cycle may adversely affect the quality of learning applied to subsequent iterations. For example, if left out of the 'assess' phase, important (or politically powerful) stakeholders will be removed from joint fact-finding efforts to collectively construct a baseline picture of existing ecological and social conditions. As a result, these stakeholders may stalemate future agreement on measurable project objectives during the 'design' phase or prohibit experiments through use of litigation during the 'implement' phase.

However, not all adaptive management projects can and will move at a desired pace, one that keeps in step with funding cycles and those agencies and organizations needing an evaluation of project success. So how can success be determined by an unfinished project, i.e. how can success in adaptive management be determined without a completed iteration of the adaptive management cycle? We suggest that a definition of success in adaptive management depends on effectively completing *individual* phases of the adaptive management cycle. Therefore, *measuring* success in adaptive management may have two distinct components: measuring progress toward explicit objectives created at the outset of an adaptive management project; and/or reviewing the success of implementing the adaptive management process through a criteria-driven assessment of individual phases of the adaptive management cycle.

Table 6.3 presents a framework to assist with a programmatic assessment of adaptive management through the review of adaptive management cycle phases. Our framework suggests three types of assessment: (1) a qualitative review, which includes questions to describe and compare details of the process, (2) a quantitative review including metrics to measure relative degree of phase completion; and (3) an assessment of how and what might be reviewed externally with the help of experts (Table 6.3). To demonstrate its usefulness, the following section will briefly analyze what is publically known about the GCDAMP in the context of the evaluation framework.

Table 6.3 A proposed framework for evaluating adaptive management processes

	Phase of AM Cycle					
	Assess	Design	Implement	Monitor	Evaluate	Adjust
Questions to ask of phase (qualitative inquiry)	Are all relevant stakeholders involved or engaged?	Are objectives explict, prioritized, shared, and measurable?	Has the program moved to action?	Who is in charge of monitoring? Is there benefit to or danger in joint-monitoring responsibility?	Was any new information learned?	Were adjustments made to experiments (management policies) in light of new information learned?
	Has clear and comprehensive baseline information been established?	Can management alternatives be posed as testable hypotheses?	Are experiments performed with the rigor of the scientific method?	Is there funding secured for the necessary and realistic monitoring timeframe?	Do the results match what was expected?	Do objectives still make sense in light of new knowledge or should adjustments be made?
Potential Metrics for measuring each phase (quantitative inquiry)	Stakeholder mapping: map the management scenario to determine individuals and groups who may be affected by a management decision; inform map through interviews and/or spatial analysis	1. Public and stakeholder surveys on objectives; 2. model management alternatives to determine potential outcomes; 3. design measurable indicators for measuring progress toward goals	Measure number and duration of completed experiments. What is the length and duration of committed funding for experimental policies?	Measure consistency of the monitoring program including overseeing agency, funding, and any interruptions during the duration of monitoring	Quantify new information learned. Integrate into models. Quantiatively compare observed vs. predicted data	Measure the number and breadth of adjustments made to management policies
External peer-review	Review of baseline ecological data for management scenario	Review of models used	Review the procedures for implementing policies as experiments- was there a control, were they replicable?	Review of monitoring procedures- were the best possible practices used for an adequate duration of time?	Review collected data to ensure accurate interpretations were made	Review adjustments to management policies as accurate interpretations of monitoring as well as reasonable iterations of the AM cycle

Testing the Framework: An Evaluation of the GCDAMP Process

The GCDAMP is one of the most discussed adaptive management projects ever envisioned and thus one of the most high profile. The Glen Canyon Dam (GCD–completed in 1963) is the largest feature of the Colorado River Storage Project[1]. The hydroelectric power generation facility boasts a capacity of 1320 megawatts and can service 1.3 million residential customers. However, when completed, the GCD significantly altered the Colorado River ecosystem both above and below its location. Above the dam, Glen Canyon created Lake Powell, inundating almost 200 miles of river habitat and important Native American cultural sites. However, Lake Powell became a highly sought after recreation destination, the use of which is protected through federal law as a National Recreation Area—Glen Canyon National Recreation Area (GCNRA). Below the dam, flows of the Colorado River enter Grand Canyon National Park (GCNP), and have been significantly altered—dramatic, silt-laden seasonal flows have been replaced by reservoir cooled, clear flows that fluctuate daily (although at relatively low volumes) in response to peaking energy production needs. As a result, important resources in the Grand Canyon have suffered including endangered species habitat of the humpback chub and Kaibab ambersnail. All told, eight endangered and three threatened species call Grand Canyon home and all are affected in some way by the controlled flows from GCD. In addition, GCD flows impact access to important Native American cultural sites in GCNP (historic dwellings, sites of origin stories, and traditional sacred lands) and the seasonal building of beaches that serve as camping locations for the over 20,000 river users who visit each year.

In the late 1980s and in response to pressures mounting from the applicability of the ESA to the operations of GCD, the Bureau of Reclamation ordered a study of GCD operations and its potential effects on GCNP and GCNRA. This process led to the passage of the Grand Canyon Protection Act (GCPA) of 1992 which provided a direction for an environmental impact statement (EIS) to be conducted with the goal to "protect, mitigate adverse impacts to and improve the values for which [GCNP] and [GCNRA] were established" (Public Law 102–575). The EIS took five years, the efforts of 15 agencies, and more than 40 individual research projects/efforts. The final Record of Decision (ROD) was signed by the Secretary of Interior in 1996, and among other things, *mandated* an AM approach to management of the resources of GCNP which included GCD operations. At the same time, Interior Secretary Bruce Babbitt created a Federal Advisory Committee, which would become the GCD Adaptive Management Working Group (AMWG).

---

[1] Information for this section was taken from the following websites and pages contained within, unless otherwise referenced: Glen Canyon Dam Adaptive Management Program Website (http://www.gcdamp.gov/); U.S. Bureau of Reclamation GCDAMP website (http://www.usbr.gov/uc/rm/amp/); and the Grand Canyon Monitoring and Research Center website (http://www.gcmrc.gov/).

The goals of the Working Group (and the greater GCDAMP) were to advise the Secretary on how to manage GCD to improve the resources of importance in GCNP and GCNRA outlined in the ROD—through the use of adaptive management. Ultimately, the AMWG guides the GCDAMP: "The AMP provides a process through which the effects of dam operations on downstream resources are monitored and assessed; and operational adjustments are recommended to the Secretary of the Interior based on those assessments" (GCDAMP 2010). The AMWG is charged with determining how the dam can be operated to provide water and power while protecting and improving natural and cultural resources downstream.

Since its official inception in 1997, the GCDAMP has built an inclusive community of affected stakeholders in the AMWG. With the assistance of the Grand Canyon Research and Monitoring Center (GCRMC—established in 1995 and operated by the U.S. Geological Survey (USGS) for the GCDAMP) and the Technical Work Group subcommittee (TWG), the AMWG has been responsible for several high-flow experiments aimed at testing hypotheses on how to rebuild sediment loads and resultant sandbar (beach) habitat in GCNP. The AMWG has also been responsible for the experimental removal of non-native trout species near the Colorado River confluence with the Little Colorado to increase survival of the endangered humpback chub. The GCDAMP has been both heralded as a successful example of adaptive management as well as an example of how the process can fail (Gunderson and Light 2006, Zellmer and Gunderson 2009, Susskind et al. 2010, Susskind et al. 2012). Since the management scenario charged to the GCDAMP appears to be ripe for adaptive management implementation (at first glance it is: high uncertainty, high controllability through protected areas, inclusive of major stakeholders, adequate funding and capacity, etc.), we can move forward with an abbreviated review of the GCDAMP in the context of our framework for assessing adaptive management success.

Assess

A review of public records and Internet resources concerning the GCDAMP, suggests that the program—at least on the surface—took the 'assess' phase of the adaptive management cycle seriously. The GCPA of 1992 provided direction for a comprehensive EIS, the components of which spanned 40 individual research projects across 15 agencies and took five years—creating a robust set of baseline data from which to build on under the direction of adaptive management. The AMWG webpage lists its stakeholder members, which include relevant state and federal agencies including the Western Area Power Administration, the states with interests in the Colorado Basin (AZ, CA, CO, NM, NV, UT, and WY), local Native American tribes, environmental and conservation groups, recreational user groups, and federal hydropower purchase contractors (power distributors). Although this appears to be a comprehensive list of major stakeholders, it is unclear whether or not any group or population has been marginalized by exclusion, weak representation, or dynamics within the working group itself (see Gunderson and Light 2006). No evidence of quantitative metrics for assessing stakeholder participation are evident, however, some measure or scoping evaluation may be contained in the original EIS studies. Evidence of peer-review on many of the baseline ecological studies exists, and one stakeholder, the Southern Paiute Consortium, commissioned a study on the effectiveness of its own participation in the AMWG (Austin et al. 2007). This is an excellent example of the use of qualitative research and peer-review to evaluate participation in an adaptive management project, and could be expanded to encompass the contributions and effectiveness of all stakeholders.

Design

According to the Bureau of Reclamation's website for the GCDAMP, management objectives for adaptive management were (and are, in the case of adjustments) created through the following procedure: the AMWG collectively determines and prioritizes broad goals and policies; a subcommittee (the Technical Working Group—TWG—made up of technical representatives from each member organization of the AMWG) determines measurable objectives, research standards, and monitoring procedures; and the Grand Canyon Research and Monitoring Center (USGS) is responsible for performing the experiments and continuing monitoring. In designing the GCDAMP, the AMWG created categories of resources central to its directives under the GCPA of 1992 and accessible through an adaptive management approach. These categories include cultural resources, endangered species, hydropower program, recreational trout fishery, native fishes, sediment, storage, and whitewater recreation. While experiments initiated in the design phase have addressed many if not all of these priority resource categories since the 1996 ROD, we will simplify this cursory review of the GCDAMP by focusing on only one resource category, sediment. According to a 1996 document entitled *Glen Canyon Dam Management Objectives*, the majority of the objectives crafted to manage sediment in the Colorado River are explicit and measurable—a critical requirement of the 'design' phase of adaptive management. For example, one objective for sediment reads as follows:

> "As a minimum, maintain the number and average size (area and thickness) of sandbars between the stages associated with flows of 8000 and 45,000 cfs and the number and average size of backwaters at 8000 cfs that existed during the 1990–1991 research flows" (Ralston et al. 1998).

This objective is targetable and measurable due to its use of numeric measurements. Stakeholder input from the AMWG was translated through the TWG into objectives, testable hypotheses, and models used to predict potential effects of "beach/habitat-building flows" on GCNRA and GCNP resources (USGS 2011a).

Implement and Monitor

Three high-flow experiments (HFEs) have been performed during the life of the GCDAMP: March 1996, November 2004, and March 2008. The GCRMC has designed and performed these HFEs, under the overarching goal "to determine whether high flows could be used to benefit important resources in [GCNRA] and [GCNP] that have been affected by the existence and operation of [GCD]" (USGS 2011b). As the lead agency in the GCRMC, the USGS has continuously and extensively monitored these events and published a comprehensive report on their findings (USGS 2011a). While an evaluation of monitoring efforts is beyond the scope of this analysis, it is apparent that adequate resources are available to monitor these experiments and it can be assumed that the USGS employs the best practices for monitoring design and data collection.

Evaluate

The three HFEs allowed researchers to uncover uncertainties about the dynamics of sediment transport in the Colorado River below GCD. Specifically, researchers found that the primary sources of sediment to the river post-GCD are the Little Colorado River and the Paria River tributaries (USGS 2011a). Prior to the HFEs, most of this tributary sediment was not stored in the river bed, but instead transported downstream as a result of typical dam operations (USGS 2011b). Through monitoring the sediment transport before, during, and after the HFEs, scientists learned that the timing of HFEs matter—strategically timed HFEs following natural tributary flood events allowed for the efficient use of recently (and naturally) added tributary sediment in rebuilding Colorado River sandbars (USGS 2011b).

Adjust

The monitoring performed in coordination with the three HFEs allowed scientists to better understand how sandbars form (sand from lower elevation bars accumulates to build higher elevation bars) as well as how the timing of HFEs corresponds with sandbar construction and status-quo dam operations increase sandbar erosion (USGS 2011b). These three findings have allowed scientists, through consultation with the AMWG and the TWG subcommittee, to adjust future flow experiments (2004, 2008, 2012, and most recently 2013) to more effectively transport sediment and build sandbars along the Colorado River in GCNP. For example, the 2004 and 2008 HFEs were shorter in duration than the 1996 event and timed to occur after a significant sediment release by tributary streams (USGS 2011b). The USGS (2011b) reports that "from February 1996 to October 2008—the span of the three HFEs—75 percent of the sandbars at long-term study sites in Grand Canyon experienced net increases in volume, despite ongoing sandbar erosion between HFEs". There is little evidence available to determine whether root objectives have been modified as a result of the findings of these adaptive management experiments, but it is obvious that the experiments themselves have been adjusted in light of information gained during the adaptive management process.

Conclusion

This analysis of the phases of adaptive management in the GCDAMP illustrates the inherent value of the adaptive management approach to management. While the use of the scientific method in a relatively short time period to directly influence flow management from GCD demonstrates an adaptive management approach to natural resource management, the infrastructure and decision-making processes inspired by the adaptive management cycle are of paramount importance. The flow information has institutionalized learning—from the members of the AMWG, through the TWG, to agency scientists performing experiments, and then fed back to influence dam operations through the policy-making discretion of the AMWG. In this case, adaptive management offers a more explicit, inclusive, and rigorous approach to managing a social-ecological system surrounded by uncertainty.

There is of course the long-standing argument that the whole is never simply equal to the sum of its parts, and that argument could be leveraged against evaluating adaptive management via process alone. Does it stand that just because each phase of the cycle is complete and accurate that adaptive management as a whole will be successful? Maybe not, but what *will* take place as a result of completing the adaptive management cycle is *learning*, and specifically, learning on a system-wide scale. If strict attention is paid to each phase of adaptive management, the process will inherently guide learning that was not previously incorporated into management or was potentially absent from the system as a whole. Learning may still take place even if some process components or phases are incomplete or weak, but results will not be as robust. Thus capacity and resources should be leveraged to address weak phases in the adaptive management cycle.

What is needed moving forward is not more lip service to adaptive management, but instead more investment in training managers to implement an adaptive management process and thus institutionalize a culture of learning as a tool for management. Instead of simply granting agencies money to apply adaptive management to stream or wetland restoration projects, grants should promote workshops to engage stakeholder leadership in the initiation of the adaptive management process and build the capacity for self-directed adaptive management. Local universities, specifically land-grant institutions with a directive of agricultural and natural resource service through extension activities, are perfectly poised to provide peer-review for adaptive management projects, providing valuable feedback during each phase of the adaptive management cycle. Not every resource management scenario ripe for adaptive management will encompass the scale or necessitate the investment that the GCDAMP boasts, but it is strikingly apparent that very few management communities currently have the capacity for implementing an adaptive management approach. Despite the lack of documented success in the literature, including the difficulty in measuring success, there is hope for adaptive management. Adaptive management represents a tool for creating science-management-policy networks that can legitimately guide natural resource management by inclusive stakeholder investment and the implementation of best available science. But a culture shift in natural resource management may be necessary—when the adaptive management process becomes as important as its outputs—only then can we hope to see success defined by the institutionalization of learning.

References

Alexander, G. G., & Allan, J. D. (2007). Ecological success in stream restoration: Case studies from the Midwestern United States. *Environmental Management, 40,* 245–255.

Allen, C. R., & Gunderson, L. (2011). Pathology and failure in the design and implementation of adaptive management. *Journal of Environmental Management, 92,* 1379–1384.

Austin, D., Phillips, A., & Seibert, D. (2007). *Southern Paiute participation in the Glen Canyon adaptive management program: A ten year review*. Tucson: University of Arizona, Bureau of Applied Research in Anthropology.

Bearlin, A. R., Schreiber, E. S. G., Nicol, S. J., Starfield, A. M., & Todd, C. R. (2002). Identifying the weakest link: Simulating adaptive management of the re-introduction of a threatened fish. *Canadian Journal of Fisheries and Aquatic Sciences, 59,* 1–8.

Broderick, K. (2008). Adaptive management for water quality improvement in the Great Barrier Reef catchments: Learning on the edge. *Geographical Research, 46,* 303–313.

Burbridge, P. R. (1997). A generic framework for measuring success in integrated coastal management. *Ocean and Coastal Management, 37,* 175–189.

Day, J. (2008). The need and practice of monitoring, evaluating and adapting marine planning and management—lessons from the Great Barrier Reef. *Marine Policy, 32,* 823–831.

Downs, P. W., & Kondolf, G. M. (2002). Post-project appraisals in adaptive management of river channel restoration. *Environmental Management, 29,* 477–496.

Duxbury, J., & Dickinson, S. (2007). Principles for sustainable governance of the coastal zone: In the context of coastal disasters. *Ecological Economics, 63,* 319–330.

Fontaine, J. J. (2011). Improving our legacy: Incorporation of adaptive management into state wildlife action plans. *Journal of Environmental Management, 92,* 1403–1408.

Fulton, E. A. (2010). Approaches to end-to-end ecosystem models. *Journal of Marine Systems, 81,* 171–183.

GCDAMP (Glen Canyon Dam Adaptive Management Program). (2010) Adaptive management program purpose and goals. Glen Canyon Dam Adaptive Management Program: http://www.gcdamp.gov/fs/amp_pg.pdf. Accessed 26 Sept 2013.

Gregory, R., Ohlson, D., & Arvai, J. (2006). Deconstructing adaptive management: Criteria for applications to environmental management. *Ecological Applications, 16,* 2411–2425.

Gunderson, L., & Light, S. S. (2006). Adaptive management and adaptive governance in the everglades ecosystem. *Policy Sciences, 39,* 323–334.

Hockings, M. (2003). Systems for assessing the effectiveness of management in protected areas. *BioScience, 53,* 823–832.

Kapos, V., Balmford, A., Aveling, R., Bubb, P., Carey, P., Entwistle, A., Hopkins, J., Mulliken, T., Safford, R., Stattersfield, A., Walpole, M., & Manica, A. (2008). Calibrating conservation: New tools for measuring success. *Conservation Letters, 1,* 155–164.

Kenney, D. S. (2001). Are community-based watershed groups really effective? Confronting the thorny issue of measuring success. In P. Brick, D. Snow, & S. Van de Wetering (Eds.), *Across the great divide: Explorations in collaborative conservation and the American West* (pp. 188–193). Washington, D.C.: Island Press.

Lee, K. (1993). *Compass and gyroscope*. Washington, D.C.: Island Press.

McCarthy, M. A., & Possingham, H. P. (2007). Active adaptive management for conservation. *Conservation Biology, 21,* 956–963.

McConnaha, W. E., & Paquet, P. J. (1996). Adaptive strategies for the management of ecosystems: The Columbia River experience. *American Fisheries Society Symposium, 16,* 410–421.

McFadden, J. E., Hiller, T. L., & Tyre, A. J. (2011). Evaluating the efficacy of adaptive management approaches: Is there a formula for success? *Journal of Environmental Management, 92,* 1354–1359.

Morghan, K. J. R., Sheley, R. L., & Svejcar, T. J. (2006). Successful adaptive management—The integration of research and management. *Rangeland Ecology and Management, 59,* 216–219.

Murray, C., & Marmorek, D. (2003). Adaptive management and ecological restoration. In P. Freiderici (Ed.), *Ecological restoration of southwestern ponderosa pine forests* (pp. 417–428). Washington, D.C: Island Press.

O'Donnell, T. K., & Galat, D. L. (2008). Evaluating success criteria and project monitoring in river enhancement within an adaptive management framework. *Environmental Management, 41,* 90–105.

Olsen, S., Tobey, J., & Kerr, M. (1997). A common framework for learning from ICM experience. *Ocean and Coastal Management, 37,* 155–174.

Pahl-Wostl, C., Sendzimir, J., Jeffrey, P., Aerts, J., Berkamp, G., & Cross, K. (2007). Managing change toward adaptive water management through social learning. *Ecology and Society, 12*(2), 30. http://www.ecologyandsociety.org/vol12/iss2/art30/

Parrish, J. D., Braun, D. P., & Unnasch, R. S. (2003). Are we conserving what we say we are? Measuring ecological integrity within protected areas. *BioScience, 53,* 851–860.

Ralston, B., Winfree, R., & Gold, B. (1998). Beach Habitat Building Flow Resource Criteria: A process document. U.S. Bureau of Reclamation. http://www.usbr.gov/uc/rm/amp/twg/mtgs/99jan11/Attach_01a.pdf. Accessed 30 Dec 2014.

Schreiber, E. S. G., Bearlin, A. R., Nicol, S. J., & Todd, C. R. (2004). Adaptive management: A synthesis of current understanding and effective application. *Ecological Management and Restoration, 5,* 177–182.

Smith, C. B. (2011). Adaptive management on the central Platte River—Science, engineering and decision analysis to assist in the recovery of four species. *Journal of Environmental Management, 92,* 1414–1419.

Stankey, G. H., Bormann, B. T., Ryan, C., Shindler, B., Sturtevant, V., Clark, R. N., & Philpot, C. (2003). Adaptive management and the Northwest forest plan: Rhetoric and reality. *Journal of Forestry, 101,* 40–46.

Stojanovic, T., Ballinger, R. C., & Lalwani, C. S. (2004). Successful integrated coastal management: Measuring it with research and contributing to wise practice. *Ocean and Coastal Management, 47,* 273–298.

Susskind, L., Camacho, A. E., & Schenk, T. (2010). Collaborative planning and adaptive management in Glen Canyon: A cautionary tale. *Colombia Journal of Environmental Law, 35,* 1–55.

Susskind, L., Camacho, A. E., & Schenk, T. (2012). A critical assessment of collaborative adaptive management in practice. *Journal of Applied Ecology, 49,* 47–51.

USGS (United States Geological Survey). (2011a). Effects of Three High-Flow Experiments on the Colorado River Ecosystem Downstream from Glen Canyon Dam, Arizona. Circular 1366. Geological Survey. Reston, VA, U.S.A.

USGS (United States Geological Survey). (2011b). Three Experimental High-Flow Releases from Glen Canyon Dam, Arizona—Effects on the Downstream Colorado River Ecosystem. Fact Sheet 2011–3012. Geological Survey Reston, U.S.A.

Weinstein, M. P., Balletto, J. H., Teal, J. M., & Ludwig, D. F. (1997). Success criteria and adaptive management for a large-scale wetland restoration project. *Wetlands Ecology and Management, 4,* 111–127.

Williams, B. K., & Brown, E. D. (2012). Adaptive management: The U.S. Department of the Interior applications guide. U.S. Department of the Interior. Adaptive Management Working Group. Washington, D.C., U.S.A.

Williams, B. K., Szaro, R. C., & Shapiro, C. D. (2009). Adaptive management: The U.S. Department of Interior technical guide. U.S. Department of the Interior. Adaptive Management Working Group. Washington, D.C., U.S.A.

Zellmer, S., & Gunderson, L. (2009). Why resilience might not always be a good thing: Lessons in ecosystem restoration from Glen Canyon and the Everglades. *Nebraska Law Review, 87,* 893–949.

Chapter 7
The Role of Bridging Organizations in Enhancing Ecosystem Services and Facilitating Adaptive Management of Social-Ecological Systems

Olivia Odom Green, Lisen Schultz, Marmar Nekoro and Ahjond S. Garmestani

Keywords Resilience · Adaptive management · Adaptive governance · Bridging organizations · Ecosystem services

Adaptive Management, Panarchy, and Environmental Policy

Adaptive management is an approach for monitoring the response of ecological systems to different policies and practices and attempts to reduce the inherent uncertainty in ecological systems via system monitoring and iterative decision making and experimentation (Holling 1978). Monitoring is an essential aspect of adaptive management, as information from the system (e.g., monitoring data) feeds back into the management process in an iterative manner that allows managers to adapt to changing circumstances (Green and Garmestani 2012). Management actions are hypotheses to be put "at risk" in an adaptive management framework, and

O. Odom Green (✉)
U.S. Environmental Protection Agency,
26 W. Martin Luther King Drive, Cincinnati, Ohio 45268, USA
e-mail: green.olivia@epa.gov

A. S. Garmestani
Office of Research and Development, U.S. Environmental Protection Agency,
26 W. Martin Luther King Drive, Cincinnati, Ohio 45268, USA
e-mail: garmestani.ahjond@epa.gov

L. Schultz
Stockholm Resilience Centre, Stockholm University,
106 91 Stockholm, Sweden
e-mail: lisen.schultz@stockholmresilience.su.se

M. Nekoro
Stockholm University, Baltic Sea Centre,
106 91 Stockholm, Sweden
e-mail: marmar.nekoro@su.se

C. R. Allen, A. S. Garmestani (eds.), *Adaptive Management of Social-Ecological Systems*,
DOI 10.1007/978-94-017-9682-8_7

information that allows for learning is generated to improve management decisions (Green and Garmestani 2012).

Scale is the critical variable in monitoring and, therefore, policy associated with linked social-ecological systems. However, organizational arrangements can act as barriers for managing at the appropriate scale (Garmestani et al. 2009). One useful model for characterizing social-ecological systems and the organizations that manage these systems is panarchy, which describes the cross-scale interactions of a social-ecological system (Gunderson and Holling 2002). For our purposes, panarchy is useful because environmental policy often falls into the "one-size fits all" trap that can lead to adverse policy outcomes, as there is no "best" scale for implementation of policy for linked social-ecological systems (Brock and Carpenter 2007).

While panarchy shares similarities with traditional hierarchical models of organization, panarchy differs in key elements. Unlike the top-down control envisioned in traditional hierarchies, connectivity between adaptive cycles in a panarchy can be from levels above or below. In a hierarchy, lower-level patterns and processes are dominated by higher levels in the hierarchy. Panarchy differs from this characterization of nesting, with respect to complex systems , in that conditions can arise that trigger bottom-up (i.e., cross-scale cascading) change in the system (Garmestani et al. 2009). This model of social-ecological systems more accurately captures the "surprise" or uncertainty inherent in such systems. For example, cumulative impacts have the capacity to "scale up" in terms of their effect (Ruhl et al. 2007). As an illustration, large scale destruction and degradation of wetlands, and the ecological services associated with those wetlands, has occurred primarily as a result of innumerable, small conversions of wetlands for agricultural and urban development - a tyranny of many small decisions (Ruhl et al. 2007). Within this context, no single conversion of a wetland appears to have much of an impact upon the delivery of ecosystem services (Ruhl et al. 2007). However, the cumulative impact of small-scale wetland transformations has produced large-scale degradation of the ecosystem services associated with those wetlands (Ruhl et al. 2007). Further, levels in a panarchy are not static states, but rather adaptive cycles that are interconnected to other adaptive cycles in the panarchy. Each cycle operates over a discrete range of scale in both time and space and is connected to adjacent levels (adaptive cycles). It is important to note that adaptive cycles do not exist in isolation. Since adaptive cycles operate over specific ranges of scale, a system's resilience is dependent upon the interactions between processes and structures at multiple scales (Gunderson and Holling 2002).

Gunderson et al. (1995) highlight the importance of panarchy for the management of social-ecological systems. Numerous case studies document the management failures that have manifested from focusing upon a single scale, resulting in crises that may have otherwise been avoided (Gunderson 1999). Since social-ecological systems are not scale-free, and can exhibit non-linear change, the generation of adaptive capacity in management is an investment or insurance policy for sustainability (Gunderson 1999). Adaptive capacity in social systems is characterized by past history and local knowledge, as well as open and frequent lines of communication between institutions at multiple scales. As a result of the realization that adaptive management of nested social-ecological systems requires collaboration across sectors and scales, the term adaptive co-management was coined in the early 2000s (Ruitenbeek and Carter 2001, Olsson et al. 2004). In 2005, the concept was further developed into adaptive governance (Folke et al. 2005), recognizing that adaptive management requires an adaptive institutional and policymaking context and the collaboration between state and non-state actors in order to develop.

Gunderson (1999) has characterized the policymaking process as an adaptive cycle in which policies are formed and implemented (r); concern manifests as policies begin to show signs of failure (K); the policies fail (Ω); and policies are reorganized (α). Societies often react slowly to problems that may not appear to be imminent but in reality demand immediate attention (Scheffer et al. 2003). This phenomenon is exacerbated in societies with strong social control, as well as situations in which the problem is downplayed by a leader or by competition for attention from other problems (Scheffer et al. 2003). Brock (2006) asserts that regime shifts may be driven by endogenous or exogenous change, thus identifying the type of driver is paramount. Brock's (2006) model of tipping points in policy shows how group pressure to conform can overcome free-riders to produce regime shifts in policy. In the model, policy can remain relatively constant, but over time, small or slow changes can result in policy regime shifts. Alternatively, a strong leader can reduce the time before there is a shift in policy to deal with the problem (Scheffer et al. 2003).

According to Kingdon (1995), there are two types of policy windows: a problem-driven window and a politically-driven window. A problem-driven window opens when a policymaker believes that a policy is necessary for a specific issue. A politically-driven window is driven by a particular theme adopted by a policymaker who looks for problems that fit within the theme. Kingdon (1995) contends that significant changes in policy occur when conditions (e.g., problems, solutions, and politics) converge at the same time, which creates the window of opportunity for change. Wood and Doan (2003) assert that defining how an issue becomes perceived as a public problem is critical to agenda setting. Their research model demonstrates that problem definition can result in regime shifts via slow or sudden change. Wood and Doan (2003) found that if most individuals accept a particular condition, negative feedback works to maintain public opinion in that particular regime. However, if the individuals in the regime develop a critical mass of distaste for a particular issue, public opinion can cross a threshold and reorganize into an alternative regime. Importantly, interest groups, the media, and other agents can have an effect on agenda setting and creating the climate necessary for a shift in public opinion (Wood and Doan 2003).

Olsson et al. (2004) studied environmental policy in Sweden and Canada and concluded that this form of adaptive co-management of ecological systems was most effective when there was: Leadership with vision for the system of interest; legislation that created the environment for adaptive management; funds for adaptive management; monitoring of the ecological system; information flow (i.e., cross-scale linkage); combination of a variety of sources of knowledge; and a venue for collaboration. Olsson et al. (2004) contend that these factors are critical to building resilience in social-ecological systems, as they help to protect the system from the failure of management decisions under uncertainty (i.e., imperfect information). Skjaerseth (2006), in a study of regime interaction between institutions governing marine pollution in the northeast Atlantic Ocean, asserts that there are two factors that explain successful environmental management in this region: (1) all of the institutions in the region worked on essentially the same environmental issue; and (2) a long history of cooperation between the institutions in the area, which allowed for adaptation in the dynamics between the institutions over time. Olsson et al. (2006) contend that leaders and shadow networks are the key components to fostering change in the management of social-ecological systems. Alternative strategies can be developed and tested by shadow networks, which enable these policies to be rapidly implemented should a window of opportunity open (Olsson et al. 2006). Shadow networks and bridging organizations composed of various interested parties result in better matches between policy and the scale of the system being managed (Olsson et al. 2004, 2006).

Bridging Organizations

One of the most critical aspects in the panarchy of ecological systems and organizations appears to be organizations that monitor the status of the system and manifest rapid change, if conditions are deteriorating (Kinzig et al. 2003). The nested nature of social-ecological systems across scales requires a multi-scale approach to such monitoring and response, and Stokke and Coffey (2006) assert that the flow of concepts and ideas between organizations at different scales is critical to facilitation of effective environmental management. However, in many cases this flow is hindered by hierarchical structures and bureaucratic procedures. Recent research therefore suggests that bridging organizations that facilitate collaboration and learning across sectors and scales are key to adaptive governance (Hahn et al. 2006, Berkes 2009, Schultz 2009, Crona and Parker 2012). It has even been argued that bridging organizations are essential for adaptive governance to occur (Schultz 2009).

Bridging organizations were first mentioned in the development literature by Brown (1991), as organizations that span the gaps among diverse constituencies to work on development problems. He suggested that bridging organizations can play key roles in building local organizations, creating horizontal linkages, increasing grassroots influence on policy, and disseminating new visions and organizational innovations (Brown 1991). In the literature on natural resource management, the concept of bridging organizations was introduced by Westley (1995) and further developed by Folke et al. (2005), Hahn et al. (2006), Cash et al. (2006), Hahn et al. (2008), Malayang et al. (2007), Olsson et al. (2007), Berkes (2009), Garmestani et al. (2009), Schultz (2009) and Crona and Parker (2012). In these papers, bridging organizations are discussed in relation to adaptive co-management and the authors describe the role of these organizations in creating arenas for trust-building , knowledge generation, collaborative learning, preference formation, and conflict resolution among actors in relation to specific environmental issues. A central network position in combination with specific leadership skills help bridging organizations facilitate vertical and horizontal collaboration, provide sense-making and identification of common interests, filter external threats and redirect such threats into opportunities (Schultz 2009).

A recent definition of bridging organizations is put forward by Crona and Parker (2012), who define them as "organizations that link diverse actors or groups through some form of strategic bridging process", adding that they differ from informal social networks in that they are organizations in their own right with personnel and resources that are distinct from the parties they work to link. As such, they provide a long-term platform for learning and collaboration, where adaptive management can be initiated, coordinated and sustained over time (Schultz 2009). In the next sections we provide a practical example of a bridging organization active in southern Sweden, and the impact that this organization has had on ecosystem services generated in this area.

The Case of Kristianstads Vattenrike

Kristianstads Vattenrike roughly translates as Kristianstad's Water Realm and represents both a geographical area and a municipal initiative for ecosystem management. Geographically, the Kristianstads Vattenrike encompasses the catchment area of the Helgeå River within the municipality of Kristianstad (Fig. 7.1). The 1100 km^2 includes Sweden's largest flooded meadows used for grazing and hay-making. Many of the unique biological and cultural values of the area are associated with these social-ecological systems, which require active management and annual flooding to be sustained. Other habitats include two shallow lakes, large beech forests, wet forests and willow bushes in the lowlands and sandy grasslands with unique flora and fauna. Much of the area is agricultural land; the sandy and clay soils around Kristianstad have been and still are important for agricultural production and the area is one of the most productive in Sweden. The area also holds the largest groundwater reserve in northern Europe and the city Kristianstad with 30,000 inhabitants.

The Kristianstads Vattenrike is known for its rich fauna and flora, and in 1975, the 35 km stretch of wetlands along the lower Helgeå River from Torsebro to the Hanö Bay in the Baltic Sea was granted protection by the Convention on Wetlands of International Importance. It became known as the Ramsar Convention Site and the County Administration Board became responsible for the management. The Ramsar Convention on Wetlands provided a framework for protecting wetland areas from further exploitation, and in its official plan from 1975, the County Administrative Board suggested that almost the whole area, 49 km^2, should become a nature reserve. However, in 1989, only three percent of the Ramsar Convention Site was protected by reserves (Magnusson et al. 1989). Ownership within the Kristianstads Vattenrike area is mixed between private and public.

During the 1980s, several biological inventories showed that values of the lower Helgeå River and the Ramsar Convention Site were decreasing, despite a number of plans, policy documents, and protection efforts. A major reason was that the area

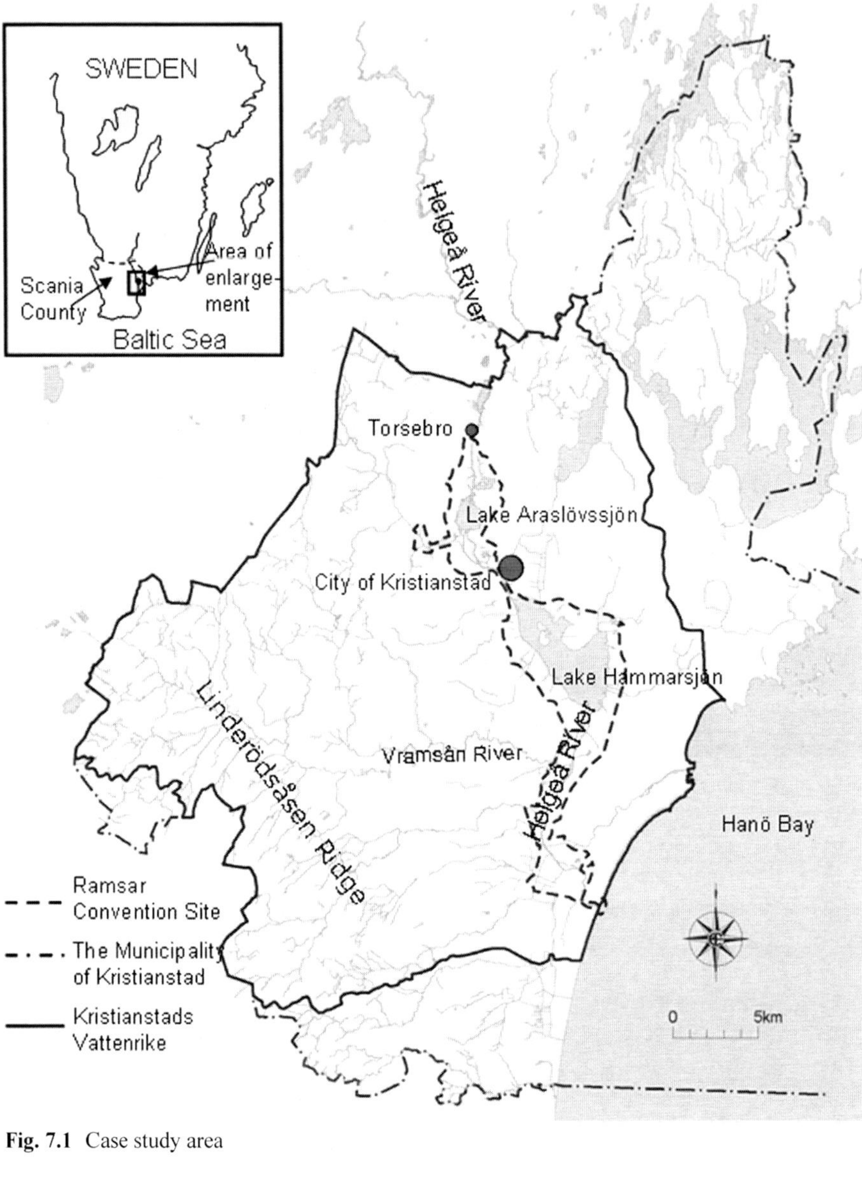

Fig. 7.1 Case study area

of flooded meadows used for haymaking and grazing had decreased dramatically (Magnusson et al. 1989). Even flooded meadows in nature reserves on state owned land were deteriorating. There was a growing concern that giving the wetlands of the lower Helgeå River Ramsar Convention Site status was not enough to sustain the natural and cultural values of the area. As a response actors started to self-organize and a multilevel governance network for the wetland landscape emerged, with the organization Ecomuseum Kristianstad Vattenrike as a key bridge between local actors and higher levels of governance. The Ecomuseum Kristianstads Vattenrike was launched in 1989 as a municipal organization co-financed by the county board administration. Their first project was to restore flooded meadows in collaboration with farmers. They also enhanced access to the wetlands for recreational and educational purposes and worked to change the perception of wetlands from "water-logged swamps" to a "water realm." Since then, they expanded their work across the landscape, and in June 2005, the area was given the status of Biosphere Reserve by UNESCO, thereby becoming a model for sustainable development. In response, the Ecomuseum Kristianstads Vattenrike changed its name to the Biosphere Office.

The Kristianstads Vattenrike case has been extensively studied by resilience researchers. Olsson et al. (2004) described the transformation of governance that led to the establishment of Ecomuseum Kristianstads Vattenrike as a bridging organization. Hahn et al. (2006, 2008) analyzed the strategies of Ecomuseum Kristianstads Vattenrike, and Schultz et al. (2007) analyzed the local steward networks that are involved in their work. This research shows that the Biosphere Office provides a crucial link between local stewards, governmental administrations, and scientists. Local stewards contribute on-site management and monitoring efforts, local ecological knowledge, and links to specialized expert networks, and governmental administrations provide access to funding resources, larger data-sets, maps, etc., and the capacity to undertake coordinated efforts (Schultz et al. 2007, Hahn et al. 2008). The bridging organization facilitates this collaboration by identifying the key actors, engaging in building personal relationships with them, and building an attractive and clear vision for the area with a flexible approach to achieving it. Furthermore, they ensure that management is in tune with the ecosystem by synthesizing knowledge from various sources, documenting local ecological knowledge as well as scientific knowledge, and facilitating learning between actors.

An important task for environmental managers is to build support and motivation for management of ecosystem services among citizens. The Biosphere Office Kristianstads Vattenrike invests in enhancing access to places where people can reconnect with local ecosystems, such as outdoor museums, nature schools and walking trails (Schultz and Lundholm 2010). Furthermore, they make an effort to identify and communicate win-win situations between ecosystem management and other private and societal goals (Hahn et al. 2006). They build on participants' emotional motivations, such as sense-of-place, identity, and the joy of contributing to something meaningful (Schultz et al. 2007) as well as the rational, such as the enhancement of ecosystem services. Whether the enhancement of ecosystem services has really been achieved through the work of the bridging organization is the theme of the rest of the chapter.

Ecosystem Services

Ecosystems play a fundamental role in supporting human activities by providing essential ecosystem goods and services. There are many definitions and classification schemes of ecosystem services. In 2005, the UN-initiated Millennium

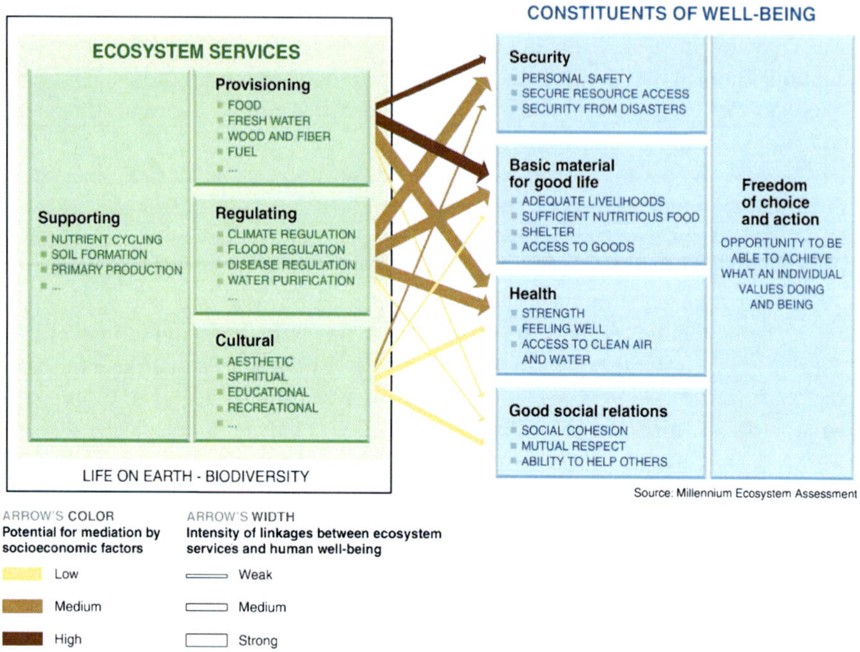

Fig. 7.2 Linkages between categories of ecosystem services and components of human well being that are commonly encountered

Ecosystem Assessment published its findings after a worldwide assessment of the consequences of ecosystem change for human wellbeing. The Millennium Ecosystem Assessment describes ecosystem services as "goods and services that people obtain from ecosystems" (i.e., the benefits that humans receive from ecosystems), and divides ecosystem services into four categories: provisioning, regulating, supporting, and cultural services, and also linked them to different constituents of wellbeing (Fig. 7.2).

Provisioning services are also termed ecosystem goods and include renewable resources such as food, timber, fuel, water, and genetic resources. Cultural services include educational, recreational, and aesthetic values. Supporting services uphold the provision of the other three categories and include primary production, nutrient cycling, production of atmospheric oxygen, soil formation, and the provision of habitat. The last category, regulating services, includes services such as climate regulation, pollination, water purification, protection against storms and floods, and regulation and control of pests and diseases (e.g., Odum 1989, Costanza et al. 1997, Daily 1997, Millennium Ecosystem Assessment 2005).

The Millennium Ecosystem Assessment concluded that human activities have resulted in a significant degradation and loss of ecosystem services worldwide, with approximately 60% of the ecosystem services examined being degraded or used unsustainably (Millennium Ecosystem Assessment 2005). This leads to a decrease in human well-being and represents a loss of natural assets, which may ultimately compromise the sustainability of humans in the biosphere. As Costanza et al. stated in their seminal paper on valuations of the world's ecosystem services and natural capital in 1997, "The economies of the Earth would grind to a halt without the services of the ecological life-support system, so in one sense their total value to the economy is infinite" (Costanza et al. 1997).

Assessing Ecosystem Services in the Flooded Meadows of Kristianstads Vattenrike

The studies in the Millennium Ecosystem Assessment were undertaken at local, regional and global scales, incorporating local knowledge and scientific research. One of the Swedish sub-global assessments during the Millennium Ecosystem Assessment was performed in Kristianstads Vattenrike. Through millennia, this area has been transformed through different agricultural practices, creating a unique cultural landscape. The distinctive morphology and geology, the interface between lakes and running water and the brackish water of the Baltic Sea, and the variations in local climates create unique conditions for a diversity of land cover types that, in turn, support a large number of ecosystems and species. The area is home to high biological diversity, including some 20 globally red-listed (i.e., according to the IUCN Red List of Threatened Species which identifies a species' risk of becoming extinct in an area) species, some 60 EU listed species, and more than 700 nationally red-listed species of flora and fauna. This figure is high compared to other areas of Sweden, with approximately 30 % of the red-listed species in the province of Skåne occurring in the biosphere reserve (Cronert and Lindblad 2004, Magnusson 2004, Magnusson et al. 2004).

The Kristianstads Vattenrike contains many different biotopes, including the largest area of managed flooded meadows in Sweden, covering over 1660 hectares. In this chapter we focus on the flooded meadows, which represent a highly dynamic ecosystem with annual flooding and fluctuating water levels. Floods generally occur in wintertime with high water levels that decrease toward summer. These flooded meadows are managed by continuous grazing and mowing by cattle, thus maintaining the open landscape (Magnusson 2004).

In their social-ecological inventory, Nekoro and Svedén (2009) mapped ecosystem services supplied by the flooded meadows in Kristianstads Vattenrike through a combination of literature studies, in-depth interviews with local stakeholders, a web-based questionnaire available for the public, and a workshop with stakeholders and experts in the area. Nekoro and Svedén found that all four groups of ecosystem services, including very rich cultural services, are represented in the flooded meadows of Kristianstads Vattenrike (Fig. 7.3). In the group of regulating services, water regulation and flood control, water purification, and air quality control were the most important services. The provisioning services identified were fodder, meat, fish, manure and freshwater. The large number of cultural services included recreational values such as bird watching, trekking/hiking, hunting, and fishing, and the value

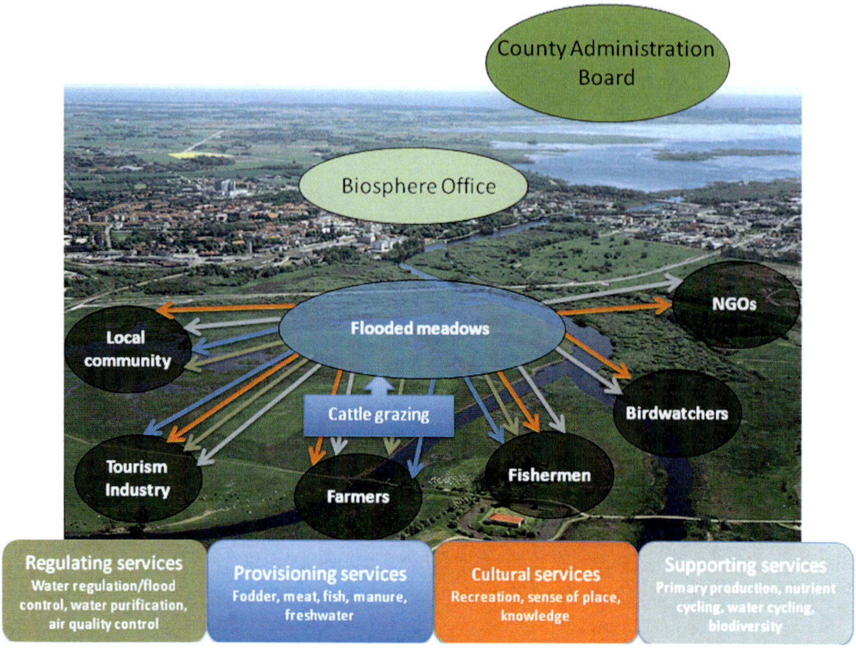

Fig. 7.3 Visual representation of linkages between ecosystem services and the parties that benefit from them in Kristianstads Vattenrike

of the flooded meadows as a symbol of Kristianstad and the local identity, as well as education, inspiration, and cultural history. The supporting services, essential for generating the other groups, included primary production, nutrient cycling and cycling of water. The flooded meadows also foster a rich biodiversity, including rare and unique species of both flora and fauna, such as the black-tailed godwit (*Limosa limosa*), white-tailed eagle (*Haliaeetus albicilla*), ruff (*Philomachus pugnax*), great raft spider (*Dolomedes plantarius*), European catfish (*Silurus glanis*), and musk orchid (*Herminium monorchis*) (Nekoro and Svedén 2009).

Effects of the Bridging Organization

When the Ecomuseum Kristianstads Vattenrike was formed in 1989 it included two nature reserves, covering approximately 190 hectares. In 2011, the area, now a biosphere reserve, comprises 21 nature reserves, with an area covering more than 3600 hectares of various biotopes (Kristianstads Vattenrike Biosphere Office 2011). This development has not only led to protection of important and endangered flora and fauna but also to the safeguarding of ecosystem services. Flooded meadows constitute an ecosystem which has decreased in large parts of Europe as a result of diminished management and development. The flooded meadows of Kristianstads Vattenrike are unique, showing an opposite trend to many other cultural landscapes. Thanks to active involvement by the Ecomuseum Kristianstads Vattenrike, which helped to increase awareness about their importance, local stakeholders were able to not only halt the loss of, but actually increase the area of flooded meadows (Magnusson et al. 2004, Walker and Salt 2006).

In 1989 flooded meadows in Kristianstads Vattenrike covered 1222 hectares, while in 2009 the meadows enclosed over 1660 hectares (Oveson 2009). The natural, cultural, and historical values and ecosystem services are related to cultivation and the annual floods and are managed by a large network of local steward associations. This management combines traditional knowledge with new methods developed to maintain the valuable ecosystem. At the end of the nineteenth century, as a result of cultivation of fodder on arable land, Swedish haymaking and grazing of flooded meadows had, with the exception of Kristianstads Vattenrike, almost ceased.

Coping with Surprise

Nekoro and Svedén (2009) identified several cross-scale threats to the flooded meadows and the ecosystem services they provide. A cease in the management of grazing and mowing would directly and negatively influence the provisioning of ecosystem services. Cease of management can have several underlying causes. A generation shift of the farmers or loss of ownership can lead to a halt or stop in management. This can also be triggered by changes in compensation systems or profitability (e.g., from national to EU level) due to policy changes. The same is true if key persons with knowledge about the importance of the management regime in the local or regional organizations or authority are lost. The establishment of Ecomuseum Kristianstads Vattenrike and its active involvement has been a crucial component in safeguarding the management of these rare and valuable ecosystems. This bridging organization has played an important role in upholding knowledge about the combination of traditional knowledge and new methods needed to maintain the flooded meadows and the ecosystem services provided by them.

Other threats include climate change-induced changes in the patterns of flooding, which can complicate or hinder proper management. As the flooded meadows of Kristianstads Vattenrike are directly dependent upon management and seasonal flooding, a change in these cyclic conditions would affect the provisioning of ecosystem services and, with great probability, influence the resilience of the ecosystem. In July 2007, the Kristianstads Vattenrike experienced unusually large summer floods, covering the flooded meadows for weeks and rendering grazing and mowing impossible. In addition, a blackish brown layer of unidentified sludge covered the grass and destroyed large parts of the vegetation in the flooded meadows, leading to decreased areas for nesting habitats. The long-term effects of this event on these sensitive areas are still unknown.

Nekoro and Svedén (2009) showed that Ecomuseum Kristianstads Vattenrike is vital in safeguarding the management of the flooded meadows. Ecomuseum Kristianstads Vattenrike's extensive knowledge and contacts with stakeholders at multiple scales, helps to combine local ecological knowledge, new methods, and adaptive management, thus helping to manage the resilience of the social-ecological system and the provisioning of ecosystem services that are directly and indirectly important on a regional and international scale.

Synthesis

Previous work on adaptive management and adaptive governance highlights the following interacting processes as crucial in relation to management of ecosystem services in social-ecological systems (Folke et al. 2005):

1. Building knowledge and understanding of resource and ecosystem dynamics and combining different knowledge systems to increase the capacity of detecting and responding to environmental feedback;
2. Feeding ecological knowledge into adaptive management practices by continuously testing, monitoring, and reevaluating to enhance adaptive responses;
3. Collaboration of a diverse set of stakeholders, operating at different levels through social networks, to gain legal, political, and financial support to ecosystem management initiatives; and
4. Developing the capacity to deal with and even create opportunities in the face of global change drivers like climate, disease outbreaks, hurricanes, global market shocks, demand, subsidies, and governmental policies.

This chapter suggests that all of these processes can be catalyzed, sustained, and protected by bridging organizations with well-developed social networks. However, in the beginning of an adaptive co-management initiative, bridging organizations often do not exist. The Kristianstads Vattenrike case illustrates that in this phase, the bridging function may be performed by individuals in various organizations who perceive a need for a new orientation of management, identify other individuals that can help formulate this new orientation, and mobilize action. Later in the adaptive co-management process, if it is successful, bridging actors may be turned into formal organizations whose main function is to facilitate learning processes and sustain adaptive collective action within the vision of ecosystem-based management. At this stage, they become bridging organizations, and the adaptive co-management process can be seen as institutionalized. Unless a (resilient) bridging organization is formed or identified, the initiative remains vulnerable to changes in key actors and staff, availability of funds, etc., and might eventually fade out.

The challenges that such formalization brings, however, are for the bridging organization to remain adaptive, innovative, and legitimate (Schultz 2009). Over time, organizations tend to refine and streamline routines, increase bureaucracy, and improve efficiency at the expense of flexibility and innovation. In some cases, the organization's own survival becomes its main purpose, and a bridging organization could potentially be used in the self-interest of a few individuals, instead of serving a common good. These are tendencies that need to be combated within bridging organizations, which act within complex adaptive systems, and hold a powerful position in networks concerned with ecosystem management. There is a need for investments in multiple-loop learning, experimentation and critical evaluation. Even though many issues can be solved through strategic collaboration, certain issues need to be negotiated in democratic forums, such as representative boards.

In Kristianstads Vattenrike, the mandate of the bridging organization was given by the municipal board on a 1-year basis. The organization has an advisory board where representatives are chosen by different stakeholder groups, not as a result of strategic selection from the bridging organization. This forum is particularly important for issues that cannot be solved in consensus (Hahn et al. 2008). A recent article that analyzed the accountability of the governance networks of Kristianstads Vattenrike shows that final decision-making processes are fully embedded in the democratic structure, which means that the governance network complements representative democracy, rather than substitutes for it (Hahn 2011).

Conclusion

To summarize, bridging organizations can facilitate cross-scale linkages, enabling formal management entities operating at discrete scales to improve communication channels and create opportunities for collaboration. These results will allow for management to set new target levels and modify policy to reach those target levels as new information is generated on scale-specific system attributes (Karkkainen 2002). The lack of communication and cooperation between organizations at even small scales further illuminates that bridging organizations may help bring about sound management of natural resources at multiple scales (Roy et al. 2008). Bridging organizations should facilitate communication between organizations, incubate new ideas for environmental management, and provide a forum for coming to agreement on contentious issues (Brown et al. 2000).

In order to manage for resilience, the goal must be to generate improved understanding of the entire system of interest, rather than specific, detailed knowledge from parts of the system (Folke et al. 2005). In the Kristianstads Vattenrike example, social and ecological change at one scale triggered cross-scale effects which resulted in a window of opportunity for the transition to adaptive governance (Olsson et al. 2006). With respect to the adaptive cycle, the critical variable is that change is the only constant. The trick is to embrace change and recognize that while change can be negative, change can also be a catalyst for new opportunities.

References

Berkes, F. (2009). Evolution of co-management: Role of knowledge generation, bridging organizations and social learning. *Journal of Environmental Management, 90,* 1692–1702.

Brock, W. A. (2006). Tipping points, abrupt opinion change, and punctuated policy change. In R. Repetto (Ed.), *Punctuated equilibrium and the dynamics of U.S. Environmental Policy* (pp. 47–77). New Haven: Yale.

Brock, W. A., & Carpenter, S. R. (2007). Panaceas and diversification of environmental policy. *Proceedings of the National Academy of Sciences of the United States of America, 104,* 15206–15211.

Brown, L. D. (1991). Bridging organizations and sustainable development. *Human Relations, 44,* 807–831.

Brown, L. D., Khagram, S., Moore, M. H., & Frumkin, P. (2000). Globalization, NGOs and multisectoral relations. In J. S. Nye & J. D. Donohue, editors. *Governance in a Globalizing World* (pp. 271–296). Washington, D.C.: Brookings Institution.

Cash, D. W., Adger, W. N., Berkes, F., Garden, P., Lebel, L., Olsson, P., Pritchard, L., & Young, O. (2006). Scale and cross-scale dynamics: Governance and information in a multilevel world. *Ecology and Society, 11*(2), 8. http://www.ecologyandsociety.org/vol11/iss2/art8/.

Costanza, R., D'Aarge, R., De Groot, R., Farber, S., Grasso, M., Hannon, B., Limburg, K., Naeem, S., O'Neill, R., Paruelo, J., Raskin, R. G., Sutton, P., & Van den Belt, M. (1997). The value of the world's ecosystem services and natural capital. *Nature, 387,* 253–260.

Crona, B., & Parker, J. (2012). Learning in support of governance: Theories, methods and findings to assess how bridging organizations contribute to adaptive resource governance. *Ecology and Society, 17*(1), 32. http://dx.doi.org/10.5751/ES-04534-170132.

Cronert, H., & Lindblad, T. (2004). Strandängsinventering längs nedre Helgeån i Kristianstads Vattenrike våren 2003. ANSER 2/04: pp. 65–78.

Daily, G. (Ed). (1997). *Nature's services: Societal dependence on natural ecosystems.* Washington, D.C.: Island Press.

Folke, C., Hahn, T., Olsson, P., & Norberg, J. (2005). Adaptive governance of social-ecological systems. *Annual Review of Environment and Resources, 30,* 441–473.

Garmestani, A. S., Allen, C. R., & Cabezas, H. (2009). Panarchy, adaptive management and governance: Policy options for building resilience. *Nebraska Law Review, 87,* 1036–1054.

Green, O. O., & Garmestani, A. S. (2012). Adaptive management to protect biodiversity, best available science and the Endangered Species Act. *Diversity, 4,* 164–178.

Gunderson, L. H. (1999). Stepping back: Assessing for understanding in complex regional systems. In K. N. Johnson, F. Swanson, M. Herring, & S. Greene (Eds), *Bioregional assessments: Science at the crossroads of management and policy* (pp. 27–40). Washington, D.C.: Island Press.

Gunderson, L. H., & Holling, C. S. (2002). *Panarchy: Understanding transformations in human and natural systems.* Washington, D.C.: Island Press.

Gunderson, L. H., Holling, C. S., & Light, S. S. (1995). *Barriers and bridges to renewal of ecosystems and institutions.* New York: Columbia University Press.

Hahn, T. (2011). Self-organized governance networks for ecosystem management: Who is accountable? *Ecology and Society, 16*(2), 18. http://www.ecologyandsociety.org/vol16/iss2/art18/.

Hahn, T., Olsson, P., Folke, C., & Johansson, K. (2006). Trust-building, knowledge generation and organizational innovations: The role of a bridging organization for adaptive co-management of a wetland landscape around Kristianstad, Sweden. *Human Ecology, 34,* 573–592.

Hahn, T., Schultz, L., Folke, C., & Olsson, P. (2008). Social networks as sources of resilience in social-ecological systems. In J. Norberg & G. Cumming (Eds), *Complexity theory for a sustainable future* (pp. 119–148). New York: Columbia University Press.

Holling, C. S. (1978). *Adaptive environmental assessment and management.* Chichester: Wiley.

Karkkainen, B. C. (2002). Collaborative ecosystem governance: Scale, complexity and dynamism. *Virginia Environmental Law Journal, 21,* 189–244.

Kingdon, J. W. (1995). *Agendas, alternatives and public policies*. New York: Harper Collins College.

Kinzig, A., Starrett, D., Arrow, K., Aniyar, S., Bolin, B., Dasgupta, P., Ehrlich, P., Folke, C., Hanemann, M., Heal, G., Hoel M., Jansson, A., Jansson, B. O., Kautsky, N., Levin, S., Lubchenco, J., Maler, K. G., Pacala, S. W., Schneider, S. H., Siniscalco, D., & Walker, B. (2003). Coping with uncertainty: A call for a new science-policy forum. *Ambio, 32,* 330–335.

Kristianstads Vattenrike Biosphere Office. (2011). http://www.vattenriket.kristianstad.se/eng/index.php. Accessed 13 Sep 2011.

Millennium Ecosystem Assessment. (2005). *Ecosystem and human well-being: Synthesis*. Washington, D.C.: Island.

Magnusson, S. E. (2004). The changing perception of the wetlands in and around Kristianstad, Sweden—from waterlogged areas toward a future Water Kingdom, Kristianstad Vattenrike Biosphere Reserve. *Annals of the New York Academy of Sciences, 1023,* 323–327.

Magnusson, S. E., Andersson, J., & Vägren, G. (1989). *Markhävdkartering 1989. Helgeåns vattenområde från Torsebro till havet [Mapping of land-use practices. Lower Helgeå River catchment from Torsebro to the sea]. Kristianstad*. Sweden: Nordöstra Skånes Fågelklubb and Kristianstads Vattenrike.

Magnusson, S. E., Magntorn, K., Wallsten, E., Cronert, H., & Thelaus, M. (2004). Kristianstads Vattenrike biosphere reserve nomination form.

Malayang, B. S. III, Hahn, T., & Kumar, P. (2007). Responses to ecosystem change and to their impacts on human well-being. Pages 203–226 in Millennium Ecosystem Assessment, Ecosystems and Human Well-being: Multiscale Assessments, Findings of the Sub-global Assessments Working Group. Washington, D.C.: Island Press.

Nekoro, M., & Svedén, J. (2009). Ekosystemtjänstanalys i Kristianstads Vattenrike 2008 - Pilotstudie strandängar [Ecosystemservices in Kristianstads Vattenrike—pilot study flooded meadows, Report in the Swedish Environmental Protection Agency series], Naturvårdsverket, Rapport 5947, May 2009. ISBN 978-91-620-5947-7pdf. ISSN 0282–7298

Odum, E. P. (1989). *Ecology and our endangered life support systems*. Sunderland: Sinauer Associated.

Olsson, P., Folke, C., & Berkes, F. (2004). Adaptive co-management for building resilience in social-ecological systems. *Environmental Management, 34,* 75–90.

Olsson, P., Gunderson, L. H., Carpenter, S. R., Ryan, P., Lebel, L., Folke, C., & Holling, C. S. (2006). Shooting the rapids: Navigating transitions to adaptive governance of social-ecological systems. *Ecology and Society, 11*(1), 18. http://www.ecologyandsociety.org/vol11/iss1/art18/.

Olsson, P., Folke, C., Galaz, V., Hahn, T., & Schultz, L. (2007). Enhancing the fit through adaptive co-management: creating and maintaining bridging functions for matching scales in the Kristianstads Vattenrike Biosphere Reserve Sweden. *Ecology and Society, 12*(1), 28. http://www.ecologyandsociety.org/vol12/iss1/art28/.

Oveson, P. (2009). Markhävdkartering 2008 hävdtillståndet på betesmarker och slåtterängar inom nedre Helgeåns våtmarksområde i Kristianstads Vattenrike, Vattenriket i fokus 2009:05 Mapping of traditional land-use 2008: current maintenance of pastures and hay meadows in the lower Helgeå wetland area in Kristianstads Vattenrike.

Roy, A. H., Wenger, S. J., Fletcher, T. D., Walsh, C. J., Ladson, A. R., Shuster, W. D., Thurston, W., & Brown, R. R. (2008). Impediments and solutions to sustainable, watershed-scale urban stormwater management: Lessons from Australia and the United States. *Environmental Management, 42,* 344–359.

Ruitenbeek, J., & Cartier, C. (2001). *The invisible wand: Adaptive co-management as an emergent strategy in complex bio-economic systems*. Bogor: Center for International Forestry Research.

Ruhl, J. B., Kraft, S. E., & Lant, C. L. (2007). *The law and policy of ecosystem services*. Washington, D.C.: Island Press.

Scheffer, M., Westley, F., & Brock, W. (2003). Slow response of societies to new problems: Causes and costs. *Ecosystems, 6,* 493–502.

Schultz, L. (2009). Nurturing resilience in social-ecological systems: Lessons learned from bridging organizations. Dissertation, Stockholm University.

Schultz, L., & Lundholm, C. (2010). Learning for resilience? Exploring learning opportunities in Biosphere Reserves. *Environmental Education Research, 16,* 5–6.
Schultz, L., Folke, C., & Olsson, P. (2007). Enhancing ecosystem management through social-ecological inventories: Lessons from Kristianstads Vattenrike, Sweden. *Environmental Conservation, 34,* 140–152.
Skjaerseth, J. B. (2006). Protecting the northeast Atlantic: One problem, three institutions. In S. Oberthur & T. Gehring, (Eds), *Institutional interaction in global environmental governance* (pp. 103–125). Cambridge: MIT Press.
Stokke, O. S., & Coffey, C. (2006). Institutional interplay and responsible fisheries: Combating subsidies, developing precaution. In S. Oberthur & T. Gehring (Eds), *Institutional interaction in global environmental governance* (pp. 127–155). Cambridge: MIT Press.
Walker, B., & Salt, D. (2006). *Resilience thinking—sustaining ecosystems and people in a changing world.* Washington, D.C.: Island Press.
Westley, F. (1995). Governing design: The management of social systems and ecosystems management. In L.H. Gunderson, C. S. Holling, & S. S. Light (Eds), *Barriers and bridges to the renewal of ecosystems and institutions* (p. 20). New York: Columbia University Press.
Wood, B. D., & Doan, A. (2003). The politics of problem definition: Applying and testing threshold models. *American Journal of Political Science, 47,* 640–653.

Chapter 8
Adaptive Management for Novel Ecosystems

Nicholas A. J. Graham and Christina C. Hicks

Keywords Adaptive management · Novel ecosystems · Coral reefs · Ecology · Uncertainty

Novel Ecosystems and Marine Ecology

Many drivers of ecosystem change, such as exploitation, climate change and the introduction of invasive species, lead to non-random species extinctions or changes in abundance (Purvis et al. 2000). For example, exploitation typically targets larger species in an assemblage first (Owens and Bennett 2000), climate change affects species based on their thermal performance windows (Pörtner and Farrell 2008), and invasive predatory species can cause local extinctions of prey species least adapted to predator evasion (Blackburn et al. 2004). Although appreciation of ecosystem degradation and differential species loss through anthropogenic drivers is not new (Carson 1962, Myers 1987), the recognition that changes are predictable and sometimes lead to persistent new ecosystem configurations (species compositions and relative abundances) has led to the emerging concept of novel ecosystems (Hobbs et al. 2006, 2013). This concept, which has similarities to the no-analog literature in paleoecology (Williams and Jackson 2007), explicitly recognizes that many ecosystems are changing and are unlikely to return to pre-impact conditions. However, these new configurations may still provide valuable goods and services to society, and consequently there is a need to understand the properties of emergent

N. A. J. Graham (✉) · C. C. Hicks
ARC Centre of Excellence for Coral Reef Studies, James Cook University, Townsville, QLD 4811, Australia
e-mail: nick.graham@jcu.edu.au

C. C. Hicks
Center for Ocean Solutions, Stanford Woods Institute for the Environment, Stanford University, Monterey, CA 93940, USA
e-mail: christina.c.hicks@gmail.com

C. R. Allen, A. S. Garmestani (eds.), *Adaptive Management of Social-Ecological Systems*,
DOI 10.1007/978-94-017-9682-8_8

novel ecosystems (Williams and Jackson 2007) and determine the most appropriate management in the new ecosystem contexts (Seastedt et al. 2008).

The majority of work on emerging novel ecosystems has been restricted to the terrestrial realm (Hobbs et al. 2013). However, given substantial recent and predicted changes for a range of coastal marine ecosystems (Polunin 2008), there is an increasing awareness of the need to understand how the novel ecosystem concept relates to the marine environment (e.g., Harborne and Mumby 2011, Doney et al. 2012). Many classic marine ecology studies assessed the dynamics and structure of communities in response to natural disturbances and gradients, with little thought to anthropogenic influences. Indeed, such research and knowledge is well founded in the study of shallow coastal marine habitats, with natural disturbances such as storms, rainfall, temperature anomalies and diseases playing key roles in the structuring and dynamic nature of many of these habitats (Dayton 1971, Connell 1978, Sousa 1979, Thistle 1981). However, there has been an increasing recognition that anthropogenic disturbances have also affected the structure and dynamics of shallow marine habitats for at least the past two centuries (Jackson et al. 2001, Pandolfi et al. 2003), and the intensity and frequency of such disturbances are increasing exponentially. Indeed, anthropogenic stressors are becoming the dominant drivers of community structure in many systems (Nyström et al. 2000, Hughes et al. 2003, Polunin 2008).

A range of anthropogenic disturbances threaten coastal ecosystems, including overfishing, nutrient input, sedimentation, land reclamation and dredging. However, climate change is rapidly emerging as perhaps the most substantial threat for many ecosystems (Walther et al. 2002, Hughes et al. 2003, Polunin 2008, Hoegh-Guldberg and Bruno 2010). The effects of these anthropogenic disturbances is leading to concerns over the long-term persistence or changing nature of a variety of ecosystems, including kelp forests (Steneck et al. 2002), seagrass beds (Duarte 2002), mangrove forests (Alongi 2002), and coral reefs (McClanahan 2002). As such, it seems particularly pertinent to apply the novel ecosystem concept to the marine environment. Coral reefs provide a powerful focal ecosystem to examine the importance of the novel ecosystem concept and implications for management, having been substantially altered by a range of direct anthropogenic drivers and global climate change (Hughes and Connell 1999, Hughes et al. 2003, Graham et al. 2006, Hoegh-Guldberg et al. 2007, Graham et al. 2013). In this chapter we examine the utility of adaptive management for novel ecosystems through the lens of coral reefs.

Coral Reefs as Emerging Novel Ecosystems

The novel ecosystem concept could be particularly powerful in coral reef science and management as there are currently two outlooks dominating the literature. The first focuses on identifying the few remaining pristine coral reef locations, characterizing the reefs, and suggesting that these locations are useful baselines and reference points for management targets (Sandin et al. 2008). The second outlook is largely resigned to the complete degradation of coral reefs unless global climate change is brought under control (Hoegh-Guldberg et al. 2007, Veron et al. 2009). However, the former sets up unrealistic, largely historic, goals that do not recognize that the majority of reefs exist in human dominated seascapes (Kittinger et al. 2012), while the latter fails to recognize that the large variation in vulnerability and recovery potential of many coral reef organisms to major threats should result in some reefs persisting with altered community compositions (e.g., Pandolfi et al. 2011, Riegl et al. 2013). A growing literature suggests a middle ground may be more likely where coral reefs change due to ongoing human pressures, including climate change, but do not disappear (Pandolfi et al. 2011, Hughes et al. 2012, Graham et al. 2013, McClanahan et al. 2014).

Much of the research around novel ecosystems in terrestrial systems has focused on land use change, restoration and in particular invasive species (Lindenmayer et al. 2008, Hobbs et al. 2013), but has also identified differential responses of species to climate change (Williams and Jackson 2007, Archer and Predick 2008, Hobbs et al. 2013). Novel ecosystems in this context have been broadly defined as changes in species composition, relative abundances, interactions, and ecosystem functions, although they have also sometimes been used to describe even more dramatic ecosystem change related to threshold-driven regime shifts (Hobbs et al. 2013). Following Graham et al. (2014), we offer a definition for coral reefs that differentiates coral-dominated reefs from those that have undergone a regime shift to an alternative non-coral state. Specifically, we define novel coral ecosystems as changes in species configurations, interactions and functions, that are historically novel, but still within the parameter space of calcifying coral-dominated reefs. In doing so, we differentiate between heavily degraded reefs that have shifted to a non-coral ecosystem (e.g., from corals to seaweed) (Hughes et al. 2010), and reefs that remain in a calcifying condition but have shifted in composition and function.

Invasive species may lead to novel ecosystem compositions on coral reefs. Two key examples are the invasive macroalgae, *Gracilaria salicornia*, that is altering the physical and compositional make up of some reefs in Hawaii (Martinez et al. 2012) and the Pacific red lionfish, *Pterois volitans*, that is proving a voracious competitor and predator on its non-native Caribbean reefs (Albins and Hixon 2013). However, in both these cases the invasive species are contributing to reef degradation and likely reinforcing pathways to non-coral dominated systems in already degraded locations. Furthermore, invasive species are not nearly as prevalent a problem on coral reefs as they are in many terrestrial systems.

Species range shifts are becoming more common at the latitudinal fringes of coral reefs (as documented on the east and west coasts of Australia and in Japan), leading to novel ecosystem compositions where tropical and temperate reef organisms co-inhabit stretches of coastline and new ecological interactions unfold (Yamano et al. 2011, Feary et al. 2014). Both invasive species and range shifts document entirely new species in locations where they have not been present in recent geological time periods, which will undoubtedly lead to changing ecosystem dynamics that may be large or small depending on the ecological dominance of the species involved. However, novel ecosystems can also emerge in situations where existing species composition and abundances change in long-term and predictable ways, for example through responses to climate change (Williams and Jackson 2007, Hobbs et al. 2013). This is true for the vast majority of coral reefs, and represents the most ubiquitous cause of novel coral reef configurations with widespread relevance.

Coral reefs are responding to climate change in a non-random fashion. For instance, there is a large body of literature demonstrating that some species of corals are far more susceptible than others to temperature induced coral bleaching and mortality (e.g., Marshall and Baird 2000, Loya et al. 2001) (Fig. 8.1). Similarly, based on experiments in aquariums, there is considerable variation in coral species specific responses to ocean acidification (Kleypas et al. 2006, Pandolfi et al. 2011). Indeed, a range of other common threats to reef corals are known to be non-random in their impact. The feeding preferences of the corallivorous crown-of-thorns starfish are specific to certain genera of corals (Pratchett et al. 2009). Strong storms influence corals differentially based on the strength of the coral attachment to the reef benthos and the coral morphological exposure to wave energy (Madin and Connolly 2006). Fishing can influence corals through trampling or gears (e.g. traps and nets) that directly damage colonies, and typically those corals with more fragile morphologies are most susceptible (Darling et al. 2013). Sedimentation from land-based catchment sources influences coral species based on their susceptibility to be smothered, or requirements for high light penetration (Fabricius 2005). Aside from responses to disturbances, differences in recovery potential among coral taxa following disturbances further contributes to changing species dominance and composition (Pandolfi et al. 2011, Riegl et al. 2013).

The influence of these benthic changes on associated reef fish assemblage structure is also predictable, with susceptibility dictated by specialization and body size (Wilson et al. 2006, Graham et al. 2011). Species with smaller body size or high levels of specialization (diet, habitat use or settlement cues) are most vulnerable to reductions in live coral cover, or shuffling of the coral species present (Pratchett et al. 2008) (Fig. 8.1). Conversely, larger bodied species and those with slower life-history traits (such as slow growth rates and late maturity) are most vulnerable to fishing impacts (Fig. 8.1). Taken together, these non-random disturbance responses and recovery patterns of corals and fishes are likely to lead to predictable ecosystem configurations composed of different species abundances, compositions, structures and functions. Evidence supporting this notion is mounting, with large scale alterations to coral reef species composition persisting over extended temporal scales (McClanahan et al. 2007, Pratchett et al. 2011, van Woesik et al. 2011). Indeed, recognizing that ocean warming is leading to asymmetrical losses of species of corals, the novel ecosystem concept has been used to frame coral reefs of the Caribbean (Yakob and Mumby 2011), and more recently, the future of coral reefs globally (Graham et al. 2014).

These changing dynamics and compositions on coral reefs highlight that the pristine reef conditions of the past have become increasingly rare through time (Fig. 8.2). Indeed, it is highly unlikely that any reefs will exist in a completely pristine state in the future. The proportion of reefs that have undergone a shift to an alternative non-coral system state has increased through time, and is likely to increase more into the future (Fig. 8.2). However, the emergence of novel coral reefs gives hope for a future with some coral-dominated ecosystems, albeit with novel compositions. Indeed, the concept provides a much more pragmatic and attainable outlook on a future with coral-dominated reefs. Furthermore, local management

Fig. 8.1 Species of corals and fishes display differential susceptibility to a range of disturbances. **A** Acropora corals are highly susceptible to the effects of ocean warming, and some other threats, such as crown-of-thorns starfish outbreaks and storm damage. **B** Porites corals are much more tolerant of warm water and can often persist on reefs when many other species are lost. **C** Many species of butterflyfish are vulnerable to coral loss or a change in the composition of coral species due to their reliance on certain species of corals for food and/or shelter. **D** Species of grouper are vulnerable to population depletions through fishing because they are targeted for their size, and have life histories that make population recovery difficult. Photos **A** and **B** taken by N. Graham. Photo **C** taken by M Pratchett. Photo **D** taken by F. Januchowski-Hartley

actions may ensure a greater number of reefs are maintained in a coral-dominated novel condition, rather than in a degraded non-coral state (Fig. 8.2).

Will Current Management Paradigms Be Appropriate?

There is mounting evidence that local management can be important for coral reef ecosystem condition in the face of the growing global threats to reefs and may therefore help ensure more reefs are maintained in novel coral dominated conditions in the future (Fig. 8.2). Although many of the major threats to coral reefs, such as temperature induced coral bleaching and tropical storms, are difficult to manage locally, two of the key threats (fishing and water quality) can be managed locally and doing so should provide substantial benefits to the ecosystem (Graham et al. 2013).

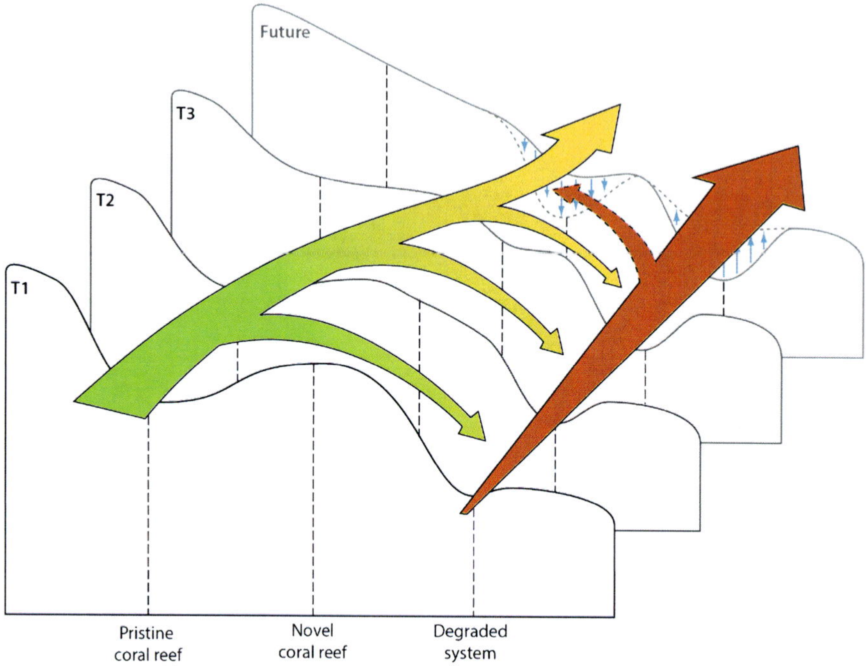

Fig. 8.2 Conceptualizing changing scenarios for coral reef ecosystems. Pristine coral reefs have become less common through time, and are unlikely to be a realistic goal for the future of coral reefs under continued human pressure and anthropogenic climate change. Degraded non-coral systems have become more common through time, and many argue will become the dominant reef condition in the future. However, a middle ground, where altered 'novel' coral reef community compositions persist may be a realistic goal for many of the world's coral reefs. This will be possible due to the differential susceptibility of many reef organisms to a range of pressures. It may also be more likely with appropriate adaptive management actions, which reduce the number of reefs in degraded conditions and increase the number of reefs persisting in a novel condition (depicted by dashed lines and blue arrows in figure)

Importantly, the functional groups of fish thought to be most important for reef recovery processes are responsive to local fisheries management policies (Graham et al. 2011). Well enforced marine reserves can lead to changes in ecosystem dynamics with increased fish herbivory, reduced fleshy algal cover and enhanced rates of coral recovery (Mumby et al. 2006, Stockwell et al. 2009, Mumby and Harborne 2010, Wilson et al. 2012). Furthermore, gradients of fish biomass highlight the importance of maintaining fish biomass above thresholds where key reef processes are reduced and the reef condition changes (McClanahan et al. 2011). Similarly, gradients or changes in water quality (sediments, pollutants and nutrients) highlight the potential for local reductions of negative terrestrial (catchment) input to enhance coral recovery and diversity (Smith et al. 1981, De'ath and Fabricius 2010).

Although compositions will be different due to the non-random reduction in occupancy, abundance, and recovery potential of species, local management should have the potential to stimulate continued recovery of hard coral communities into the future. If stewardship of reefs is improved and becomes adaptive to changing threats and social contexts, it should be possible to ensure a greater number of coral reefs are maintained in a novel ecosystem composition (Fig. 8.2). Critically, however, the unfolding novel composition of reefs highlights, more than ever, that we should not be complacent with our current management approaches. Instead, we need to re-evaluate how effective they are in novel coral reef contexts, and become more innovative in developing new adaptive management approaches that may yield new solutions, or alternative pathways to beneficial ecosystem configurations.

Marine protected areas have largely dominated the space for coral reef management. They are seen to be useful, partly because they represent measurable management outcomes based on the amount of area protected. Furthermore, implementing and managing marine protected areas can appear straightforward because they can be sited close to shore making monitoring feasible and are also amenable to top down enforcement and control. However, there is considerable debate as to the extent to which marine protected areas deliver conservation and fishery benefits (Sale et al 2005, Agardy et al. 2011, Mora and Sale 2011). In most countries no-take marine protected areas typically only represent a small proportion of the total coral reef area (McClanahan et al. 2008a), and compliance with the rules is often weak (Mora et al. 2006, Pollnac et al. 2010, Daw et al. 2011, Campbell et al. 2012). If compliance is not a large issue, marine protected areas can serve a vital role in the broader management of coral reef systems. However, less clear, and seldom articulated, is what that role is; and how, when, and where marine protected areas should be implemented. Marine protected area implementation would benefit from progressing beyond the precautionary principal approach that they currently fill (i.e., setting areas aside as an insurance) (Lauck 1998), to hypothesis driven implementation that is monitored and enables managers to learn what works and better anticipate realistic coral reef futures (McCook et al. 2010). Critically, although frameworks have been suggested (Grafton and Kompas 2005), marine protected area management on coral reefs is rarely adaptive, with rigid legislation locking an area of the ocean aside for goals that may end up changing though time, and against threats that may not be mitigated by closures. This highlights the need to manage coral reefs across wider reefscapes, reducing the negative effects of overfishing and other disturbances at larger scales (Steneck et al. 2009). The solution space needs to be more innovative, adaptive, and act at larger scales.

The Need to Manage for Uncertainty: Adaptive Management

The emergence of novel coral reef ecosystems and the recognition that species will have differential responses to a changing climate brings with it a growing urgency to manage effectively, for and with, uncertainty. However, the desire for management to be based on the best available science paradoxically hinders the very process it tries to advance. This is because science tends to be produced with laboratory style precision, leaving scientists reluctant to make recommendations based on partial information, and chastised when they do so (Willis et al. 2003). Management actions are therefore often delayed or developed in an ad hoc manner. This ad hoc management tends to be based on anecdotal evidence or advocacy and developed in response to emerging challenges and opportunities (Lee 1999, Willis et al 2003). Adaptive management offers a middle ground between laboratory style precision that tends not to be feasible and ad hoc trial-and-error that seems risky (Lee 1999).

Adaptive management is grounded in the admission that we do not know enough about managing ecosystems while simultaneously recognizing the need for action. Adaptive management further recognizes that ecosystems are naturally dynamic; and because they are inextricably linked to the human communities that interact and depend on them, they are also complex. Uncertainty is therefore inherent in any management decisions that are made regarding these dynamic and complex systems. Because a lack of information should not delay action, adaptive management explicitly addresses the uncertainties to present a structured process of learning by doing, and adapting based on what's learned (Walters and Holling 1990). Implicit is an acknowledgement that the most important uncertainties should be tested early. By recognizing the value in tracking ecosystem responses to management, this approach enables management to proceed with impartial knowledge (Lee 1999). Strategies are developed as experiments to provide information about the system that is being managed, and this information is then used to refine, or develop, future strategies (Holling 1978, Walters 1986). Because adaptive management is designed to employ a diversity of testable actions, it affords a system a greater capacity to manage for resilience than the current tendency to employ single policies across ecological space or time that could potentially lead to disaster (Tompkins and Adger 2004, Bodin and Norberg 2005).

Adaptive Management and Coral Reefs

Effectively managing single marine fish stocks for economic return has proved challenging (Roughgarden and Smith 1996). Managing complex systems with multiple benefits, as is the case with most coral reef settings, is recognized as far more challenging. Although adaptive management has been suggested as an appropriate way to address these challenges, there are relatively few examples of adaptive coral reef management. The re-zoning of the Great Barrier Reef Marine Park in 2004 to incorporate a 33 % no-take area, rather than the previous 5 %, and protect a wider range of habitat types, is one of the few examples of an adaptive process (Hughes et al. 2007). This re-zoning was based on a huge amount of learning, including experimental fishing experiments in existing no-take zones (Mapstone et al. 2004, Sale et al. 2005), and a greater understanding of the functioning of the system. On-going understanding of the drivers of ecosystem change in the Great Barrier Reef is leading to renewed attempts to control water quality problems from adjacent catchments (Queensland Government 2009), and a great deal of science and management attention is currently trying to understand how to control outbreaks of predatory crown-of-thorns starfish (Rivera-Posada et al. 2013). Although this learning and management is clearly dynamic, whether changes to the zoning plan or water quality objectives will be possible at appropriate time scales is yet to be seen.

Customary coral reef management in parts of the Pacific has developed through processes of learning, experimentation and adaptation (Johannes 1982). For example, the use of periodic closures, where a section of the reef is closed to fishing to build up stocks and influence fish behavior (Feary et al. 2011) to ultimately enhance catch during important ceremonies or events, has been studied as an adaptive management cycle (Cinner et al. 2006). Indeed, the strong traditional ecological knowledge in parts of Melanesia has been harnessed to develop effective adaptive co-management arrangements between science NGOs and local communities (Weeks and Jupiter 2013). However, the huge changes occurring on coral reefs in response to climate change and other escalating drivers requires the adoption of adaptive management more widely and in a much more experimental way.

Experimentation, Monitoring and Learning

The core principles of adaptive management are that it is experimental, multi-scalar, and place-based (Norton 2005). Experimental recognizes the ongoing search for knowledge that is necessary to set and achieve goals. Multi-scalar stresses the need for managers to model and monitor management on multiple scales of space and time. Place-based accepts that ecosystems are occupied by people, therefore, the way people are likely to perceive and respond to management will reflect their social and geographic orientations (Norton 2005). The experimental component of adaptive management stresses the importance of not delaying action for a lack of knowledge and the opportunity to take advantage of the situation and learn from it. Management actions are therefore designed in a structured way such that lessons can be rigorously inferred, similar to the way in which experiments are designed. Management actions should therefore be designed so that they have clear hypotheses, a way of controlling factors that are (thought to be) extraneous to the hypotheses, and with opportunities to replicate the management "experiment" to check its reliability (Lee 1999). These management actions should be monitored so that the data provides a way of appraising the hypotheses, and can be continuously fed back into decision-making in a forward thinking way, rather than simply tracking toward some hypothesized ecological baseline. The emergence of novel coral reef ecosystems requires a great deal of experimentation and learning to elucidate appropriate management strategies to deal with changing ecosystem compositions and processes.

A great deal of scientific work is needed to fully understand the range of important ecological processes on coral reefs, how the relative importance of these processes may change in different contexts and with changes in reef compositions, and how we can effectively influence ecosystem processes in management. Most current attention is given to the process of herbivory, because herbivorous fish and urchins can maintain cropped algae that encourages successful coral recruitment and recovery processes (Bellwood et al. 2004, Mumby et al. 2007). Knowledge of herbivory on coral reefs is rapidly improving, with an appreciation of the non-linear relationships between body size and bite volume (Lokrantz et al. 2008) and foraging range (Nash et al. 2013). Furthermore, assessments of the feeding preferences of herbivores are identifying groups of herbivores that target turf algae to feed upon, and others that eat mature fleshy macroalgae (Bellwood et al. 2006, Hoey and Bellwood 2011). These different types of herbivore functions are key to influencing different reef trajectories, will vary according to reef condition (Graham et al. 2013), and may require different forms of management experimentation.

Although outright bans on catching herbivores have been enacted in some countries (Steneck et al. 2009), in other places innovative management experiments are being attempted such as escape gaps that allow key species of herbivorous fish to exit fishing traps (Johnson 2010, Mbaru and McClanahan 2013). Importantly, modifying fishing technologies in these ways can promote important ecosystem processes on coral reefs while maintaining the income of fishers (Johnson 2010). In other locations, phasing out certain types of fishing gear may reduce impacts on key groups of fish, as each gear is typically selective toward a certain portion of the fish community (Cinner et al. 2009a). Such restrictions on the types of gears in use may be used in experimental ways, or in response to specific impacts to reefs, in an adaptive management framework (McClanahan and Cinner 2008). For example, phasing out certain gears that target species of fish known to promote coral recovery post disturbance, may be an appropriate way to prevent reefs transitioning to potentially undesirable states (Cinner et al. 2009a, Graham et al. 2013). However, reducing catch of herbivores may not always be appropriate or ethical; in some heavily fished locations the dominance of herbivores is sustaining the catch of the fishery (McClanahan et al. 2008b, Hicks and McClanahan 2012).

Although the importance of the process of herbivory is well established in the ecology of coral reefs, other processes that have been largely overlooked may become increasingly important in understanding and managing reefs in the future. One example is the realization that excess sediments in algal turfs can exclude many species of herbivore (Bellwood and Fulton 2008). Moreover, these longer turf algal states with sediment loads can be a barrier to successful coral settlement and establishment (Arnold et al. 2010). As such, the process of sediment removal from reefs is likely to be important in many locations, but poorly understood. There are likely to be a range of other processes, such as excavation of dead unstable carbonate structures, which are poorly understood, and thus not incorporated into management actions. Improving this understanding will allow management experimentation to establish options for enhancing certain reef processes on reefs dependent on the condition and dominant processes required to improve reef trajectories.

Changes in reef community composition are likely to cause shifts in species dominance patterns, which resource users and managers should carefully evaluate. For example, lightly fished coral reefs where coral communities are not too heavily impacted may be able to sustain a fishery on reef predatory species, such as the coral trout fishery in Australia (Russ and Williams 1994). However, in locations where the coral reef community has shifted substantially, and the fishing is heavier, the dominant fish species on the reef and those that sustain the catch of local fisheries can be quite different. In these contexts, slow growing species of predators have been over-exploited long ago, and the catch composition represents a suite of species with faster life history traits (McClanahan and Hicks 2011). This may also interact with the condition of the reef. For example, in Kenya fishing pressure is high and corals in the lagoon area have become scarce, leading to a more seagrass and macroalgae dominated system. Over 60 % of the biomass of the catch along this stretch of coastline is made up by two species of fish: the rabbitfish *Siganus sutor*, and the parrotfish *Leptoscarus vaigiensis* (Hicks and McClanahan 2012). Both of these species feed on seagrass and mature macroalgae (Chong-Seng et al. 2014). Indeed, before-after studies in the region assessing changes in reef condition and fish assemblages in response to the 1998 coral bleaching event, showed that macroalgal feeding species of rabbitfishes and parrotfishes like *Siganus argenteus*, *Leptoscarus vaigiensis* and *Calotomus carolinus* were among the few species that responded positively to the disturbance (Graham et al. 2007). Importantly, some of these species have very fast life history traits, with the ability to grow to plate size and reproduce within 12 months (Hicks and McClanahan 2012). The opportunities for emerging, potentially sustainable, fisheries dominated by "new" species needs to be carefully monitored. Importantly, rather than management attempting to maintain fisheries of species that are being depleted due to local fisheries and slow life history strategies, or factors external to the fishery (e.g., loss of recruitment habitat), management that is adaptive can take advantage of more sustainable fishing opportunities when different species become abundant. Clearly, such decision making will need to navigate trade-offs between biodiversity conservation and food security/livelihood needs, and may require mixed management approaches across the seascape (McClanahan et al. 2011, White et al. 2012).

Although managing reef processes and adapting to changing dominance patterns in reef resources may be preferred and in some way optimal, in many situations it will be difficult to achieve. Setting more achievable management targets, based on scientific underpinnings with the flexibility to use a range of adaptive management approaches to achieve those targets may be one solution. Although fisheries have a history of single species targeting and associated population assessments for management, recent studies have suggested that a more balanced approach, where the whole fish community is exploited in a non-selective way, may be more appropriate (Zhou et al. 2010). Indeed, this premise has recently been applied to coral reefs, with the idea of biomass based fisheries management, which provides simple guidelines for how much fish biomass should be maintained on a reef to try to maximize both fisheries sustainability and ecosystem condition. Using a large ecological data set across the Indian Ocean, McClanahan et al. (2011) showed that as biomass is reduced a series of clear and abrupt changes occur in various reef processes and state variables, such as predation, switches in dominant herbivores and increases in fleshy algae. Using these ecological switchpoints, and principals of multispecies maximum sustainable yield estimates, they proposed a biomass window that resource users and managers in the region should strive for in order to promote functioning reefs and sustain fisheries. Importantly, a wide range of approaches may be employed to meet these targets, such as community closed areas, restrictions on the types of fishing gears that can be used, or enhancing the capacity to access pelagic resources, rather than reef-based resources. Management experimentation may be important in many settings to explore ways to achieve these biomass targets.

Moving toward a Social-Ecological Systems Understanding

Key to managing adaptively is having an explicit vision or model of the ecosystem one is trying to guide (Walters 1986). That explicit vision provides a baseline for defining surprise and thus expanding the boundaries of what is known (Lee 1999). A clear articulation of what novel ecosystems are likely to emerge is therefore critical to this process. Indeed, this has been put forward as a key research frontier for novel coral reef futures, with a range of methods available to achieve the goal (Graham et al. 2014). Understanding how to guide specific novel ecosystem emergence and enable desirable novel ecosystems to persist will be key to the adaptive learning process. Integral to the persistence of certain ecosystem configurations is the perceived desirability of the ecosystem goods and services that are delivered to the stakeholders that interact and ultimately benefit from the ecosystem. Understanding who benefits, how they benefit, and how they perceive those benefits is therefore also critical to the success of any management system (Lebel et al. 2006, Hicks 2011).

Changes in coral reef compositions are likely to affect the generation of ecosystem services, however few studies have explicitly investigated these changes (Graham et al. 2013). Dive tourists often desire high coral cover and lots of big fish (Williams and Polunin 2000), but less dramatic reef conditions, and even reefs dominated by soft corals, can still attract tourists. Reefs in different conditions will also likely offer very different fishing opportunities (Mumby et al. 2008). Although lightly exploited reefs, such as those in Australia, can support fisheries on high level predators such as coral trout (Russ and Williams 1994), in more heavily exploited and altered reef locations, lower trophic level species, such as herbivorous rabbitfishes and parrotfishes, can largely sustain the fishery (McClanahan et al. 2008b, Hicks and McClanahan 2012). Importantly, different stakeholders tend to value and prioritize ecosystem services in different ways (Hicks et al. 2013). Therefore, understanding how novel coral reef ecosystems may affect people's wellbeing will not only require quantifying the ecosystem service production of alternative reef communities but also developing a better understanding of how these services are valued by different segments of society. For example, although near-pristine coral reefs may contribute to a multi-billion dollar tourism industry, profits from tourism, which are generally higher than those from fisheries, often do not benefit poor fishing communities (Hicks et al. 2009). In contrast, heavily fished reefs may support fewer recreational services, but provide important food security for low-income households. A strong understanding of how different stakeholders are likely to perceive changes in the delivery of ecosystem services for any given management action is therefore integral to long-term success.

Rather than presenting an obstacle to management, adaptive management turns a lack of evidence into an opportunity to simultaneously manage, monitor, and learn about a system (Williams 2011). Indeed, there will be a great deal of learning necessary as novel ecosystems and social-ecological interactions emerge. This learning will be in the form of new science, but also learning by doing. Learning is a process that occurs over time; with information on whether management actions succeed in effecting ecosystem outcomes, or not, used to continuously appraise and refine future strategies (Lee 1999). Importantly, management is a social-ecological process, with learning outcomes ultimately aiming to inform public policy and collective choices; to do so effectively, social perspectives should be part of what is learned (Heclo 1974, Parson and Clark 1995). There are benefits to experiment-based learning becoming a social process; learning is information-intensive and requires active participation, so involving people in the learning process means this will only need to happen once. In addition, social learning is more likely to lead to the acceptance of strategies that build social and ecological resilience (Tompkins and Adger 2004). So, if the outcomes of adaptive management and learning are to influence the ways in which people behave, then the involvement, early on, of those required to change their behavior will clearly be advantageous (see Margoluis and Salafsky 1998).

Successful adaptive management needs to be situated in institutions that have the capacity to interpret, learn from, and respond to changes in resource conditions and the response of stakeholders to these changes. Stakeholder involvement is recognized as a key component of adaptive management (Norton 2005). This is partly because the involvement of stakeholders can aid in the learning process (Tompkins and Adger 2004). Moreover, management systems that are designed locally are more likely to be adaptive and responsive to successes and failure (Cinner et al. 2006). Indeed many traditional resource management systems have evolved in an adaptive manner (e.g., Johannes 1982). Part of what makes traditional resource systems amenable to management, and stakeholder involvement critical to success, is that that local or traditional ecological knowledge plays an important role in decision making. This knowledge enables people to make sense of and react to changes in their natural resource systems (Johannes 1982, Cinner et al. 2006, Tompkins and Adger 2004). A better understanding of local knowledge and desirability opens up a wider range of avenues for management experimentation.

Interdisciplinary research identifying key linkages and feedbacks between societies and ecosystems has critical implications for sustainable resource use and governance. Indeed, thinking about societies and ecosystems as linked social-ecological systems highlights numerous social, economic, and political issues that can help identify a broader suite of potential management and policy actions (Berkes and Folke 1998, Ostrom 2007). For example, utilizing the social-ecological systems framework developed by Elinor Ostrom (Ostrom 2007, 2009), a wide range of potential actions can be taken in different components of the social-ecological system that could be used in the management experimentation of novel ecosystems (Graham et al. 2013). The governance system is comprised of the organizations with management jurisdictions, the formal and informal rules that govern resource use, and the processes through which people participate in decisions about resource use. Various options are available here that may improve system outcomes, for example providing property rights to local resource users, experimenting with centralized versus decentralized management, and developing co-management arrangements (Cinner et al. 2012). Options regarding the resource users include altering the technologies they use and how dependent they are on the resource (Cinner et al. 2009a). The resource characteristics include aspects of target species' life histories, handling and storage capabilities, and economic values (which can be profoundly influenced by subsidies) (Brewer et al. 2012). The broader social, economic, and political setting, offers opportunities to influence consumer awareness and demand, the prevalence of corruption, and levels of socioeconomic development (Cinner et al. 2009b, Fabinyi 2012). Finally, related ecosystems, such as mangroves and seagrass provide habitat mosaics for many targeted and functionally important species, and management of these ecosystems can have important flow-on implications for adjacent coral reefs (Mumby et al. 2004).

The most appropriate governance or management options will clearly be context specific, and dependent on both the vulnerability of the ecological system to change and the capacity of the human society to adapt and embrace new opportunities and management scenarios (Cinner et al. 2013). This was the basis of a framework put forward by McClanahan et al. (2008a), that proposed a bivariate plot of environmental vulnerability (i.e., exposure to ecosystem stress such as climate anomalies, storms, land based pollution and sedimentation) against social adaptive capacity (incorporating factors such as livelihood diversity, assets, and social capital) that enables site specific environmental management decisions to be made that reflect the specific social-ecological system context (McClanahan et al. 2008a). This framework is particularly useful when conceptualizing adaptive management responses to novel coral reef ecosystems. If environmental vulnerability is very high, and social adaptive capacity very low, many forms of management are likely to fail due to continued and strong disturbances and a lack of capacity for people to change. Here the most appropriate action is more likely donor aid and relief to try to decouple people's dependence on a dwindling resource (Fig. 8.3). Where both environmental vulnerability and social adaptive capacity is low, an opportunity exists to build capacity and move those social-ecological systems towards the bottom right quadrant of the framework to commence adaptive management approaches (Fig. 8.3). Indeed, in the bottom right quadrant, adaptive management and experimentation aimed at different approaches to preserve the ecosystem may be successful (Fig. 8.3). Protected areas, and a range of other larger spatial scale approaches, as outlined above, may be appropriate here. Finally, if environmental vulnerability and social adaptive capacity are both high, opportunities for diverse experimentation and adaptive management approaches to transform the system and drive different novel system outcomes exist (Fig. 8.3). Embracing social-ecological systems thinking and recognizing that specific approaches are only appropriate in certain contexts will greatly enhance the likelihood of management success and equitable outcomes.

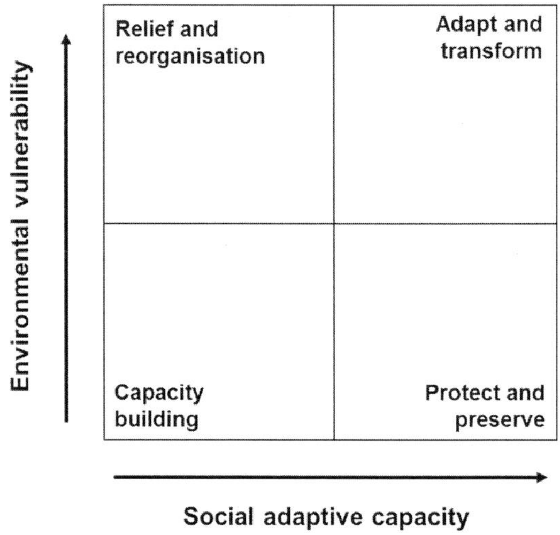

Fig. 8.3 Framework for governance and management decision making that incorporates social-ecological context. By assessing how vulnerable the ecosystem at a location is to predicted disturbances, and how much social adaptive capacity exists at a site, specific options are available (as highlighted in the 4 quadrants). Sites with low social adaptive capacity require donor relief and/or capacity building depending on the environmental vulnerability. Sites with high social adaptive capacity are more appropriate for management experimentation, with the goals reflecting how vulnerable the sites are to disturbances. Adapted from McClanahan et al. (2008a)

Challenges

Adaptive management has been clearly influential as an idea. Although the popularity of adaptive management has grown considerably since some of its earliest articulations (e.g., Beverton and Holt 1957, Holling 1978, Walters and Hilborn 1978) there is considerable ambiguity about what adaptive management is, which has limited the practical influence it has been able to have on conservation or natural resource management (Lee 1999). Furthermore, the resources available for management are often scarce. Over time, this has created a push to develop more efficient and cost effective management approaches. These approaches have significant information requirements which are sometimes lacking and often fail to incorporate uncertainty. Adaptive management, as an alternative, specifically manages for uncertainty, incorporating redundancy, but is likely to be costly and in many situations slow (e.g., Walters et al. 1993). Justifying expenditures on actions that will, in some instances, prove wrong where the tendency is to streamline and cut costs will be tricky. A mental shift is needed, from one in which the costs and benefits are weighed up at time zero, to one that accepts the uncertainties and the need for failures to ensure success. Adaptive managers will need to strive to achieve cost-effective testing of hypotheses and develop monitoring strategies that emerge from a skeptical appraisal of what kinds of information one can afford to collect (Rogers 1998).

The process of social learning, can also bring with it challenges. Although social engagement and learning is clearly beneficial for communicating and spreading adaptive management successes, there should not be a filter on what is communicated. This means that stakeholders will be equally aware of successes and failures. Such mixed outcomes are likely to impact public confidence and therefore support for future management strategies, unless they are carefully communicated. Communicating the goals of adaptive management, and the very real potential for mixed success will be key to long term support and success.

The majority of countries with coral reefs are low income, with local communities that have a high dependence on their natural resources, and where governance structures are weak (Donner and Potere 2007, Allison et al. 2009). Such settings create challenges for management, highlighting the need to tailor actions to local social-ecological system contexts (McClanahan et al. 2008a). Adaptive management can take many forms, and targeting locally appropriate actions will be key to successes (Cinner et al. 2012). Challenges will also exist for coral reefs due to the complexity of the system, which is incredibly hard to predict. A large conceptual leap will be needed from various stakeholders, from ideas of returning reefs to baselines and targets of previous conditions, to managing for uncertainty and future novel compositions. Although this may take time, there is an increasing recognition of the changing composition of coral reefs, and realistic goals for their management in this context will be imperative.

Conclusion

Coral reef ecosystems are amongst the most influenced by anthropogenic climate change and a range of other drivers, leading to changing compositions and the development of novel coral reef ecosystems. With these changes comes a great deal of uncertainty, and a need to adapt our management approaches and targets. Assessing how effective existing management approaches are in a novel coral reef ecosystem context, and trialing new approaches will inherently require adaptive management that embraces experimentation and learning. Critically, successful and equitable approaches will need to be tailored to local social-ecological contexts and involve multiple stakeholders. Although many challenges exist, embracing novel futures will enable the development of innovative management approaches by moving away from the notion of trying to return to pristine ecosystem configurations that are no longer achievable. A failure to recognize the changing compositions, species dominance, and related changes to ecological processes and functions will likely lead to unrealistic management expectations and could create ecological surprises that we are not well prepared to deal with. Embracing change and experimenting with adaptive management will be key to understanding and sustainably using novel coral reef futures.

Acknowledgements Many thanks to Morgan Pratchett and Fraser Januchowski-Hartley for contributing photographs for Fig. 8.1, and to David Bellwood for helping to develop an earlier version of Fig. 8.2. Funding was provided by the Australian Research Council.

References

Agardy, T., di Sciara, G. N., & Christie, P. (2011). Mind the gap: Addressing the shortcomings of marine protected areas through large scale marine spatial planning. *Marine Policy, 35,* 226–232.

Albins, M., & Hixon, M. (2013). Worst case scenario: Potential long-term effects of invasive predatory lionfish (*Pterois volitans*) on Atlantic and Caribbean coral-reef communities. *Environmental Biology of Fishes, 96,* 1151–1157.

Allison, E. H., Perry, A. L., Badjeck, M. C., Adger, W. N., Brown, K., Conway, D., Halls, A. S., Pilling, G. M., Reynolds, J. D., Andrew, N. L., & Dulvy, N. K. (2009). Vulnerability of national economies to the impacts of climate change on fisheries. *Fish and Fisheries, 10,* 173–196.

Alongi, D. M. (2002). Present state and future of the world's mangrove forests. *Environmental Conservation, 29,* 331–349.

Archer, S. R., & Predick, K. I. (2008). Climate change and ecosystems of the Southwestern United States. *Rangelands, 30,* 23–28.

Arnold, S. N., Steneck, R. S., & Mumby, P. J. (2010). Running the gauntlet: Inhibitory effects of algal turfs on the processes of coral recruitment. *Marine Ecology Progress Series, 414,* 91–105.

Bellwood, D. R., & Fulton, C. J. (2008). Sediment-mediated suppression of herbivory on coral reefs: Decreasing resilience to rising sea levels and climate change? *Limnology and Oceanography, 53,* 2695–2701.

Bellwood, D. R., Hughes, T. P., Folke, C., & Nystrom, M. (2004). Confronting the coral reef crisis. *Nature, 429,* 827–832.

Bellwood, D. R., Hughes, R. H., & Hoey, A. S. (2006). Sleeping functional group drives coral-reef recovery. *Current Biology, 16,* 2434–2439.

Berkes, F., & Folke, C. (Eds.). (1998). *Linking social and ecological systems*. Cambridge: Cambridge University Press.

Beverton, R. J. H., & Holt, S. J. (1957). *On the dynamics of exploited fish populations*. London: Chapman and Hall.

Blackburn, T. M., Cassey, P., Duncan, R. P., Evans, K. L., & Gaston, K. J. (2004). Avian extinction and mammalian introductions on oceanic islands. *Science, 305,* 1955–1958.

Bodin, O., & Norberg, J. (2005). Information network topologies for enhanced local adaptive management. *Environmental Management, 36,* 772–773.

Brewer, T. D., Cinner, J. E., Fisher, R., Green, A., & Wilson, S. K. (2012). Market access, population density, and socioeconomic development explain diversity and functional group biomass of coral reef fish assemblages. *Global Environmental Change, 22,* 399–406.

Campbell, S. J., Hoey, A. S., Maynard, J., Kartawijaya, T., Cinner, J., Graham, N. A. J., & Baird, A. H. (2012). Weak compliance undermines the success of no-take zones in a large government-controlled marine protected area. *PLOS One, 7,* e50074.

Carson, R. (1962). *Silent spring*. Boston: Houghton Mifflin Company.

Chong-Seng, K. M., Nash, K. L., Bellwood, D. R., & Graham, N. A. J. (2014). Macroalgal herbivory on degraded versus recovering coral reefs. *Coral Reefs, 33,* 409–419.

Cinner, J., Marnane, M. J., McClanahan, T. R., & Almany, G. R. (2006). Periodic closures as adaptive coral reef management in the Indo-Pacific. *Ecology and Society, 11,* 31.

Cinner, J. E., McClanahan, T. R., Daw, T. M., Graham, N. A. J., Maina, J., Wilson, S. K., & Hughes, T. P. (2009a). Linking social and ecological systems to sustain coral reef fisheries. *Current Biology, 19,* 206–212.

Cinner, J. E., McClanahan, T. R., Graham, N. A. J., Pratchett, M. S., Wilson, S. K., & Raina, J. B. (2009b). Gear-based fisheries management as a potential adaptive response to climate change and coral mortality. *Journal of Applied Ecology, 46,* 724–732.

Cinner, J. E., McClanahan, T. R., MacNeil, M. A., Graham, N. A. J., Daw, T. M., Mukminin, A., Feary, D. A., Rabearisoa, A. L., Wamukota, A., & Jiddawi, N. (2012). Comanagement of coral reef social-ecological systems. *Proceedings of the National Academy of Sciences of the United States of America, 109,* 5219–5222.

Cinner, J. E., Huchery, C., Darling, E. S., Humphries, A. T., Graham, N. A. J., Hicks, C. C., Marshall, N., & McClanahan, T. R. (2013). Evaluating social and ecological vulnerability of coral reef fisheries to climate change. *PLOS One, 8,* e74321.

Connell, J. H. (1978). Diversity in tropical rain forests and coral reefs. *Science, 199,* 1302–1310.

Darling, E. S., McClanahan, T. R., & Côté, I. M. (2013). Life histories predict coral community disassembly under multiple stressors. *Global Change Biology, 19,* 1930–1940.

Daw, T., Cinner, J. E., McClanahan, T. R., Graham, N. A. J., & Wilson, S. K. (2011). Design factors and socioeconomic variables associated with ecological responses to fishery closures in the western Indian Ocean. *Coastal Management, 39,* 412–424.

Dayton, P. K. (1971). Competition, disturbance, and community organization: The provision and subsequent utilization of space in a rocky intertidal community. *Ecological Monographs, 41,* 351–389.

De'ath, G., & Fabricius, K. (2010). Water quality as a regional driver of coral biodiversity and macroalgae on the Great Barrier Reef. *Ecological Applications, 20,* 840–850.

Doney, S. C., Ruckelshaus, M., Emmett Duffy, J., Barry, J. P., Chan, F., English, C. A., Galindo, H. M., Grebmeier, J. M., Hollowed, A. B., Knowlton, N., Polovina, J., Rabalais, N. N., Sydeman, W. J., & Talley, L. D. (2012). Climate change impacts on marine ecosystems. *Annual Review Marine Science, 4,* 11–37.

Donner, S., & Potere, D. (2007). The inequity of global threat to coral reefs. *BioScience, 57,* 214–215.

Duarte, C. M. (2002). The future of seagrass meadows. *Environmental Conservation, 29,* 192–206.

Fabinyi, M. (2012). Historical, cultural and social perspectives on luxury seafood consumption in China. *Environmental Conservation, 39,* 83–92.

Fabricius, K., De'ath, G., McCook, L., Turak, E., & Williams, D. M. (2005). Changes in algal, coral and fish assemblages along water quality gradients on the inshore Great Barrier Reef. *Marine Pollution Bulletin, 51,* 384–398.

Feary, D. A., Cinner, J. E., Graham, N. A. J., & Januchowski-Hartley, F. A. (2011). Effects of customary marine closures on fish behavior, spear-fishing success, and underwater visual surveys. *Conservation Biology, 25,* 341–349.

Feary, D. A., Pratchett, M. S., Emslie, M., Fowler, A. M., Figueira, W. F., Luiz, O. J., Nakamura, Y., & Booth, D. J. (2014). Latitudinal shifts in coral reef fishes: Why some species do and others do not shift. *Fish and Fisheries, 15,* 593–615.

Grafton, R. Q., & Kompas, T. (2005). Uncertainty and the active adaptive management of marine reserves. *Marine Policy, 29,* 471–479.

Graham, N. A. J., Wilson, S. K., Jennings, S., Polunin, N. V. C., Bijoux, J., & Robinson, J. (2006). Dynamic fragility of oceanic coral reef ecosystems. *Proceedings of the National Academy of Sciences of the United States of America, 103,* 8425–8429.

Graham, N. A. J., Wilson, S. K., Jennings, S., Polunin, N. V. C., Robinson, J., & Daw, T. M. (2007). Lag effects in the impacts of mass coral bleaching on coral reef fish, fisheries, and ecosystems. *Conservation Biology, 21,* 1291–1300.

Graham, N. A. J., Chabanet, P., Evans, R. D., Jennings, S., Letourneur, Y., MacNeil, M. A., McClanahan, T. R., Öhman, M. C., Polunin, N. V. C., & Wilson, S. K. (2011). Extinction vulnerability of coral reef fishes. *Ecology Letters, 14,* 341–348.

Graham, N. A. J., Bellwood, D. R., Cinner, J. E., Hughes, T. P., Norström, A. V., & Nyström, M. (2013). Managing resilience to reverse phase shifts in coral reefs. *Frontiers in Ecology and the Environment, 11,* 541–548.

Graham, N. A. J., Cinner, J. E., Nörstrom, A. V., & Nyström, M. (2014). Coral reefs as novel ecosystems: embracing new futures. *Current Opinions Environmental Sustainability, 7,* 9–14.

Harborne, A. R., & Mumby, P. J. (2011). Novel ecosystems: Altering fish assemblages in warming waters. *Current Biology, 21,* R822–R824.

Heclo, H. (1974). *Modern social politics in Britain and Sweden.* New Haven: Yale University Press.

Hicks, C. C. (2011). How do we value our reefs? Risks and tradeoffs across scales in "biomass-based" economies. *Coastal Management, 39,* 358–376.

Hicks, C. C., & McClanahan, T. R. (2012). Assessing gear modification needed to optimize yields in a heavily exploited, multi-species, seagrass and coral reef fishery. *PLOS One, 7,* e36022.

Hicks, C. C., McClanahan, T. R., Cinner, J. E., & Hills, J. M. (2009). Trade-offs in values assigned to ecological goods and services associated with different coral reef management strategies. *Ecology and Society, 14,* 10.

Hicks, C. C., Graham, N. A. J., & Cinner, J. E. (2013). Synergies and tradeoffs in how managers, scientists, and fishers value coral reef ecosystem services. *Global Environmental Change, 23,* 1444–1453.

Hobbs, R. J., Arico, S., Aronson, J., Baron, J. S., Bridgewater, P., Cramer, V. A., Epstein, P. R., Ewel, J. J., Klink, C. A., Lugo, A. E., Norton, D., Ojima, D., Richardson, D. M., Sanderson, E. W., Valladares, F., Vilà, M., Zamora, R., & Zobel, M. (2006). Novel ecosystems: Theoretical and management aspects of the new ecological world order. *Global Ecology and Biogeography, 15,* 1–7.

Hobbs, R. J., Higgs, E. S., & Hall, C. (2013). *Novel ecosystems: Intervening in the new ecological world order*. Chichester: Wiley.

Hoegh-Guldberg, O., & Bruno, J. F. (2010). The impact of climate change on the world's marine ecosystems. *Science, 328,* 1423–1428.

Hoegh-Guldberg, O., Mumby, P. J., Hooten, A. J., Steneck, R. S., Greenfield, P., Gomez E, E., Harvell, C. D., Sale, P. F., Edwards, A. J., Caldeira, K., Knowlton, N., Eakin, C. M., Iglesias-Prieto, R., Muthiga, N., Bradbury, R. H., Dubi, A., & Hatziolos, M. E. (2007). Coral reefs under rapid climate change and ocean acidification. *Science, 318,* 1737–1742.

Hoey, A. S., & Bellwood, D. R. (2011). Suppression of herbivory by macroalgal density: A critical feedback on coral reefs? *Ecology Letters, 14,* 267–273.

Holling, C. S. (1978). *Adaptive environmental assessment and management.* New York: John Wiley and Sons.

Hughes, T. P., & Connell, J. H. (1999). Multiple stressors on coral reefs: A long-term perspective. *Limnology Oceanography, 44,* 932–940.

Hughes, T. P., Baird, A. H., Bellwood, D. R., Card, M., Connolly, S. R., Folke, C., Grosberg, R., Hoegh-Guldberg, O., Jackson, J. B. C., Kleypas, J., Lough, J. M., Marshall, P., Nystrom, M., Palumbi, S. R., Pandolfi, J. M., Rosen, B., & Roughgarden, J. (2003). Climate change, human impacts, and the resilience of coral reefs. *Science, 301,* 929–933.

Hughes, T. P., Gunderson, L. H., Folke, C., Baird, A. H., Bellwood, D., Berkes, F., Crona, B., Helfgott, A., Leslie, H., Norberg, J., Nyström, M., Olsson, P., Österblom, H., Scheffer, M., Schuttenberg, H., Steneck, R. S., Tengö, M., Troell, M., Walker, B., Wilson, J., & Worm, B. (2007). Adaptive management of the Great Barrier Reef and the Grand Canyon World Heritage Areas. *Ambio, 36,* 586–592.

Hughes, T. P., Graham, N. A. J., Jackson, J. B. C., Mumby, P. J., & Steneck, R. S. (2010). Rising to the challenge of sustaining coral reef resilience. *Trends in Ecology and Evolution, 25,* 633–642.

Hughes, T. P., Baird, A. H., Dinsdale, E. A., Moltschaniwskyj, N. A., Pratchett, M. S., Tanner, J. E., & Willis, B. L. (2012). Assembly rules of reef corals are flexible along a steep climatic gradient. *Current Biology, 22,* 736–741.

Jackson, J. B. C., Kirby, M. X., Berger, W. A., Bjorndal, K. A., Botsford, L. W., Bourque, B. J., Bradbury, R. H., Cooke, R., Erlandson, J., Estes, J. A., Hughes, T. P., Kidwell, S., Lange, C. B., Lenihan, H. S., Pandolfi, J. M., Peterson, C. H., Steneck, R. S., Tegner, M. J., & Warner, R. R. (2001). Historical overfishing and the recent collapse of coastal ecosystems. *Science, 293,* 629–638.

Johannes, R. E. (1982). Traditional conservation methods and protected marine areas in Oceania. *Ambio, 11,* 258–261.

Johnson, A. E. (2010). Reducing bycatch in coral reef trap fisheries: Escape gaps as a step towards sustainability. *Marine Ecology Progress Series, 415,* 201–209.

Kittinger, J. N., Finkbeiner, E. M., Glazier, E. W., & Crowder, L. B. (2012). Human dimensions of coral reef social-ecological systems. *Ecology and Society, 17,* 17.

Kleypas, J., Feely, R. A., Fabry, V. J., Langdon, C., Sabine, C. L., & Robbins, L. L. (2006). Impacts of ocean acidification on coral reefs and other marine calcifiers: a guide for future research. NSF, NOAA, USGS.

Lauck, T., Clark, C. W., Mangel, M., & Munro, G. R. (1998). Implementing the precautionary principle in fisheries management through marine reserve. *Ecological Applications, 8,* S72–S78.

Lebel, L., Anderies, J. M., Campbell, B., Folke, C., Hatfield-Dodds, S., Hughes, T. P., & Wilson, J. (2006). Governance and the capacity to manage resilience in regional social-ecological systems. *Ecology and Society, 11,* 19.

Lee, K. N. (1999). Appraising adaptive management. *Conservation Ecology, 3,* 3.

Lindenmayer, D. B., Fischer, J., Felton, A., Crane, M., Michael, D., Macgregor, C., Montague-Drake, R., Manning, A., & Hobbs, R. J. (2008). Novel ecosystems resulting from landscape transformation create dilemmas for modern conservation practice. *Conservation Letters, 1,* 129–135.

Lokrantz, J., Nystrom, M., Thyresson, M., & Johansson, C. (2008). The non-linear relationship between body size and function in parrotfishes. *Coral Reefs, 27,* 967–974.

Loya, Y., Sakai, K., Yamazato, K., Nakano, Y., Sambali, H., & van Woesik, R. (2001). Coral bleaching: The winners and the losers. *Ecology Letters, 4,* 122–131.

Madin, J. S., & Connolly, S. R. (2006). Ecological consequences of major hydrodynamic disturbances on coral reefs. *Nature, 444,* 477–480.

Mapstone, B. D., Davies, C. R., Little, L. R., Punt, A. E., Smith, A. D. M., Pantus, F., Lou, D. C., Williams, A. J., Jones, A., Ayling, A. M., Russ, G. R., & McDonald, A. D. (2004). The effects of line fishing on the Great Barrier Reef and evaluations of alternative potential management strategies. CRC Reef Research Centre Technical Report, Book 52. CRC Reef Research Centre, Townsville, Australia.

Margoluis, R., & Salafsky, N. (1998). *Measures of success*. Washington, D.C. Island Press.

Marshall, P. A., & Baird, A. H. (2000). Bleaching of corals on the Great Barrier Reef: Differential susceptibilities among taxa. *Coral Reefs, 19,* 155–163.

Martinez, J. A., Smith, C. M., & Richmond, R. H. (2012). Invasive algal mats degrade coral reef physical habitat quality. Estuary. *Coastal and Shelf Science, 99,* 42–49.

Mbaru, E., & McClanahan, T. R. (2013). Escape gaps in African basket traps reduce bycatch while increasing body sizes and incomes in a heavily fished reef lagoon. *Fisheries Research, 148,* 90–99.

McClanahan, T. R. (2002). The near future of coral reefs. *Environmental Conservation, 29,* 460–483.

McClanahan, T. R., & Cinner, J. (2008). A framework for adaptive gear and ecosystem-based management in the artisanal coral reef fishery of Papua New Guinea. *Aquatic Conservation: Marine and Freshwater Ecosystems, 18,* 493–507.

McClanahan, T. R., & Hicks, C. C. (2011). Changes in life history and ecological characteristics of coral reef fish catch composition with increasing fishery management. *Fisheries Management and Ecology, 18,* 50–60.

McClanahan, T. R., Ateweberhan, M., Graham, N. A. J., Wilson, S. K., Ruiz C, S. C., Guillaume, M. M. M., & Bruggemann, J. H. (2007). Western Indian Ocean coral communities: Bleaching responses and susceptibility to extinction. *Marine Ecology Progress Series, 337,* 1–13.

McClanahan, R. T., Hicks, C. C., & Darling, S. E. (2008a). Malthusian overfishing and efforts to overcome it on Kenyan coral reefs. *Ecological Applications, 18,* 1516–1529.

McClanahan, T. R., Cinner, J. E., Maina, J., Graham, N. A. J., Daw, T. M., Stead, S. M., Wamukota, A., Brown, K., Ateweberhan, M., & Venus, V. (2008b). Conservation action in a changing climate. *Conservation Letters, 1,* 53–59.

McClanahan, T. R., Graham, N. A. J., MacNeil, M. A., Muthiga, N. A., Cinner, J. E., Bruggemann, J. H., & Wilson, S. K. (2011). Critical thresholds and tangible targets for ecosystem-based management of coral reef fisheries. *Proceedings of the National Academy of Sciences of the United States of America, 108,* 17230–17233.

McClanahan, T. R., Graham, N. A. J., & Darling, E. S. (2014). Coral reefs in a crystal ball: predicting the future from the vulnerability of corals and reef fishes to multiple stressors. *Current Opinions in Environmental Sustainability, 7,* 59–64.

McCook, L., Ayling, T., Cappo, M., Choat, J. H., Evans, R. D., De Freitas, D. M., Heupel, M., Hughes, T. P., Jones, G. P., Mapstone, B., Marsh, H., Mills, M., Molloy, F. J., Pitcher, C. R., Pressey, R. L., Russ, G. R., Sutton, S., Sweatman, H., Tobin, R., Wachenfeld, D. R., & Williamson, D. H. (2010). Adaptive management of the Great Barrier Reef: A globally significant demonstration of the benefits of networks of marine reserves. *Proceedings of the National Academy of Sciences of the United States of America, 43,* 18278–18285.

Mora, C., & Sale, P. F. (2011). Ongoing global biodiversity loss and the need to move beyond protected areas: A review of the technical and practical shortcomings of protected areas on land and sea. *Marine Ecology Progress Series, 434,* 251–266.

Mora, C., Andrefouet, S., Costello, M. J., Kranenburg, C., Rollo, A., Veron, J., Gaston, K. J., & Myers, R. A. (2006). Coral reefs and the global network of marine protected areas. *Science, 312,* 1750–1751.

Mumby, P. J., & Harborne, A. R. (2010). Marine reserves enhance the recovery of corals on Caribbean reefs. *PLOS One, 5,* 1–7.

Mumby, P. J., Edwards, A. J., Arias-Gonzalez, J. E., Lindeman, K. C., Blackwell, P. G., Gall, A., Gorczynska, M. I., Harborne, A. R., Pescod, C. L., Renken, H., Wabnitz, C. C. C., & Llewenyn, G. (2004). Mangroves enhance the biomass of coral reef fish communities in the Caribbean. *Nature, 427,* 533–536.

Mumby, P. J., Dahlgren, C. P., Harborne, A. R., Kappel, C. V., Micheli, F., Brumbaugh, D. R., Holmes, K. E., Mendes, J. M., Broad, K., Sanchirico, J. N., Buch, K., Box, S., Stoffle, R. W., & Gill, A. B. (2006). Fishing, trophic cascades, and the process of grazing on coral reefs. *Science, 311,* 98–101.

Mumby, P. J., Harborne, A. R., Williams, J., Kappel, C. V., Brumbaugh, D. R., Micheli, F., Holmes, K. E., Dahlgren, C. P., Paris, C. B., & Blackwell, P. G. (2007). Trophic cascade facilitates coral recruitment in a marine reserve. *Proceedings of the National Academy of Sciences of the United States of America, 104,* 8362–8367.

Mumby, P. J., Broad, K., Brumbaugh, D. R., Dahlgren, C. P., Harborne, A. R., Hastings, A., Holmes, K. E., Kappel, C. V., Micheli, F., & Sanchirico, J. N. (2008). Coral reef habitats as surrogates of species, ecological functions, and ecosystem services. *Conservation Biology, 22,* 941–951.

Myers, N. (1987). The extinction spasm impending: Synergisms at work. *Conservation Biology, 1,* 14–21.

Nash, K. L., Graham, N. A. J., & Bellwood, D. R. (2013). Fish foraging patterns, vulnerability to fishing, and implications for the management of ecosystem function across scales. *Ecological Applications, 23,* 1632–1644.

Norton, B. G. (2005). *Sustainability: A Philosophy of Adaptive Ecosystem Management.* Chicago: University of Chicago Press.

Nyström, M., Folke, C., & Moberg, F. (2000). Coral reef disturbance and resilience in a human-dominated environment. *Trends in Ecology Evolution, 15,* 413–417.

Ostrom, E. (2007). A diagnostic approach for going beyond panaceas. *Proceedings of the National Academy of Sciences of the United States of America, 104,* 15181–15187.

Ostrom, E. (2009). A general framework for analyzing sustainability of social-ecological systems. *Science, 325,* 419–422.

Owens, I. P. F., & Bennett, P. M. (2000). Ecological basis of extinction risk in birds: Habitat loss versus human persecution and introduced predators. *Proceedings of the National Academy of Sciences of the United States of America, 97,* 12144–12148.

Pandolfi, J. M., Bradbury, R. H., Sala, E., Hughes, T. P., Bjorndal, K. A., Cooke, R. G., McArdle, D., McClenachan, L., Newman, M. J. H., Paredes, G., Warner, R. R., & Jackson, J. B. C. (2003). Global trajectories of the long-term decline of coral reef ecosystems. *Science, 301,* 955–958.

Pandolfi, J. M., Connolly, S. R., Marshall, D. J., & Cohen, A. L. (2011). Projecting coral reef futures under global warming and ocean acidification. *Science, 333,* 418–422.

Parson, E., & Clark, W. (1995). Sustainable development as social learning: Theoretical perspectives and practical challenges for the design of a research program. In L. H. Gunderson CSH, & S. S. Light (Eds.), *Barriers and Bridges to the Renewal of Ecosystems and Institutions*. New York: Columbia University Press.

Pollnac, R., Christie, P., Cinner, J. E., Dalton, T., Daw, T. M., Forrester, G. E., Graham, N. A. J., & McClanahan, T. R. (2010). Marine reserves as linked social-ecological systems. *Proceedings of the National Academy of Sciences of the United States of America, 107,* 18262–18265.

Polunin, N. V. C. (Ed.) (2008). *Aquatic ecosystems: trends and global prospects*. Cambridge: Cambridge University Press.

Pörtner, H. O., & Farrell, A. P. (2008). Physiology and climate change. *Science, 322,* 690–692.

Pratchett, M. S., Munday, P. L., Wilson, S. K., Graham, N. A. J., Cinner, J. E., Bellwood, D. R., Jones, G. P., Polunin, N. V. C., & McClanahan, T. R. (2008). Effects of climate-induced coral bleaching on coral-reef fishes: ecological and economic consequences. *Oceanography and Marine Biology: an Annual Review, 464,* 251–296.

Pratchett, M. S., Schenk, T. J., Baine, M., Syms, C., & Baird, A. H. (2009). Selective coral mortality associated with outbreaks of Acanthaster planci L. in Bootless Bay, Papua New Guinea. *Marine Environmental Research, 67,* 230–236.

Pratchett, M. S., Trapon, M., Berumen, M. L., & Chong-Seng, K. (2011). Recent disturbances augment community shifts in coral assemblages in Moorea, French Polynesia. *Coral Reefs, 30,* 183–193.

Purvis, A., Gittleman, J. L., Cowlishaw, G., & Mace, G. M. (2000). Predicting extinction risk in declining species. *Proceedings of the Royal Society of London Series B: Biological Sciences, 267,* 1947–1952.

Queensland Government. (2009). Reef water quality protection plan for the Great Barrier Reef World Heritage Area and adjacent catchments. The State of Queensland, Brisbane, Australia.

Riegl, B., Berumen, M., & Bruckner, A. (2013). Coral population trajectories, increased disturbance and management intervention: A sensitivity analysis. *Ecological Evolution, 3,* 1050–1064.

Rivera-Posada, J., Caballes, C. F., & Pratchett, M. S. (2013). Lethal doses of oxbile, peptones and thiosulfate-citrate-bile-sucrose agar (TCBS) for *Acanthaster planci*; exploring alternative population control options. *Marine Pollution Bulletin, 75,* 133–139.

Rogers, K. (1998). Managing science/management partnerships: A challenge of adaptive management. *Conservation Ecology, 2,* R1.

Roughgarden, J., & Smith, F. (1996). Why fisheries collapse and what to do about it. *Proceedings of the National Academy of Sciences of the United States of America, 93,* 5078–5083.

Russ, G. R., & Williams, D. M. (1994). Review of data on fishes of commercial and recreational fishing interest in the Great Barrier Reef Book 1. Great Barrier Reef Marine Park Authority, Townsville, Australia.

Sale, P. F., Cowen, R. K., Danilowicz, B. S., Jones, G. P., Kritzer, J. P., Lindeman, K. C., Planes, S., Polunin, N. V. C., Russ, G. R., Sadovy, Y. J., & Steneck, R. S. (2005). Critical science gaps impede use of no-take fishery reserves. *Trends in Ecology Evolution, 20,* 74–80.

Sandin, S. A., Smith, J. E., DeMartini, E. E., Dinsdale, E. A., Donner, S. D., Freidlander, A. M., Konotchick, T., Malay, M., Maragos, J. E., Obura, D., Pantos, O., Paulay, G., Richie, M., Rohwer, F., Schroeder, R. E., Walsh, S., Jackson, J. B. C., Knowlton, N., & Sala, E. (2008). Baselines and degradation of coral reefs in the Northern Line Islands. *PLOS One, 3,* e1548.

Seastedt, T. R., Hobbs, R. J., & Suding, K. N. (2008). Management of novel ecosystems: Are novel approaches required? *Frontiers in Ecology and the Environment, 6,* 547–553.

Smith, S. V., Kimmerer, W. J., Laws, E. A., Brock, R. E., & Walsh, T. W. (1981). Kaneohe Bay sewage diversion experiment: Perspectives on ecosystem responses to nutritional perturbation. *Pacific Science, 35,* 279–402.
Sousa, W. P. (1979). Disturbance in marine intertidal boulder fields: The nonequilibrium maintenance of species diversity. *Ecology, 60,* 1225–1239.
Steneck, R. S., Graham, M. H., Bourque, B. J., Corbett, D., Erlandson, J. M., Estes, J. A., & Tegner, M. J. (2002). Kelp forest ecosystems: Biodiversity, stability, resilience and future. *Environmental Conservation, 29,* 436–459.
Steneck, R. S., Paris, C. B., Arnold, S. N., Ablan-Lagman, M. C., Alcala, A. C., Butler, M. J., McCook, L. J., Russ, G. R., & Sale, P. F. (2009). Thinking and managing outside the box: Coalescing connectivity networks to build region-wide resilience in coral reef ecosystems. *Coral Reefs, 28,* 367–378.
Stockwell, B., Claro Renato, J. L., Abesamis, R. A., Alcala, A. C., & Russ, G. R. (2009). Trophic and benthic responses to no-take marine reserve protection in the Philippines. *Marine Ecology Progress Series, 389,* 1–15.
Thistle, D. (1981). Natural physical disturbances and communities of marine soft bottoms. *Marine Ecology Progress Series, 6,* 223–228.
Tompkins, E. L., & Adger, W. N. (2004). Does adaptive management of natural resources enhance resilience to climate change? *Ecology and Society, 9,* 10.
van Woesik, R., Sakai, K., Ganase, A., & Loya, Y. (2011). Revisiting the winners and the losers a decade after coral bleaching. *Marine Ecology Progress Series, 434,* 67–76.
Veron, J. E. N., Hoegh-Guldberg, O., Lenton, T. M., Lough, J. M., Obura, D. O., Pearce-Kelly, P., Sheppard, C. R. C., Spalding, M., Stafford-Smith, M. G., & Rogers, A. D. (2009). The coral reef crisis: The critical importance of 350 ppm CO2. *Marine Pollution Bulletin, 58,* 1428–1436.
Walters, C. (1986). *Adaptive management of renewable resources.* New York: Mamillan Publishing Co.
Walters, C., & Hilborn, R. (1978). Ecological optimization and adaptive management. *Annual Review of Ecology Systematics, 9,* 157–188.
Walters, C. J., & Holling, C. S. (1990). Large-scale management experiments and learning by doing. *Ecology, 71,* 2060–2068.
Walters, C., Goruk, R. D., & Radford, D. (1993). Rivers inlet sockeye salmon: An experiment in adaptive management. *North American Journal of Fisheries Management, 13,* 253–262.
Walther, G., Post, E., Convey, P., Menzel, A., Parmesan, C., Beebee, T. J. C., Fromentin, J., Hoegh-Guldberg, O., & Bairlein, F. (2002). Ecological responses to recent climate change. *Nature, 416,* 389–395.
Weeks, R., & Jupiter, S. D. (2013). Adaptive comanagement of a marine protected area network in Fiji. *Conservation Biology, 27,* 1234–1244.
White, C., Halpern, B. S., & Kappel, C. V. (2012). Ecosystem service tradeoff analysis reveals the value of marine spatial planning for multiple ocean uses. *Proceedings of the National Academy of Sciences of the United States of America, 109,* 4696–4701.
Williams, B. K. (2011). Adaptive management of natural resources—framework and issues. *Journal of Environmental Management, 92,* 1346–1353.
Williams, J. W., & Jackson, S. T. (2007). Novel climates, no-analog communities, and ecological surprises. *Frontiers in Ecology and the Environment, 5,* 475–482.
Williams, I. D., & Polunin, N. V. C. (2000). Differences between protected and unprotected reefs of the western Caribbean in attributes preferred by dive tourists. *Environmental Conservation, 27,* 382–391.
Willis, T. J., Millar, R. B., Babcock, R. C., & Tolimieri, N. (2003). Burdens of evidence and the benefits of marine reserves: Putting Descartes before des horse? *Environmental Conservation, 30,* 97–103.
Wilson, S. K., Graham, N. A. J., Pratchett, M. S., Jones, G. P., & Polunin, N. V. C. (2006). Multiple disturbances and the global degradation of coral reefs: Are reef fishes at risk or resilient? *Global Change Biology, 12,* 2220–2234.

Wilson, S. K., Graham, N. A. J., Fisher, R., Robinson, J., Nash, K., Chong-Seng, K., Polunin, N. V. C., Aumeeruddy, R., & Quatre, R. (2012). Effect of macroalgal expansion and marine protected areas on coral recovery following a climatic disturbance. *Conservation Biology, 26,* 995–1004.

Yakob, L., & Mumby, P. J. (2011). Climate change induces demographic resistance to disease in novel coral assemblages. *Proceedings of the National Academy of Sciences of the United States of America, 108,* 1967–1969.

Yamano, H., Sugihara, K., & Nomura, K. (2011). Rapid poleward range expansion of tropical reef corals in response to rising sea surface temperatures. *Geophysical Research Letters, 38,* L04601.

Zhou, S., Smith, A. D. M., Punt, A. E., Richardson, A. J., Gibbs, M., Fulton, E. A., Pascoe, S., Bulman, C., Bayliss, P., & Sainsbury, K. (2010). Ecosystem-based fisheries management requires a change to the selective fishing philosophy. *Proceedings of the National Academy of Sciences of the United States of America, 107,* 9485–9489.

Chapter 9
Adaptive Co-Management

Christo Fabricius and Bianca Currie

Keywords Adaptive management · Adaptive co-management · Uncertainty · Systems · Resilience

Introduction

Adaptive co-management has its foundations in the convergence of two independently evolved concepts, adaptive management and co-management (Berkes 2009, Plummer 2009). In practice these two approaches frequently blend into adaptive co-management when successive cycles of participation, learning and doing occur (Berkes 2009, Huitema et al. 2009). Berkes (2009) has highlighted the similarities and differences in linkages, temporal scope, organizational level and capacity building focus between co-management, adaptive management and adaptive co-management (Table 9.1).

Core Elements

The adaptive co-management process has four cornerstones:

1. An **enabling environment** through institutional arrangements (norms and rules), leadership, policies and legislation (e.g. incentives) (Armitage et al. 2009, Berkes 2009).

C. Fabricius (✉) · B. Currie
Sustainability Research Unit, Nelson Mandela Metropolitan University, P/Bag X6531, 6530 George, South Africa
e-mail: christo.fabricius@nmmu.ac.za

B. Currie
e-mail: Bianca.Currie@nmmu.ac.za

C. R. Allen, A. S. Garmestani (eds.), *Adaptive Management of Social-Ecological Systems*,
DOI 10.1007/978-94-017-9682-8_9

Table 9.1 Comparison of adaptive management, co-management and adaptive co-management concepts (Source: Adapted from Berkes 2009) and incorporating Plummer (2009) and Holling (1978)

	Adaptive management	Co-management	Adaptive co-management
Concept	Learning by doing process	Joint or shared decision making, conflict resolution or management process	Joint management through learning by doing collaboratively
Designed for…	Designed to continually improve management policies and practices by learning from the outcomes of previously employed policies and practices	Designed as an alternative approach from the top down to a consensus based and decentralized approach	Designed to enhance resilience and manage complex systems which transcend multiple levels and scales
Emphasis on…	Learning and experimentation through implementing monitoring and adjusting in real space and time	Sharing of rights, responsibilities and power across a range of relevant stakeholders	Joint management and learning by doing, (local and scientific knowledge), sharing of rights, responsibilities and power by relevant stakeholders at multiple scales
Linkages	Science and management for learning by doing	Vertical institutional linkages for the inclusion of diverse knowledge types and equity in resource sharing and decision making	Horizontal and vertical for joint learning by doing
Temporal scope	Medium to long term, multiple cycles of learning and adapting	Short to medium term, creates snapshots	Medium to long term, multiple cycles of learning and adapting
Organizational level	Managers needs and relationships	Bridging between local and government levels	Multi-level, with self-organized networks
Capacity building focus	Resource managers and decision makers	Community and resource users	All stakeholders

2. **Learning** through experimentation, monitoring and evaluation in a real world setting (Armitage et al. 2008, Armitage et al. 2009, Berkes 2009, Cundill and Fabricius 2009).
3. **Collaboration** across a diversity of stakeholders sharing a resource, rights and responsibilities at multiple levels and scales (Ruitenbeek and Cartier 2001, Armitage et al. 2009, Berkes 2009).
4. In a **cyclical** iterative process (Plummer 2009).

Adaptive co-management thus refers to an ongoing process that allows stakeholders to share responsibility within a system where they can explore their objectives, find common ground, learn from their institutions and practices, and adapt and modify them for subsequent cycles. While, as with adaptive management, the focus remains learning by doing, it takes into account a diversity of knowledge systems. This allows for the inclusion of informal, local and traditional knowledge, formal scientific knowledge and the sharing of rights, responsibilities and power among the diverse range of relevant stakeholders (Ruitenbeek and Cartier 2001). These interactions may occur at multiple levels, e.g., at the local community, provincial, national and even international levels. Stakeholders could include resource users, local stewardship associations, government agencies, and non-governmental organizations (NGOs) (Olsson et al. 2004a). The degree of collaboration can occur on a continuum of involvement which can also vary during different phases of the adaptive management cycle (Ruitenbeek and Cartier 2001).

Adaptive co-management is in essence an approach to ecosystem governance, as a partnership between the state or regulating authority, scientific and media institutions, resource users and "other civil society groups" (Adger 2005). Adaptive governance principles (Clark and Clarke 2011) include (1) the degree of cross-scale interaction between project participants and other governance levels; (2) the "learning and adaptation processes" that have occurred; (3) the extent of shared understandings about the goals and vision for the initiative. The result is integration of human and social "capitals" through cross-scale interactions which provide learning opportunities, with intermediaries and "bridging organizations" playing key roles. This leads to the emergence of a common vision and understanding. Clark and Clarke (2011) used these principles to assess good practices in adaptive governance in five case studies in English national parks. They found a positive correlation between indicators of local sustainability, and the application of adaptive governance principles. They conclude that national park authorities are important bridging organizations that promote sustainability through capacity development and trust building.

Despite being very well defined and its core elements explicit, adaptive co-management is sometimes misused or incorrectly used. In some papers a project or case study called adaptive management often contains a collaborative or community engagement component and should actually be called adaptive co-management. The "co" in the term is also interchangeably used as communal or collaborative, implying varied partnerships between public and private actors (Carlsson and Berkes 2005). The most widely used definition of adaptive co-management is: a process by which institutional arrangements and ecological knowledge are tested and revised in a dynamic, ongoing, self-organized process of learning by doing (Folke et al. 2002, Armitage et al. 2007).

Where is Adaptive Co-management Applicable?

Adaptive co-management is not the answer to all management challenges (Plummer 2009), but has been useful in dealing with complexity (Carlsson and Berkes 2005), when decentralization of management is desirable (Bulkeley and Betsill 2005) or there is a need to legitimize decision making (Carlsson and Berkes 2005), in conflict resolution (Pomeroy and Berkes 1997, Singleton 1998), and problem solving (Carlsson and Berkes 2005).

Dealing with Complexity

Adaptive co-management is useful in dealing with complex multi- scale and level systems and with problems where a need to link a diversity of types and levels of organizations is required through partnerships (Carlsson and Berkes 2005, Armitage et al. 2009). This may be due to conflict, or a high level of interdependence (Ansell and Gash 2007), risk sharing, or exchange of resources (Carlsson and Berkes 2005) e.g., herbicide being provided by a conservation authority to private landowners to clear alien invasive species on private property, or provision of free advice and assistance with infrastructure by conservation bodies to communities involved in integrated conservation and development projects (Garnett et al. 2007). No single agency or local user group governing the system alone will work (Carlsson and Berkes 2005, Berkes 2009); the problem requires ongoing committed cooperation and collaboration by a range of stakeholders, which is where adaptive co-management comes into its own.

Decentralization of Management

Adaptive co-management has also commonly been associated with the decentralization of natural resource management of a commons, landscape or small scale resource (e.g., urban ecosystems, Bulkeley and Betsill 2005) where local level knowledge is required; or when distinct rights to resources are shared. Decentralization should be to an appropriate level, congruent with the scale of the ecosystem being managed (Bohensky and Lynam 2005). In the same way as it does not make sense to devolve the authority to manage a regional catchment to a local community, it also would not be appropriate for central government to take sole responsibility for the management of a locally used stream, wetland or community forest (Murphree 2004).

Legitimization of Decision Making

Government agencies can also use adaptive co-management to legitimize decision making, to delegate functions too costly to manage, or to allocate tasks where divisions of labor across levels and capacities can improve efficiency, for example when a group of local producers gains access to new markets through increasing economies of scale (Carlsson and Berkes 2005).

Conflict Resolution and Problem Solving

Conflict can be a barrier to collaboration but can also be the inspiration for collaboration when stakeholders realize that the issue at hand cannot be solved without working together. Adaptive co-management can be applied as a conflict resolution mechanism between stakeholders and government (Pomeroy and Berkes 1997, Singleton 1998, Armitage et al. 2009) or where a policy deadlock occurs (Ansel and Gash 2007) and even evolve into a problem solving mechanism with time. The process can include negotiation and bargaining where the rights and responsibilities of the relevant stakeholders can become explicit, formalized and entrenched, potentially resolving conflicts over resources (Carlsson and Berkes 2005).

Conceptual Frameworks

Plummer and Baird (2013) offer one abstract model which represents the adaptive co-management process as an evolving process consisting of three stages: "inchoate", where actors with a shared resource or issue exist but have yet to begin interaction; "formulation" where the actors engage and begin interacting and deliberating; and "conjoined" where the actors continue to interact and deliberate and implement actions, monitoring and adapting or modifying approaches and practices (Fig. 9.1). They acknowledge that evolution is seldom as linear as depicted in the model and that the process is "dynamic and ongoing".

Research on the design principles for adaptive co-management centers around five core elements (Plummer and Baird 2013, Pratt Miles 2013, Wallis et al. 2013):

1. Establish an institutional platform for collaboration and stakeholder interaction. Because adaptive co-management typically addresses a series of 'wicked problems' (*sensu* Rittel and Webber 1973, Ludwig 2001), stakeholder opinions

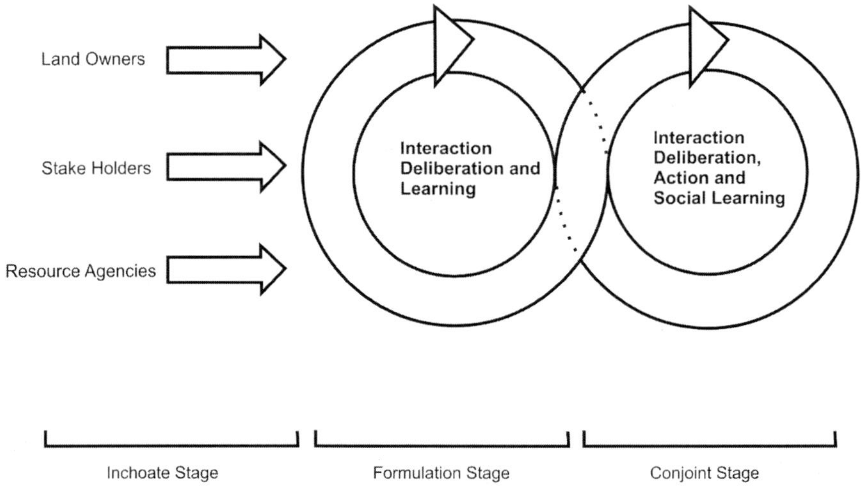

Fig. 9.1 Adaptive co-management analytical model (reproduced from Plummer and Baird 2013)

should be canvassed early and frequently. Appropriate forums for this type of interaction are essential.
2. Obtain input from key stakeholders at critical points in the adaptive management process. It is particularly important to embrace rather than avoid diversity of opinions and to adopt a 'soft systems' approach (Checkland 2000) which allows objectives to shift and be re-defined as stakeholders become more aware of the complexities, their common interests and their differences (Cundill et al. 2012).
3. Communicate and share data and information. This can be done during face to face meetings or by using technology, social media and web-based platforms to promote citizen science (Tulloch et al. 2013).
4. Define the thresholds that will initiate change and adaptation, understanding that this will vary between stakeholders, and that overcoming the inertia or "action paralysis" associated with social learning can be a huge challenge (Allen and Gunderson 2011, Fabricius and Cundill 2014).
5. Adapt, adjust and even re-design the decision-making structures and governance systems when new information becomes available (Wallis et al. 2013).

Attributes that promote good governance include "participation, representation, deliberation, accountability, empowerment, social justice, and organizational features such as being multilayered and polycentric". These attributes need to be coupled with system resilience attributes such as appropriateness of scale, the capacity to adapt to uncertainties, the capacity to maintain diversity, and to combine different sources of knowledge, and, importantly, the ability to detect looming thresholds (Lebel et al. 2006). The feedbacks between governance attributes and resilience attributes ultimately affect the system's capacity to "self-organize", "learn" and "adapt", which are measures of system resilience (Fig. 9.2).

One of the outcomes of adaptive co-management is that all actors will ultimately, and to varying degrees, perceive the system as integrated, complex and adaptive, with interactions and feedbacks between resource users, resources, ecosystems, governance and public infrastructure (Anderies et al. 2004). Learning, constantly assessing and understanding interactions, not only at the focal scale but also vertically across scales, is central to adaptive co-management. This is graphically depicted in Ostrom's (2009) social-ecological systems framework which emphasizes that the interactions between system components are less well understood than the components themselves. Ostrom and Cox (2010) emphasize that institutional complexity is the rule rather than the exception in social-ecological systems and warn against "panaceas" consisting of single solutions and diagnostic frameworks. They advocate a multi-tiered approach and suggest the use of multiple frameworks. In their IAD (Institutional Analysis and Development) framework, seven components: the institutional actors; their roles or 'positions'; the actions that they are allowed; their 'levels of control'; the outcomes for each actor; the information each actor has at their disposal; and the incentives, costs and benefits associated with their actions, are viewed as important (Fig. 9.3; from Ostrom and Cox 2010).

Fabricius et al. (2007) developed a design framework, based on the community-based assessments of the Millennium Ecosystem Assessment (Folke et al. 2005). The framework (Fig. 9.4) could also be used for diagnostic purposes to differentiate

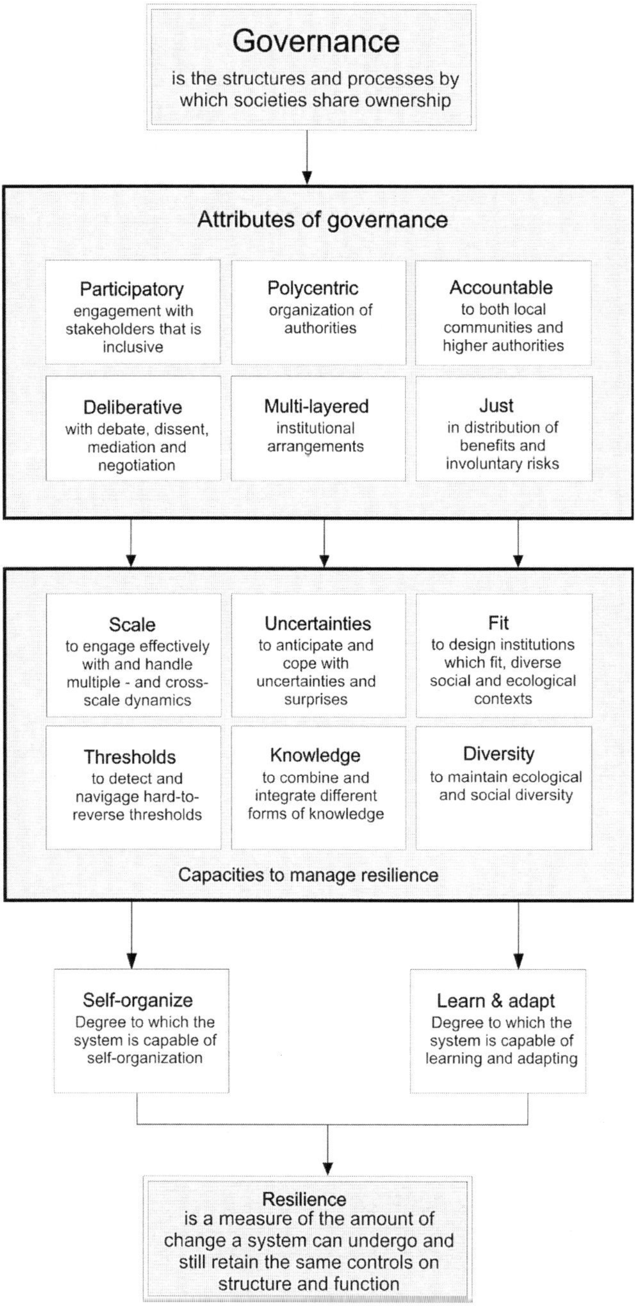

Fig. 9.2 Associations between selected attributes of governance systems and the capacity to manage resilience (reproduced from Lebel et al. 2006)

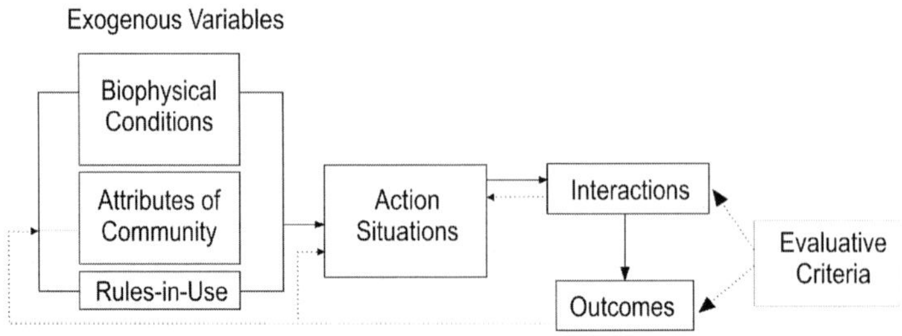

Fig. 9.3 The IAD framework (reproduced from Ostrom and Cox 2010)

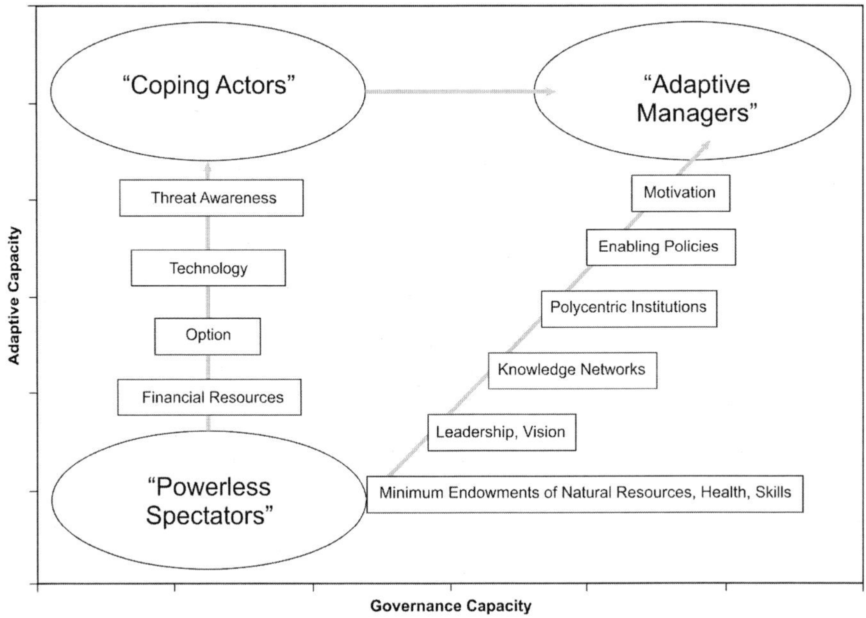

Fig. 9.4 Three types of adaptive communities along gradients of adaptive capacity and governance capacity, respectively. "Powerless spectator" communities have a low adaptive capacity and weak capacity to govern, do not have financial or technological options, and lack natural resources, skills, institutions, and networks. "Coping actor" communities have the capacity to adapt, but are not managing social–ecological systems. They lack the capacity for governance because of lack of leadership, of vision, and of motivation, and their responses are typically short term. "Adaptive manager" communities have both the capacity to adapt and the governance capacity to sustain and internalize this adaptation. They invest in the long-term management of ecosystem services. Such communities are not only aware of the threats, but also take appropriate action for long-term sustainability. Adaptive co-management becomes possible through leadership and vision, the formation of knowledge networks, the existence or development of polycentric institutions, establishing or maintaining links between culture and management, the existence of enabling policies, and high levels of motivation in all role players (adapted from Fabricius et al. 2007)

between "powerless spectators"; "coping actors" and "adaptive co-managers". The framework hinges on two factors: stakeholder's capacity for governance; and their adaptive capacity. When governance capacity and adaptive capacity are low, stakeholders are mere spectators who powerlessly observe events unfold and affect them. When adaptive capacity is high but governance capacity low, stakeholders merely cope without taking part in management. Only when both adaptive capacity and governance capacity are strong enough can adaptive co-management be effective. Fabricius et al. (2007) conclude that 'adaptive manager' communities have both adaptive capacity and governance capacity to sustain and internalize this adaptation. They invest in the long-term management of ecosystem services. Such communities are not only aware of the threats, but also take appropriate action for long-term sustainability. The enablers of adaptive co-management include having adequate endowments of ecosystem services; leadership; vision; knowledge networks; polycentric institutions; enabling policies; and motivation (including incentives) to take action (Fig. 9.4).

Contextual factors, despite being extremely important, are often overlooked when attempting to understand adaptive co-management. No two cases are alike and the same case at one point in time might have a different context at a different point in time. Plummer and Hashimoto (2011) describe two fishery case studies, in Canada and Japan, with different contexts and different outcomes. Physical geography, e.g., connectedness of the fishery to fish stocks, culture and traditions of cooperation, and the "embeddedness" of co-management in national and local cultures are important contextual factors to consider. They offer a framework to help understand the relevance of context in adaptive co-management (Fig. 9.5).

Contextual factors often provide the backdrop for participation, an intrinsic component of adaptive co-management. Stringer et al. (2006) outline the reasons why participation is important and provide a framework for designing a participatory learning process at different stages in the adaptive co-management cycle (Fig. 9.6).

An Enabling Policy Environment

For adaptive co-management to take place it is essential that a facilitating or enabling institutional policy and legislative environment is created or provided (Olsson et al. 2004a). This either enables or disables real power and responsibility sharing (Armitage et al. 2009). To allow for the necessary experimentation to take place and for dealing with the complex nature of the systems, the policy and legislative environment should be flexible and adaptive (Berkes 2009); transcend multiple spatial levels and time scales horizontally (e.g. across a landscape) and vertically (e.g., across different levels of governance) (Huitema et al. 2009, Armitage et al. 2009); be innovative and offer incentives (Rouse 2008); include the provision of funds, resources and capacity for monitoring (Barthel et al. 2005); facilitate collaboration (Barthel et al. 2005); enable self-organization, i.e., growth of formal and informal networks for information flow that incorporate a diversity of knowledge

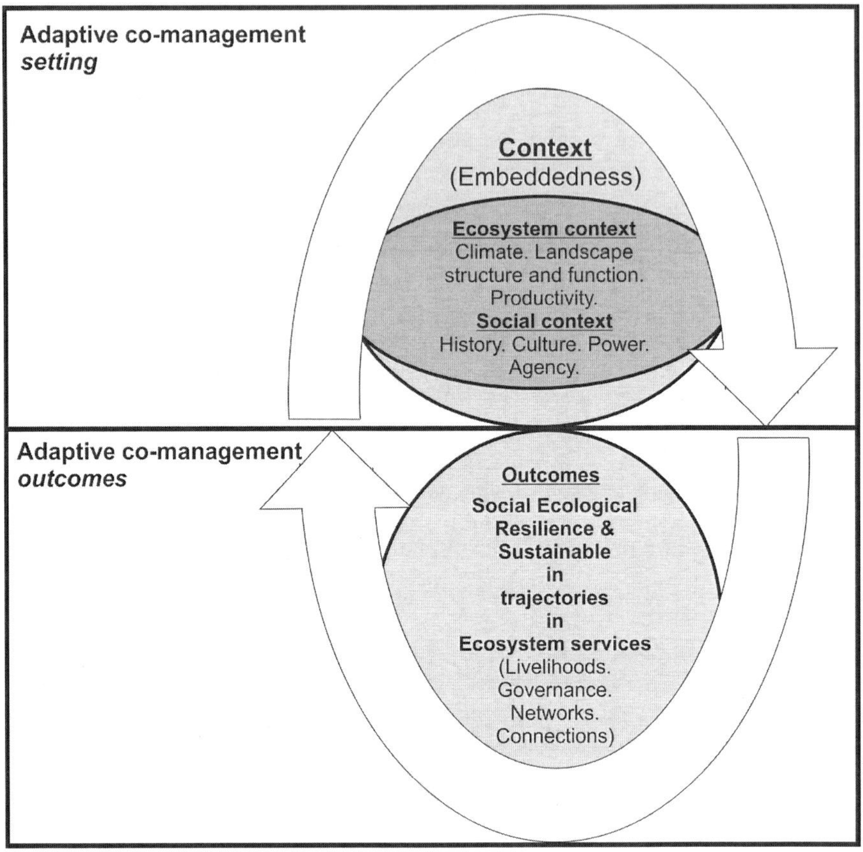

Fig. 9.5 Framework for tailoring adaptive co-management and enhancing adaptability (reproduced from Plummer and Hashimoto 2011)

types and develop formal and informal rules and norms (e.g. constitutions, laws, policies, behaviors and conventions) (Barthel et al. 2005, Berkes 2009, Armitage et al. 2009); play a role in empowering and improving capacity for collaboration by promoting repeated interaction, preferably through face to face dialogue (Armitage et al. 2009).

Social Learning in Adaptive Co-management

Social learning is integral to adaptive co-management (Cundill and Fabricius 2009, Roux et al. 2011). Armitage et al. (2008) assessed the learning theories in adult education, organizational development and business management that inform adaptive co-management. According to Armitage et al. (2008) three types of learning are possible in adaptive co-management:

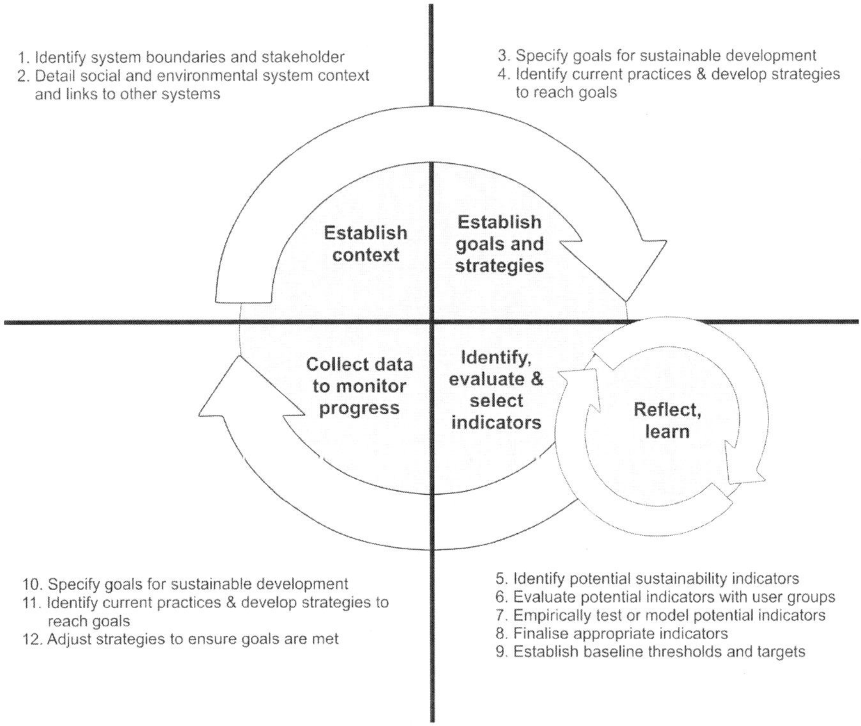

Fig. 9.6 Framework for participation at different stages in the adaptive co-management cycle (adapted from Stringer et al. 2006; adapted from Reed et al. 2006)

a. Experiential learning: learning through experience or doing which occurs in a continuous learning cycle of four phases: (1) concrete experience, (2) reflective observation, (3) abstract conceptualization (4) active experimentation (Kolb 1984).
b. Transformative learning: learning for change through a process of reflection and critical engagement (Mezirow 2000, 2009).
c. Social learning: learning with others in a continuous cycle of reflection through shared experiences, ideas and environment. Social learning includes single, double and triple loop learning (Armitage et al 2009).

When the conditions for social learning are present, sustainable outcomes are more likely to be achieved than when the conditions are absent (Clark and Clarke 2011). Aspects that need to be incorporated into the design of social learning experiments include (Bos et al. 2013): (a) a "shared learning agenda" that guides the operational environment and provides meaning; (b) political legitimacy that endorses activities; (c) dedicated financial and other resources to enable implementation of innovative experiments; (d) projects that are locally driven and which draw on a wide range of knowledge systems; (e) multi-organizational learning groups that serve as learning platforms; (f) partnerships between research and management; (g) multiple responsibilities for facilitation, distributed across a variety of actors and organizations;

(h) room for adaptation and innovation; and (i) enough time to develop trust. The Cooks River Sustainability Initiative in Sydney (Bos et al. 2013) provides an example of the application of these principles in practice (Box 1).

Box 1. A governance experiment in social learning in the Cooks River Sustainability Initiative, Australia (Bos et al. 2013)

To promote social learning in the Cooks River catchment in Sydney, Bos et al. (2013) initiated a governance experiment aimed at transforming water planning and governance in the catchment. The "OurRiver" Cooks River Sustainability Initiative was designed to develop stakeholder-driven management plans for six sub-catchments in the river system. Stakeholders were encouraged to explore ways to find synergies between their agendas and initiatives, and the researchers' roles were to promote interaction and communication. Several official and civil society groups operating across the municipality were identified, and these interacted during steering committee meetings, technical working groups and workshop sessions.

Their innovations included using a fine-grained, sub-catchment perspective, working and cooperating across disciplines, working closely with local communities in seeking solutions, and taking a regional approach to governance. This resulted in inter-organizational cooperation and the adoption of a systems perspective, with technical specialists learning to engage with local communities and everyone exploring new technologies.

Examples of the application of adaptive co-management design principles applied in this governance experiment included:

Shared learning agenda—In the Cooks River Sustainability Initiative the project's 'shared learning agenda' was to explore "mutual interdependencies" between officials and citizens to develop "context-specific solutions".

Legitimacy—They received political endorsement from the respective mayors.

Resources—They received grant funding for proposal development and follow-up funding from municipal budgets.

Focus—Their focus was on sub-catchment planning that drew on a variety of disciplines.

Multi-organizational peer groups—They appointed a variety of steering and expert committees as learning platforms.

Distributed facilitation—The project included numerous facilitators from the project team, university partners and steering committee members.

Science and research—Scientists provided expert advice throughout the initiative, but especially in the early stages.

Adaptability—The facilitators and the process needed to be flexible due to the culturally and geographically diverse sub-catchments involved.

Time—It took time to build trust and to appreciate the complexity of the challenges being addressed.

Collaboration

Collaboration can be conceptualized as: (1) an exchange system between separate spheres where information, goods and or services are exchanged; or (2) a joint organization with overlapping sectors for formal cooperation; (3) governmental/state imbedded systems where the state holds the legal rights and private stakeholders are entrusted with responsibilities; (4) a community nested system where the resource users have the rights; and (5) a network, where the state consists of numerous authorities and agencies (Carlsson and Berkes 2005). The stakeholders or resource users (Ostrom 2009) have a role in committing to and buying into long term open and transparent communication processes, where trust and social capital can be built (Armitage et al. 2009, Olsson et al. 2004a,b). In this context, leadership is essential to create an enabling environment. All stakeholders have the responsibility to ensure inclusivity and the development of a shared understanding and vision (Brink et al. 2010). The development of a shared vision, trust and commitment develops over time, with facilitators and bridging organizations playing a crucial role to 'soften' interactions and promote open dialogue (Childs et al. 2013). Over time self-organization, shared responsibility and empowerment will develop. Empowerment is not usually a starting point but rather an outcome of the process of adaptive co-management which will enhance equity, efficiency in decision making and legitimization of actions taken by authorities and institutions (Fabricius et al. 2007).

Actors Involved in Adaptive Co-management

The governance partnership that constitutes adaptive co-management incorporates local resource users, technical experts (including academics), regulating authorities and, in some cases, private business. Donor agencies and their advisors and consultants may also play a supporting role (Fabricius and Collins 2007). Local resource users are the primary stakeholders and their contributions are important. In adaptive co-management all actors should have the right to participate in decision making, share information and contribute to solutions.

1. *Leaders* are the legitimate representatives of a constituency playing an important role because it is not always possible for every individual in a group to participate (Olsson et al. 2004a). They have personal power and are crucial to the success of adaptive co-management. The ideal leader should act in the interest of the group and society and should adopt a broader strategic long term approach. But many leaders focus solely on representing the interests of their own constituencies, or promote their own interests which creates new vulnerabilities and conflicts (Ezzine De Blas et al. 2011).
2. *Intermediaries* take the form of bridging organizations or 'boundary organizations' facilitating communication and knowledge building, and trust and cooperative behavior. Scientists and facilitators are ideally suited for the role, but it does require time commitments beyond the typical three-year project cycle (Clark and Clarke 2011, Roux et al. 2011).

3. *Resource users* are the primary stakeholders who depend on the resource base for their well-being, either by directly using the resource, depending on it for regulating or supporting services, or by relying on it for cultural purposes.
4. *Regulators or public infrastructure providers* (*sensu* Anderies et al. 2004) are officials and their organizations tasked with implementing government or other regulations. They benefit from adaptive co-management through a reduction in the transaction costs of resource management, learning from local people, and reduced conflicts (Redpath et al. 2013).

In adaptive co-management, the vertical linkages between actors are vital for the flow of information and knowledge, and to ameliorate or mediate the interactions between actors (Plummer and Baird 2013). Adger et al. (2005) identified five types of linkage functions of adaptive co-management: a forum for participatory management; a provider of vertical linkages for participation; a connector of local resource users to similar users elsewhere; a conduit that connects all actors to scientific information; and an influencer of policies and regulations. These linkages are instrumental in the co-production of knowledge (Berkes 2009), especially if bridging organizations are involved (Clark and Clarke 2011). The relevance of cross-scale linkages is not unique to adaptive co-management and is crucial in all resource management systems (Berkes 2009). It is the failure to acknowledge such interactions that often results in unexpected or undesirable outcomes or conflicts (Adger et al. 2005).

Processes of knowledge gathering, knowledge sharing, interpretation of complex information, and application of knowledge requires trust and mutual respect (Dale and Armitage 2011). When there is investment in the "intermediate steps" of trust building and communication, actors become more reasonable and accommodating of each other's views during the early stages if there are skilled intermediaries (Monroe et al. 2013). The awareness of actors of the issues forms the foundation for a common vision, motivation, including financial and non-financial incentives to co-create solutions, and capacity to act (Plummer et al. 2013). This often leads to motivation and capacity which enable role players to constructively interact to determine future actions (Lambin 2005) (Table 9.2).

The Relevance of Stakeholder Diversity

Understanding the plurality of stakeholders in an area requires careful and in-depth analysis–merely guessing who is involved almost certainly results in significant stakeholders, especially disempowered ones, to be overlooked (Plummer and Baird 2013). Stakeholder analysis is useful to identify who does what; categorize them; and investigate linkages between them. Reed et al. (2009) assessed the methods used in stakeholder analysis, the resources required to implement them and their strengths and weakness (Table 9.3). They developed a useful method with six steps: "(1) Identify the focus, (2) Identify the system boundaries, (3) Identify stakeholders and their stake, (4) Differentiate between and identify stakeholders, (5) Investigate stakeholder relationships, (6) Recommend future activities and stakeholder engagement."

Table 9.2 Stakeholders, their roles and benefits derived from adaptive co-management

Stakeholder	Role	Benefits derived
Leaders	Representation of a constituency	Intermediary and regulator knowledge sharing
	Creating an enabling environment	
Intermediaries (scientists and facilitators)	Facilitate communication	New knowledge
	Facilitate linkages and networking	
	Build knowledge, trust and cooperative behavior	
Resource User	Committing to and buying into the process	Opportunity and platform to participate in decision making
	Sharing local knowledge	Knowledge sharing
Regulators/public infrastructure providers	Implementing government or other regulations	Reduction of transaction costs
		Local knowledge
	Creating an enabling environment	Reduction in conflict
		Legitimization of decision making

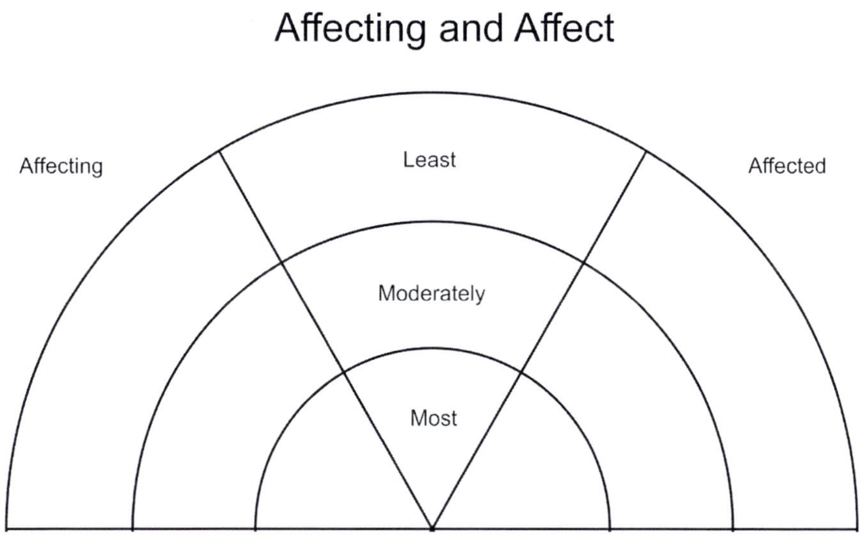

Fig. 9.7 Rainbow diagram for stakeholder analysis (reproduced from Reed et al. 2009)

The 'rainbow diagram' method (Fig. 9.7) is a useful participatory action research method which classifies individuals who are least and most affecting, or are affected by an issue (Reed et al. 2009). Social-ecological inventories are used to describe and map relationships between role players and ecosystems (Rico et al. 2012, Plummer and Baird 2013) and "should be employed in any attempt to develop and implement ecosystem management" (Schultz et al. 2007).

Table 9.3 Summary of stakeholder analysis methods, resources required and their strengths and weaknesses (modified from Reed et al. 2009)

Method	Description	Strengths	Weaknesses
Focus groups	Carefully selected groups categorize stakeholders, using techniques such as 'who counts' matrices and rainbow diagrams	Quick, flexible and cheap. Group discussions are useful in finding common interests and developing trust	Selection of focus group members is crucial, especially in multi-cultural contexts
Semi-structured interviews	Open ended questions to select individuals, using a check-list or interview schedule	Useful for in-depth insights and can be used to triangulate information from other sources	Time-consuming, costly and challenging to code or categorize. Transcription of recorded interviews can be costly and should be planned well in advance
Snow-ball sampling	A small group of initial stakeholders cyclically identify new stakeholder groups and individuals	Useful to increase sample sizes when initial knowledge of stakeholders is incomplete	Biased by the social networks of the first individual or group, especially in multi-cultural contexts
Interest–Influence matrices	Stakeholders cross-tabulated based on their comparative interests and influence	Priorities can be assigned and power dynamics assessed	Depends on the cultural identities of the focus group. Assumptions about the relevance of interests should be tested
Q methodology	Statements are written on cards. Stakeholders sort the cards according to their agreement or disagreement with them	Useful to get a 'feel' for prevalent social discourses	May be incomplete due to stakeholder biases
Actor-linkage matrices	Stakeholders are tabulated in a two-dimensional matrix and their relationships described using codes	Relatively easy, requiring few resources	Can become confusing and difficult to use if many linkages are described
Social network analysis	Used to identify the network of stakeholders and measuring relational ties between stakeholders through use of structured interview/questionnaire	Gain insight into the boundary of stakeholder network; the structure of the network; identifies influential stakeholders and peripheral stakeholders	Time-consuming; questionnaire is a bit tedious for respondents; need specialist in the method

Table 9.3 (continued)

Method	Description	Strengths	Weaknesses
Knowledge mapping	Used in conjunction with SNA; involves semi-structured interviews to identify interactions and knowledge	Identifies stakeholders that would work well together as well as those with power balances	Knowledge needs may still not be met due to differences in the types of knowledge held and needed by different stakeholders.
Radical transactiveness	Snow-ball sampling to identify fringe stakeholders; development of strategies to address their concerns	Identifies stakeholders and issues that might otherwise be missed and minimizes risks to future of project	Time-consuming and hence costly

Challenges to Adaptive Co-management

Adaptive co-management is a multi-actor collaborative governance process which faces a number of distinct challenges such as: how to deal with power differentials to avoid win-lose scenarios where more powerful stakeholders (typically regulators and scientists) dominate (Adger et al. 2005); how to facilitate vertical and horizontal communication; how to reconcile the interests of diverse actors whose interests are often at odds with one another (Plummer and Hashimoto 2011) and who are often unwilling to share power (Plummer and Armitage 2007a, Isyaku et al. 2011). One of key challenges of adaptive co-management is for powerful role players with formal scientific knowledge to respect informal knowledge (Armitage 2005). The "Holy Grail" of adaptive co-management is to motivate and capacitate a diverse range of actors to not only acknowledge each others' rights but also to share responsibilities (Plummer et al. 2013). The time constraints of formal project cycles often prevent an incremental process of trust and capacity building needed to overcome power differentials and find common ground between actors. Especially in developing countries the capacity of all actors, not only local resource users, to engage, build trust and co-create new futures is crucial (Fabricius and Cundill 2010, Lemos et al. 2013).

While decentralization is often, and quite correctly, advocated by advocates of adaptive co-management e.g., Berkes (2010), decentralization can sometimes entrench local elites where the empowered stakeholders can manipulate the incapacitated stakeholders jeopardizing the collaboration (Ansell and Gash 2007, Berkes 2009). Power imbalances can negatively influence incentives to participate (Imperial 2005) and may affect social learning. It can dictate who learns, what they learn, how they learn and how they measure their learning (Armitage et al. 2009, Fabricius and Cundill 2014).

Marginalization of stakeholders will negatively impact on equity and community welfare (Berkes 2009) and exclusivity can influence successful collaboration. If stakeholders feel excluded, other collaborations or platforms will be developed as alternative strategies. The development of multiple independent collaborative agreements may compete with or undermine adaptive co-management (Johnson 1999, Mikalsen et al. 2007).

Stakeholders need to be open to transparent collaboration and participation where a diversity of knowledge systems can be drawn from (Armitage et al. 2009). If an elitist environment is fostered where little sharing occurs learning will be severely inhibited. Face to face dialogue can provide an opportunity for sharing but only if the stakeholders buy into the process and approach it with open minds. Conflict and divisions between stakeholders with 'us and them' attitudes may be catalysts for adaptive co-management but can also complicate the process. Historical conflict results in low levels of trust and commitment and underhanded and manipulative behaviors. Effective conflict resolution, on the other hand, may promote the emergence of deeply entrenched trust and social capital (Ansell and Gash 2007). Table 9.4 highlights the bridges and barriers to adaptive co-management.

Table 9.4 Bridges and barriers to adaptive co-management

Bridges	Barriers
A complex adaptive systems view or 'wicked problem' (Rittel and Webber 1973) perspective	Viewing systems and problems in isolation or as static
An enabling policy environment which encourages multi-level collaboration, and experimentation allowing for flexibility so that innovative management can occur	A stifling and rigid policy environment will prevent any attempt at an adaptive co-management approach
Clearly identifiable social entities can help to facilitate collaboration	Stakeholder sets are hard to identify and where no organized organization with which to work with
Interdependent stakeholders sharing a resource	Independent stakeholders where no clearly defined and shared resource is obvious
Well-connected social and governance networks with functioning vertical and horizontal linkages	Multi-layered institutions where duplication, conflicts and or competition and jealousy for fiscal resources and recognition are found
Key leaders are required to drive the process	Lack of leadership or key individuals willing to drive the process
Stakeholders with a sense of place and a process involving capacity building at multiple levels to empower stakeholders to participate	A system where stakeholders lack capacity (skills, expertise, time, liberty, sense of place) which marginalizes stakeholders
Implementation of a transparent process where a diversity of knowledge systems contribute	An elitist process which does not allow for a diversity or new knowledge to be considered
Enough time for building of relationships, trust and networks as well for collaboration and learning to take place	Situations where authorities are time pressured into decision making
Face to face dialogue where stakeholders can learn about and understand each other	A lack of communication between stakeholders
Incentives for participation brought on by a crisis, policy of financial benefit	Little opportunity or incentive to participate

Conditions for Adaptive Co-management

The consensus amongst scholars is that for adaptive co-management to 'work', the following essential elements should to varying degrees be in place:

1. *Systems perspective*—As a departure point, the key actors need to conceptualize and approach adaptive co-management as a **complex adaptive process** or 'wicked problem' (Rittel and Webber 1973). Wicked problems have no clear end point and require inter-agency collaboration, with local people at the centre of the process (Patterson et al. 2013). Ideally the system should have a **well-defined shared resource system** where property rights to the resources are clear (Dietz et al. 2003) and **identifiable social groups or entities** can help to facilitate collaboration.
2. *Interdependence* among stakeholders is also an enhancing factor. If stakeholders are linked through some kind of interdependence and understand that in order to achieve their goals they need buy in and commitment from the other stakeholders, then the likelihood of sustainable adaptive co-management is increased (Logsdon 1991).
3. *An enabling environment* which encourages multi-level collaboration and experimentation is key for adaptive co-management (Wallis et al. 2013). The legislative and policy environment needs to allow for flexibility so that innovative management can occur and with no defined place-based entities involved in the management of a shared resource, adaptive co-management will be impossible. Having access to a **range of adaptable management measures** can enhance learning and experimentation (Armitage et al. 2009).
4. *Networks and linkages*—Resilient adaptive co-management requires a set of governance attributes that promote participation through **social and governance networks** with functioning vertical and horizontal linkages (Bodin and Crona 2009). Such attributes include polycentric institutions with multiple but well-connected layers of authority and decision making, to enable different governance responses that are tailored to the scale of the resource being managed and the context (Lebel et al. 2006). Multi-layered institutions may however result in duplication and even conflicts and competition. Therefore effective and regular communication and trust building between the different 'layers' is critically important. Competitive and conflicting behaviors often come to the fore when there is competition for fiscal resources or when jealousies emerge when individual role players clamber for recognition.
5. *Incentives* for participation can be motivating (Armitage et al. 2009) and may come about as result of a crisis (Olsson et al. 2004a, 2004b, 2006) or policy change. Incentives for participation are also linked to the creation of a common vision (Barthel et al. 2005), cultural factors, ethics and values (Trosper 2003, Fabricius et al. 2007). The feedbacks between awareness, information, motivation, capacity and agency are on-going and cyclical (Lambin 2005) and all actors should be made aware of these mutually dependent relationships. Motivation without knowledge may lead to maladaptations, and motivation in the absence of capacity often results in frustration.

6. *Capacity* is a multi-faceted element of adaptive co-management and can involve skills and expertise, time, sense of place, liberty and even commitment. Anyone of these facets can promote or inhibit participation or result in marginalization of stakeholders. Two factors: 'agency', i.e. "the capacity of individuals to act independently and to make their own free choices", and 'structure', i.e., factors outside the individual such as social class, religion, gender, ethnicity, customs, etc., are key determinants of the capacity of individuals and communities to make decisions and act on them (Berkes and Ross 2012, Davidson 2012). Key **leaders** are required to drive the process, as in the Kristianstad case study (Box 2), and developing the capacity of stakeholders at multiple levels can empower them to participate (Olsson et al. 2004b). It is unlikely, however, that all stakeholders (local, regional and national) will have all the skills and expertise required for meaningful participation. Often technical knowledge, local knowledge and expert knowledge need to be combined to manage complex systems (Armitage et al. 2009).
7. *Sense of place*—Collaboration can be inhibited when stakeholders **lack a sense of place** or connection to the system. Sense of place can however be a dual-edged sword, resulting in group thinking and unwillingness to innovate or involve newcomers. Peanut farmers in Queensland who had a strong sense of place and identity displayed an unwillingness to change, despite their awareness of the threats of climate change to existing practices (Marshall et al. 2012).
8. *Time*—Allocating **time** to allow for all the necessary steps in the process can enhance the collaboration and learning that takes place (Bos et al. 2013). The adaptive co-management process does not happen overnight or in one workshop. It involves building relationships and networks as well as establishing trust which is a time consuming exercise (Roussos and Fawcett 2000, Till and Meyer 2001, Margerum 2002, Imperial 2005). In some cases it can take up to a decade or more for institutional arrangements, trust and social capital to develop (Armitage et al. 2009). In situations where authorities are time pressured into decision making adaptive co-management will not be an appropriate strategy (Ansell and Gash 2007).
9. *Commitment*—Stakeholder **commitment** can be an important determinant of the success of adaptive co-management (Margerum 2002). The stakeholders need to commit to support a long term inclusive collaborative and institutional building process (Armitage et al. 2009), and this is where **incentives** for participation can assist.
10. *Power balances*—Lack of capacity (Murdock et al. 2005), lack of organizational infrastructure in the absence of civil society groups, status (Rogers et al. 1993, Buanes et al 2004) or agency (Yaffee and Wondolleck 2003) can prevent some stakeholders from collaborating fully and effectively and result in **power imbalances** distrust and a lack of commitment (Short and Winter 1999, Ansell and Gash 2007).

To demonstrate the influence of enablers and pitfalls in adaptive co-management we present two case studies (Box 2). The Kristianstad case study in the lower Helgea River in Sweden and the Goulburn Broken Catchment in Australia, both of which are comprehensively documented case studies.

Monitoring

Learning and experimentation is one of the principles of enhancing resilience (Biggs et al. 2012) and monitoring is essential to reflect on learning. In the 'resilience-experimental' school of thought of adaptive management (McFadden et al. 2011), stakeholder participation lies at the heart of the entire process, as does objective-setting, using multiple competing hypotheses and models to make predictions, acknowledging uncertainty, experimentation, evaluation and active learning. In adaptive co-management, monitoring is an essential catalyst for knowledge sharing, adaptive learning (Standa-Gunda et al. 2003) and "collective sense-making" (Prabhu et al. 2007). This raises three major challenges: dealing with the complexities of multiple stakeholders, resources and agendas; agreeing on practicable and appropriate spatial and temporal scales; and persistence of monitoring programs after scientists and NGOs have moved on (Cundill and Fabricius 2009).

Participatory monitoring is seldom sustainable when interventions and research projects are terminated, with reasons ranging from lack of funds, lack of time, and general lack of incentives to continue. The reality is that local people are almost always too busy with their own activities to continue with monitoring programs initiated by agencies or research teams, and do not see this as essential to their well-being. Researchers should not be surprised when such programs lack sustainability when left entirely in the hands of communities, and should endeavor to be part of collaborative monitoring from beginning to end and not merely as initiators, intermediaries or facilitators (Van Rijsoort et al. 2010). Scientists who initiate participatory monitoring should understand that they are in it for the long haul, unless such programs are based on activities that local people do anyway, such as e.g., livestock dipping, timber purchases, or subjective assessments of the time it takes to harvest a sufficient supply of a particular resource.

In adaptive co-management the goals of participants constantly shift, and assumptions about common objectives should be frequently revised. Using soft systems thinking, Cundill et al. (2012) identified five ways to re-assess stakeholder goals: situate the problem in its social and ecological context; raise awareness about alternative views of a problem and encourage enquiry and deconstruction of frames of reference; undertake collaborative actions; and reflect on learning. It is essential for local people to be involved in the co-creation of indicators and deliberations around thresholds of potential concern (Reed et al. 2006).

Box 2. Examples of adaptive co-management

Example 1: Kristianstad, Sweden

The Helgea River Catchment in Sweden is an example of a complex social-ecological system where adaptive co-management has been successful. Various enablers have contributed to the success. These include interdependent stakeholders, a shared resource, attractive incentives, leadership, and an enabling environment with multi level networks where learning and collaboration occurred.

Interdependent stakeholders and a shared resource: One of the towns in the area is Kristianstad (73,000 people) managed by the Kristianstad municipality. The towns' people have been deriving ecosystem benefits from the wetland in terms of a historical defense, flood abatement, habitat supply, biodiversity, and as a recreational and cultural site of significance to the community. The current dominant land use in the area historically and currently is highly productive agriculture where the annually flooded meadows are used for grazing and harvesting of hay and which subsequently maintain the meadows. Following agriculture the area also has beech and wet forests and is rich in fauna and flora. Furthermore the area contains the biggest ground water aquifers in northern Europe.

Incentive: Since the town's establishment the wetlands and the benefits derived have come under pressure and after the signing of the RAMSAR convention on wetlands (aimed at conserving the biodiversity of the area) an innovative policy and management solution was needed. The Scania County Administrative Board was eager to manage the area as a nature reserve which partially materialized with three percent of the RAMSAR site becoming a reserve (Magnusson et al. 1989) and several plans and polices developed for the area's protection. A new municipal organization was established, namely the Ecomuseum Kristianstads Vatteenrike (EKV) in 1989. Despite these developments the system continued to degrade due to neglect and inappropriate management practices.

Leadership: Through the concerted efforts of an emergent leader who drove the development and focus of the EKV.

Enabling environment: The EKV provided an enabling environment for adaptive co-management to take place and the funding support from the municipality, national Cultural Advisory Board and the World Wildlife Fund allowed for self-organization and the flexibility to monitor using inventories, experiment, learn and adapt.

Multi-level networks: The EKV took on a facilitator and coordinator role in establishing collaboration processes across international, national, regional and local stakeholders relevant to the Helgea River Catchment including researchers, nonprofit organizations, and private landowners to develop policy, design projects, conflict resolution, restorative actions and the development of goals for the catchment area, producing plans, agreements and reports.

Learning and collaboration: Knowledge of the ecosystem dynamic developed through collaboration. The EKV developed into a flexible collaborative approach to managing the ecosystem taking into account a diversity of stakeholders and knowledge systems at multiple levels. The collaboration improved the capacity to cope with changes in the system transforming it into a new social-ecological trajectory which enhanced the management of the system and the derived benefits. For more examples of adaptive co-management consult Gondo (2011).

Example 2: Goulburn-Broken Catchment, Australia

The catchments of the Goulburn and Broken Rivers form the Goulburn Broken Catchment in Australia which is a contrasting case study where the system is resilient to change. The dominant land use in the catchment includes intensive irrigated (horticulture) and dryland (dairy) agriculture as well as urban settlements (CSIRO 2003) and an emerging tourism and recreation sector. The area has come under threat due to a long history of native vegetation clearing resulting in soil erosion and dryland salinity (Lebel et al. 2006) depleting the natural capital and threatening human well-being. This example did not result in a successful case of adaptive co-management due to: severely degraded resources, the inability of leaders to foment change, the lack of an enabling environment, and the lack of a diversity of knowledge.

Lack of capacity and resources: Despite a Landcare Program which was established in 1989 and a developed stewardship ethic as well as networks facilitating the sharing of information resources, there was inadequate capacity to cope with the severely degraded locations (Lebel et al. 2006).

Incentive and leadership: As the crises became more apparent groups focused on water related impacts such as flooding and drainage became established and amalgamated in a broader network in which a trusted community leader emerged. The sharing of knowledge was encouraged. Interestingly though the emergent leadership did not facilitate adaptive governance (Olsson et al. 2006).

Lack of an enabling environment: The current land use is reflective of past values and has become entrenched in the rules and laws (e.g. property rights and the constitution) of the area making it very stable and resistant to change (Walker et al. 2009).

Lack of a diversity of knowledge: The decision-making body consisted mainly of farmers and local business owners with a vested interest which resulted in biased decision making (Olsson et al. 2006).

Setting up a Monitoring System for Adaptive Co-management

In essence, four criteria should be considered when designing monitoring programs in adaptive co-management (Cundill and Fabricius 2009): recognition of complexity and non-linearity, implying the incorporation of multiple spatial and temporal scales; the integration of both social and ecological variables; focusing on thresholds, and seeking surrogates to be able to predict their proximity; and not only monitoring the outcomes of adaptive co-management, but also the process. Clark and Clarke (2011) developed a monitoring system for local sustainability linked to adaptive governance and found that the criteria for adaptive governance and indicators of local sustainability were interdependent. Indicators included cross-scale interactions; the type of learning processes involved; and the level of shared understanding about the use of shared resources. Jacobson et al. (2009) and Clark and Clarke (2011) identified a number of more detailed process-related questions that can assist managers in establishing monitoring and feedback programs:

Monitoring questions

- *Focus:* Is monitoring conducted systematically and in relation to hypotheses? Are appropriate criteria used in indicator selection?
- *Scale:* Are short- and long-term responses monitored? Are cross-scale interactions and cooperation being considered?
- *Power:* Have stakeholders been given an opportunity to be involved? Is there a balance of power, with no clear winners and losers?
- *Learning:* Has data been collected so that management processes can be evaluated? Are enough time and resources available for reflection and adaptation?

Feedback questions

- Is a shared understanding gradually being developed?
- Are both outcome and process lessons documented?
- Is the process transparent? Are people sharing information and other resources?
- Is the process iterative and experimental? Is there room for both single and double-loop learning, i.e., are the goals being periodically assessed?
- Are both social and ecological uncertainties evaluated?
- Are management and learning processes evaluated?

Indicators of Success in Adaptive Co-management

Based on criteria put forward by Cundill and Fabricius (2010), Clark and Clarke (2011), and Bos et al. (2013), we developed a 10-point 'dashboard' to monitor the participatory process in adaptive co-management (Table 9.5).

It is, however, important to bear in mind that adaptive co-management needs to be viewed through a 'soft systems' lens. Role players should therefore constantly re-assess their goals, monitoring plans and methods of assessment (Cundill et al.

Table 9.5 A 10-point dashboard to assess adaptive co-management

Criterion	Rating			Commentary	Agreed adaptive action
	Green Satisfactory	Amber Early warning signs that things aren't going well	Red There is a problem here	What is good about this? Where are the areas for improvement? What is the justification for the rating?	
1. Is there balanced participation in meetings and discussions?	All role players have an opportunity to air their views and be heard	Some key role players are not participating	General lack of participation, key role players are excluded, looming dissatisfaction		
2. Is information being shared and understood?	Information is easily available, accessible and understandable	Some important information is unavailable, accessible and/or difficult to understand	Information is frequently unavailable or inaccessible		
3. Are people able and willing to learn from each other?	People share information, make compromises & are willing to adapt	Some information is being withheld or ignored, role players cling to strongly held views	Most role players keep their cards close to their chests, polarization and stereoptying dominates		
4. Is there progress in the development of mutual trust and respect between stakeholders?	Acceptance of diversity in opinions, needs and aspirations, sense of togetherness	Early indications of disregard for some views	Stereotyping, enemy or "*us and them*" perceptions, name-calling, judgementalism		
5. Are there signs of progress towards shared goals and a shared vision?	Common goals are articulated and agreed upon. Compromises, reconciliation	Uncompromising goal setting, signs of irreconcilable or divergent goals	Widely divergent and conflicting goals dominate		
6. Are there obvious winners and losers?	Acknowledgement of power differences. Efforts to close the gap between winners and losers. Balanced participation	Power differences are not readily acknowledged. Some interest groups are more vocal and dominant than others	Powerful interest groups take over and dominate processes and inputs		
7. Is the process flexible and adaptive?	Responsiveness, adaptation and reflection characterizes the process	Early signs of rigidity and unwillingness or inability to be flexible in the process	Most of the processes and methods are cast in stone, little adaptation despite early warning signs		
8. Is conflict being resolved?	Conflicts are acknowledged and actively managed, facilitators have good conflict management skills	Conflicts are swept under the carpet or unrecognized	Conflicts are left to fester		
9. Are enough resources and time allocated to the public participation part of the initiative?	It is possible to slow down where necessary to enable all key interest groups to make input	Some processes move too fast for role players to stay abreast, some role players lag behind	The process proceeds too fast for most role players to participate, public participation is 'quick and dirty'		
10. Are there linkages to provincial, national and global processes and role players?	Efforts are made to ensure compatibility with provincial and national plans and global trends. Provincial and national role players are consulted and make input	Some incompatibilities are evident, provincial and national role players are acknowledged but are not involved	There is little or no evidence of vertical linkages		

2012). Adaptive co-management invariably results in challenges to fundamental assumptions in true double-loop learning fashion (Huntjens et al. 2011), and this has implications for the adaptability of monitoring systems.

Knowledge and Research Gaps

Adaptive co-management isn't fully understood and several knowledge gaps exist. Areas in particular that need attention include the enabling environment, including the institutional policy arrangements which are an important core element of the process and are ultimately what results in shared responsibility and decision making. In particular what policy conditions are required for providing policy experimentation and what incentives are required? The questions are how to design flexible multi-level governance systems which facilitate learning and collaboration, organize participation and allow for experiments. Determining effectiveness or ineffectiveness of different institutional prescriptions is also required where case studies and action research are recommended as appropriate research methods (Berkes 2009, Huitema et al. 2009).

Understanding the interactions and feedbacks between users, governance systems and resource units and systems (Ostrom 2009) and a greater understanding of networks and their elements (nodes and connections and bridging organizations) is required. We have little knowledge on the role of redundancy of function in partners and the part they play when trying to create a resilient system (Berkes 2009). This element of learning in adaptive co-management requires further investigation. The learning in collaborative approaches can sometimes be superficial on the one side or can result in long term change. Little work has been done to develop best practices or learning assessments in adaptive co-management. A greater understanding of what strategies support learning, key lessons and models are required to develop best practices. The learning and collaboration nexus, evaluating learning outcomes linked to learning types, approaches and challenges (Armitage et al. 2008), and understanding the effects of learning in terms of environmental outcomes (Schultz and Lundholm 2010) are potential research questions. Berkes (2009) suggests we need to understand how stakeholder learning can transition from single, to double and triple loop learning.

Even though tools for systematic monitoring and evaluation exist, much room for refinement and development of generic parameters for ecological sustainability, livelihoods and processes in adaptive co-management are required (Plummer and Armitage 2007b). Furthermore, the development of diagnostic questions and tools is a promising area of research in the adaptive co-management field where it can then be adapted to specific cases. This aspect of research in the field will however require many case studies across different resource types and geographic areas (Berkes 2009). Calls for systematic and cooperative analysis of the adaptive co-management process in a diversity of contexts are also found in the literature (Plummer and Armitage 2007b, Plummer 2009). Plummer (2009) calls for the development of a succinct model with distinctive phases and identification of tipping points in the process. Also a better understanding of the complexity of the variables involved in the process and asking: (1) to what extent can variables be traded off; (2) which variables are essential; and (3) what are the non-essential but enabling conditions. Armitage et al. (2009) argue that insight into the expected benefits policy makers expect from adaptive co-management is also needed.

Conclusion

Adaptive co-management is no panacea and may be inappropriate in contexts where stakeholder capacity is lacking, governance is weak, problems are 'tame', solutions are urgent and trust is low. It is, however, one of the few workable options where co-management is a necessity and current knowledge is incomplete. Proponents of adaptive co-management should therefore be cautious to brashly sell the approach as a 'silver bullet' to sustainability challenges. A more appropriate approach would be to modestly experiment, acknowledge errors and shortcomings, reflect and learn. This might just prevent adaptive co-management from becoming yet another failed and self-serving academic dogma.

Acknowledgements The Nelson Mandela Metropolitan University, National Research Foundation and Resilience Alliance are acknowledged for their on-going support and funding of our work.

References

Adger, W. N., Brown, K., & Tompkins, E. L. (2005). The political economy of cross-scale networks in resource co-management. *Ecology and Society, 10,* 9. http://www.ecologyandsociety.org/vol10/iss2/art9/.

Alexander, J. A., Comfort, M. E., Weiner, B. J., & Bogue, R. (1998). Leadership in collaborative community health partnerships. *Non Profit Management and Leadership, 12*(2), 159–175. New York: Wiley.

Allen, C. R., & Gunderson, L. (2011). Pathology and failure in the design and implementation of adaptive management. *Journal of Environmental Management, 92,* 1379–1384.

Anderies, J. M., Jansson, M., & Ostrom, E. (2004). A framework to analyse the robustness of social-ecological systems from an institutional perspective. *Ecology and Society, 9,* 18. http://www.ecologyandsociety.org/vol9/iss1/art18/.

Ansell, C., & Gash, A. (2007). Collaborative governance in theory and practice. *Journal of Public Administration Research and Theory, 18,* 543–571.

Armitage, D. R. (2005). Community—based Narwhal management in Nunavut, Canada: Change, uncertainty and adaption. *Society and Natural Resources, 18,* 715–731.

Armitage, D., Berkes, F., & Doubleday, N. (2007). Introduction: Moving beyond co management. In D. Armitage, F. Berkes, & N. Doubleday (Eds.), *Adaptive co-management: Collaboration, learning, and multi-level governance* (pp. 1–18). Vancouver: UBC Press.

Armitage, D., Marschke, M., & Plummer, R. (2008). Adaptive co-management and the paradox of learning. *Global Environmental Change, 18,* 86–98.

Armitage, D., Plummer, R., Berkes, F., Arthur, R. I., Charles, A. T., Davidson Hunt, I. K., Diduck, A. P., Doubleday, N. C., Johnson, D. S., Marschke, M., McConney, P., Pinkerton, E. W., & Wollenberg, E. (2009). Adaptive co-management for social ecological complexity. *Frontiers in Ecology and the Environment, 7,* 95–102.

Barthel, S., Colding, J., Elmqvis, T., & Folke, C. (2005). History and local management of a biodiversity-rich, urban cultural landscape. *Ecology and Society, 10*(2), 10. http://www.ecologyandsociety.org/vol10/iss2/art10/.

Berkes, F. (2009). Evolution of co management: Role of knowledge generation, bridging organizations and social learning. *Journal of Environmental Management, 29,* 1692–1702.

Berkes, F. (2010). Devolution of environment and resources governance: Trends and future. *Environmental Conservation, 37,* 489–500.

Berkes, F., & Ross, H. (2012). Community resilience: Toward an integrated approach. *Society and Natural Resources, 26,* 5–20.

Biggs, R., Schlüter, M., Biggs, D., Bohensky, E. L., BurnSilver, S., Cundill, G., Dakos, V., Daw, T. M., Evans, L. S., & Kotschy, K. (2012). Toward principles for enhancing the resilience of ecosystem services. *Annual Review of Environment and Resources, 37,* 421–448.

Bodin, Ö, & Crona, B. I. (2009). The role of social networks in natural resource governance: What relational patterns make a difference? *Global Environmental Change, 19,* 366–374.

Bohensky, E., & Lynam, T. (2005). Evaluating responses in complex adaptive systems: Insights on water management from the Southern African Millennium Ecosystem Assessment (SAfMA). *Ecology and Society, 10,* 11. http://www.ecologyandsociety.org/vol10/iss1/art11/.

Bos, J. J., Brown, R. R., & Farrelly, M. A. (2013). A design framework for creating social learning situations. *Global Environmental Change, 23,* 398–412.

Brink, M., Cameron, M., Coetzee, K., Currie, B., Fabricius, C., Hattingh, S., Schmidt, A., & Watson, L. (2010). Sustainable management through improved governance in the game industry. *South African Journal of Wildlife Research, 41,* 110–119.

Buanes, A., Jentoft, S., Karlsen, G. R., Maurstadt, A., & Søreng, S. (2004). In whose interest? An exploratory analysis of stakeholders in Norwegian coastal zone planning. *Ocean and Coastal Management, 47,* 207–223.

Bulkeley, H., & Betsill, M. (2005). Rethinking sustainable cities: Multilevel governance and the 'urban' politics of climate change. *Environmental Politics, 14,* 42–63.

Carlsson, L., & Berkes, F. (2005). Co-management: Concepts and methodological implications. *Journal of Environmental Management, 75,* 65–76.

Checkland, P. (2000). Soft systems methodology: A thirty year retrospective. *Systems Research and Behavioural Science, 17,* 11–58.

Childs, C., York, A. M., White, D., Schoon, M. L., & Bodner, G. S. (2013). Navigating a murky adaptive comanagement governance network: Agua Fria Watersh, Arizona, USA. *Ecology and Society, 8*(4), 11. http://dx.doi.org/10.5751/ES-05636–180411

Clark, J. R. A., & Clarke, R. (2011). Local sustainability initiatives in English National Parks: What role for adaptive governance? *Land Use Policy, 28,* 314–324.

Cundill, G., & Fabricius, C. (2009). Monitoring in adaptive co-management: Toward a learning based approach. *Journal of Environmental Management, 90,* 3205–3211.

Cundill, G., & Fabricius, C. (2010). Monitoring the governance dimension of natural resource co-management. *Ecology and Society, 15*(1), 15. http://www.ecologyandsociety.org/vol15/iss1/art15/.

Cundill, G., Cumming, G., Biggs, D., & Fabricius, C. (2012). Soft systems thinking and social learning for adaptive management. *Conservation Biology, 26,* 13–20.

Dale, A., & Armitage, D. (2011). Marine mammal co-management in Canada's Arctic: Knowledge co-production for learning and adaptive capacity. *Marine Policy, 35,* 440–449.

Davidson, D. J. (2012). We still have a long way to go, and a short time to get there: A response to Fikret Berkes and Helen Ross. *Society and Natural Resources, 26,* 21–24.

Dietz, T., Ostrom, E., & Stern, P. C. (2003). The struggle to govern the commons. *Science, 302,* 1907–1912.

Ezzine de Blas, D., Ruiz-Pérez, M., & Vermeulen, C. (2011). Management conflicts in Cameroonian community forests. *Ecology and Society, 16*(1), 8. http://www.ecologyandsociety.org/vol16/iss1/art8/.

Fabricius, C., & Collins, S. (2007). Community-based natural resource management: Governing the commons. *Water Policy, 9,* 83–97.

Fabricius, C., & Cundill, G. (2010). Building adaptive capacity in systems beyond the threshold: The story of Macubeni, South Africa. In D. Armitage & R. Plummer (Eds.), *Adaptive capacity and environmental governance* (pp. 43–68). Berlin: Springer.

Fabricius, C., & Cundill, G. (in press, 2014). Learning in adaptive management: Insights from published practice. *Ecology and Society, 19*(1), 29. http://dx.doi.org/10.5751/ES-06263-190129.

Fabricius, C., Folke, C., Cundill, G., & Schultz, L. (2007). Powerless spectators, coping actors, and adaptive co-managers: A synthesis of the role of communities in ecosystem management. *Ecology and Society, 12*(1), 29. http://www.ecologyandsociety.org/vol12/iss1/art29/.

Folke, C., Carpenter, S., Elmqvist, T., Gunderson, L., Holling, C. S., Walker, B., Bengtsson, J., Berkes, F., Colding, J., Danell, K., Falkenmark, M., Gordon, L., Kasperson, R., Kautsky, N., Kinzig, A., Levin, S., Mäler, K. G., Moberg, F., Ohlsson, L., Olsson, P., Ostrom, E., Reid, W., Rockström, J., Savenije, H., & Svedin, U. (2002). Resilience and sustainable development: Building adaptive capacity in a world of transformation. Scientific Background Paper on Resilience for the process of The World Summit on Sustainable Development on behalf of The Environmental Advisory Council to the Swedish Government.

Folke, C., Fabricius, C., Cundill, G., Schultz, L. (2005). Communities, ecosystems and livelihoods. *Ecosystems and human well-being: multiscale assessments, 4,* 261–277.

Garnett, S. T., Sayer, J. A., & Toit, J. du. (2007). Improving the effectiveness of interventions to balance conservation and development: A conceptual framework. *Ecology and Society, 12,* 2. http://www.ecologyandsociety.org/vol12/iss1/art2/.

Gondo, T. (2011). Adaptive co-management of natural resource: A solution or problem? *Journal of Human Ecology, 33,* 119–131.

Holling, C. S. (1978). *Adaptive environmental assessment and management.* London: John Wiley and Sons.

Huitema, D., Mostert, E., Egas, W., Moellenkamp, S., Pahl Wostl, C., Yalcin, R. (2009). Adaptive water governance: Assessing the institutional prescriptions of adaptive co-management from a governance perspective and defining a research agenda. *Ecology and Society, 14,* 1. http://www.ecologyandsociety.org/vol14/iss1/art26/.

Huntjens, P., Pahl-Wostl, C., Rihoux, B., Schlüter, M., Flachner, Z., Neto, S., Koskova, R., Dickens, C., & Nabide Kiti, I. (2011). Adaptive water management and policy learning in a changing climate: A formal comparative analysis of eight water management regimes in Europe, Africa and Asia. *Environmental Policy and Governance, 21,* 145–163.

Imperial, M. T. (2005). Using collaboration as a governance strategy: Lessons from six watershed management programs. *Administration and Society, 37,* 281–320.

Isyaku, U., Chindo, M., & Ibrahim, M. (2011). Assessing community-based natural resources management at Lake Naivasha, Kenya. *Environment and Natural Resources Research, 1,* 106–116.

Johnson, B. L. (1999). Introduction to the special feature: Adaptive management— scientifically sound, socially challenged? *Conservation Ecology, 3*(1), 10. http://www.consecol.org/vol3/iss1/art10/.

Jacobson, C., Hughey, K. F. D., Allen, W. J., Rixecker, S., & Carter, R. W. (2009). Toward more reflexive use of adaptive management. *Society and Natural Resources, 22,* 484–495.

Kolb, D. A. (1984). *Experiential learning: Experience as the source of learning and development.* Upper Saddle River: Prentice-Hall.

Lambin, E. F. (2005). Conditions for sustainability of human-environment systems: Information, motivation, and capacity. *Global Environmental Change, 15,* 177–180.

Lebel, L., Anderies, J. M., Campbell, B., Folke, C., Hatfield-Dodds, S., Hughes, T. P., & Wilson, J. (2006). Governance and the capacity to manage resilience in regional social-ecological systems. *Ecology and Society, 11*(1), 19. http://www.ecologyandsociety.org/vol11/iss1/art19/.

Lemos, M., Agrawal, A., Eakin, H., Nelson, D., Engle, N., & Johns, O. (2013). Building adaptive capacity to climate change in less developed countries. In G. R. Asrar & J. W. Hurrell (Eds.), *Climate science for serving society* (pp. 437–457). Netherlands: Springer.

Logsdon, J. (1991). Interests and interdependence in the formation of social problem-solving collaborations. *Journal of Applied Behavioral Science, 27,* 23–37.

Ludwig, D. (2001). The era of management is over. *Ecosystems, 4,* 758–764.

Margerum, R. D. (2002). Collaborative planning: Building consensus and building a distinct model for practice. *Journal of Planning Education and Research, 21,* 237–253.

Marshall, N. A., Park, S. E., Adger, W. N., Brown, K., & Howden, S. M. (2012). Transformational capacity and the influence of place and identity. *Environmental Research Letters, 7,* 034022.

McFadden, J. E., Hiller, T. L., & Tyre, A. J. (2011). Evaluating the efficacy of adaptive management approaches: Is there a formula for success? *Journal of Environmental Management, 92,* 1354–1359.

Mezirow, J. (2000). Learning to think like an adult. Core concepts of transformation theory. In J. Mezirow (Ed.), *Learning as transformation. Critical perspectives on a theory in progress* (pp. 125–148). San Francisco: Jossey-Bass.

Mezirow, J. (2009). An overview on transformative learning. In K. Illeris (Ed.), *Contemporary theories of learning. Learning theorists.... In their own words* (pp. 90–105). London: Routledge.

Mikalsen, K. H., Hernes, H. K., & Jentoft, S. (2007). Learning on user-groups: The role of civil society in fisheries governance. *Marine Policy, 31,* 201–209.

Monroe, M. C., Plate, R., & Oxarart, A. 2013. Intermediate collaborative adaptive management strategies build stakeholder capacity. *Ecology and Society, 18*(2), 24. http://dx.doi.org/10.5751/ES-05444-180224.

Murdock, B., Carol, W., & Sexton, K. 2005. Stakeholder participation in voluntary environmental agreements: Analysis of 10 Project XL case studies. *Science, Technology and Human Values, 30,* 223–250.

Murphree, M. W. (2004). Communal approaches to natural resource management in Africa: From whence and to where? *Journal of International Wildlife Law and Policy, 7,* 203–216.

Olsson, P., Folke, C., & Berkes, F. (2004a). Adaptive co-management for building resilience in social ecological systems. *Environmental Management, 34,* 75–90.

Olsson, P., Folke, C., & Hahn, T. (2004b). Social-ecological transformation for ecosystem management: The development of adaptive co-management of a wetland landscape in southern Sweden. *Ecology and Society, 9*(4), 2. http://www.ecologyandsociety.org/vol9/iss4/art2/.

Olsson, P., Gunderson, L. H., Carpenter, S. R., Ryan, P., Lebel, L., Folke, C., & Holling, C. S. (2006). Shooting the rapids: Navigating transitions to adaptive governance of social-ecological systems. *Ecology and Society, 11*(1), 18. http://www.ecologyandsociety.org/vol11/iss1/art18/.

Ostrom, E. (2009). A general framework for analyzing sustainability of social-ecological systems. *Science, 325,* 419–422.

Ostrom, E., & Cox, M. (2010). Moving beyond panaceas: A multi tiered approach for social ecological analysis. *Environmental Conservation, 37,* 451–463.

Patterson, J. J., Smith, C., & Bellamy, J. (2013). Understanding enabling capacities for managing the 'wicked problem' of nonpoint source water pollution in catchments: A conceptual framework. *Journal of Environmental Management, 128,* 441–452.

Plummer, R. (2009). The adaptive co-management process: An initial synthesis of representative models and influential variables. *Ecology and Society, 14*(2), 24. http://www.ecologyandsociety.org/vol14/iss2/art24/.

Plummer, R., & Armitage, D. R. (2007a). Charting the new territory of adaptive co-management: A Delphi study. *Ecology and Society, 12*(2), 10. http://www.ecologyandsociety.org/vol12/iss2/art10/.

Plummer, R., & Armitage, D. (2007b). A resilience based framework for evaluating adaptive co management: Linking ecology, economics and society in a complex world. *Ecological Economics, 61,* 62–74.

Plummer, R., & Baird, J. (2013). Adaptive co-management for climate change adaptation: Considerations for the Barents Region. *Sustainability, 5,* 629–642.

Plummer, R., & Hashimoto, A. (2011). Adaptive co-management and the need for situated thinking in collaborative conservation. *Human Dimensions of Wildlife, 16,* 222–235.

Plummer, R., Armitage, D. R., & de Löe, R. C. (2013). Adaptive comanagement and its relationship to environmental governance. *Ecology and Society, 18*(1), 21. http://dx.doi.org/10.5751/ES-05383-180121.

Pomeroy, R. S., & Berkes, F. (1997). Two to tango: The role of government in fisheries co-management. *Marine Policy, 21,* 465–480.

Prabhu, R., McDougall, C., & Fisher, R. (2007). Adaptive collaborative management: A conceptual model. Adaptive collaborative management of community forests in Asia: Experiences from Nepal, Indonesia and the Philippines. CIFOR, Bogor, Indonesia.

Pratt Miles, J. D. (2013). Designing collaborative processes for adaptive management: Four structures for multistakeholder collaboration. *Ecology and Society, 18*(4), 5. http://dx.doi.org/10.5751/ES-05709-180405.

Redpath, S. M., Young, J., Evely, A., Adams, W. M., Sutherland, W. J., Whitehouse, A., Amar, A., Lambert, R. A., Linnell, J. D. C., Watt, A., & Gutiérrez, R. J. (2013). Understanding and managing conservation conflicts. *Trends in Ecology and Evolution, 28,* 100–109.

Reed, M. S., Fraser, E. D. G., & Dougill, A. J. (2006). An adaptive learning process for developing and applying sustainability indicators with local communities. *Ecological Economics, 59,* 406–418.

Reed, M. S., Graves, A., Dandy, N., Posthumus, H., Hubacek, K., Morris, J., Prell, C., Quinn, H., & Stringer, L. C. (2009). Who's in and why? A typology of stakeholder analysis methods for natural resource management. *Journal of Environmental Management, 90,* 1933–1949.

Rico, G., a-Amado, L., Ruiz Pérez, M., Iniesta-Arandia, I., Dahringer, G., Reyes, F., & Barrasa, S. (2012). Building ties: Social capital network analysis of a forest community in a biosphere reserve in Chiapas, Mexico. *Ecology and Society, 17*(3), 41. http://dx.doi.org/10.5751/ES-04855-170303.

Rittel, H., & Webber, M. (1973). Dilemmas in a general theory of planning. *Policy Sciences, 4,* 155–169.

Rogers, T., Beh Howard-Pitney, B., Feighery, E. C., Altman, D. G., Endres, J. M., & Roeseler, A. G. (1993). Characteristics and participant perceptions of tobacco control coalitions in California. *Health Education Research, Theory and Practice, 8,* 345–357.

Rouse, W. B. (2008). Health care as a complex adaptive system: Implication for management. *The Bridge, 38,* 17–25.

Roussos, S. T., &Fawcett, S. B. (2000). A review of collaborative partnerships as a strategy for improving community health. *Annual Review of Public Health, 21,* 369–402.

Roux, D. J., Murray, K., Nel, J. L., Hill, L., Roux, H., Driver, A. (2011). From scorecard to social learning: A reflective coassessment approach for promoting multiagency cooperation in natural resource management. *Ecology and Society, 16*(1), 24. http://www.ecologyandsociety.org/vol16/iss1/art24/.

Ruitenbeek, J., & Cartier, C. (2001). The invisible wand: Adaptive co-management as an emergent strategy in complex bio economic systems. Occasional Paper No. 54. Centre for International Forestry Research (CIFOR). ISSN 0854–9818.

Schultz, L., Folke, C., & Olsson, P. (2007). Enhancing ecosystem management through social-ecological inventories: Lessons from Kristianstads Vattenrike, Sweden. *Environmental Conservation, 34,* 140–152.

Short, C., & Winter, M. (1999). The problem of common land: Towards stakeholder governance. *Journal of Environmental Planning and Management, 42,* 613–630.

Shultz, L., & Lundholm, C. (2010). Learning for resilience? Exploring learning opportunities in biosphere reserves. *Environmental Education Research, 16,* 645–663.

Singleton, S. (1998). *Constructing Cooperation: The Evolution of Institutions of Comanagement.* Ann Arbor: University of Michigan Press.

Standa-Gunda, W., Mutimukuru, T., Nyirenda, R., Prabhu, R., Haggith, M., & Vanclay, J. (2003). Participatory modelling to enhance social learning, collective action and mobilization among users of the mafungautsi forest, Zimbabwe. *Small-scale Forest Economics, Management and Policy, 2,* 313–326.

Stringer, L. C., Dougill, A. J., Fraser, E., Hubacek, K., Prell, C., & Reed, M. S. (2006). Unpacking "participation" in the adaptive management of social ecological systems: A critical review. *Ecology and Society, 11*(2), 39. http://www.ecologyandsociety.org/vol11/iss2/art39/.

Till, J. E., & Meyer, K. R. (2001). Public involvement in science and decision making. *Health Physics, 80,* 370–378.

Trosper, R. L. (2003). Resilience in pre-contact Pacific Northwest social–ecological systems. *Ecology and Society, 7*(3), 6. http://www.ecologyandsociety.org/vol7/iss3/art6/.

Tulloch, A. I. T., Possingham, H. P., Joseph, L. N., Szabo, J., & Martin, T. G. (2013). Realising the full potential of citizen science monitoring programs. *Biological Conservation, 165,* 128–138.

Van Rijsoort, J., Jinfeng, Z., & Ten, M. (2010). Participatory resources monitoring in SW China: Lessons after five years. In A. Lawrence (Ed.), *Taking stock of nature: participatory biodiversity assessment for policy* (pp. 142–165). Cambridge: Cambridge University Press.

Walker, B. H., Abel, N., Anderies, J. M., & Ryan, P. (2009). Resilience, adaptability, and transformability in the Goulburn-Broken Catchment, Australia. *Ecology and Society, 14*(1), 12. http://www.ecologyandsociety.org/vol14/iss1/art12/.

Wallis, P. J., Ison, R. L., & Samson, K. (2013). Identifying the conditions for social learning in water governance in regional Australia. *Land Use Policy, 31,* 412–421.

Yaffee, S. L., & Wondolleck, J. (2003). Collaborative ecosystem planning processes in the United States: Evolution and challenges. *Environments, 31*, 59–72.

Chapter 10
Adaptive Management Today: A Practitioners' Perspective

Carol L. Murray, David R. Marmorek and Lorne A. Greig

Keywords Adaptive management · Uncertainty · Practitioner · Environmental management · Future

Introduction

It has been more than 30 years since Crawford S. (Buzz) Holling and his colleagues at the University of British Columbia and the Institute for Applied Systems Analysis in Vienna introduced the world to the concept of adaptive management (AM), then referred to as Adaptive Environmental Assessment and Management (Holling 1978). In their work to develop and demonstrate the concepts of adaptive management they encountered early on many of the social and organizational issues that have continued to challenge effective implementation of adaptive management.

Today, the practice of adaptive management reflects an evolution that has played out across the globe in a wide variety of resource management contexts. The current practice of adaptive management reflects a dynamic mix of valiant efforts, which as discussed in other chapters in this book includes both notable successes and dismal failures. In our three decades of experience with adaptive management we have learned significant lessons pointing the way toward improved and expanded practice.

Notwithstanding the lessons learned by practitioners about how to do adaptive management well, we see a future with intensified social and institutional challenges inhibiting effective adaptive management, continued misunderstanding of

---

C. L. Murray (✉) · D. R. Marmorek · L. A. Greig
ESSA Technologies Ltd., 2695 Granville Street, Suite 600, Vancouver, BC V6H 3H4, Canada
e-mail: cmurray@essa.com

D. R. Marmorek
e-mail: dmarmorek@essa.com

L. A. Greig
e-mail: lgreig@essa.com

C. R. Allen, A. S. Garmestani (eds.), *Adaptive Management of Social-Ecological Systems*,
DOI 10.1007/978-94-017-9682-8_10

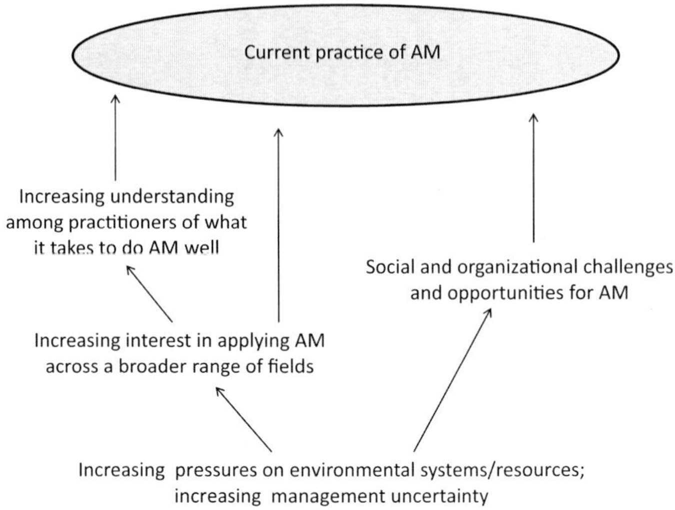

Fig. 10.1 Major influences on the practice of adaptive management today

what true adaptive management really is, and ever increasing environmental management challenges that cry out for rigorous adaptive management (Fig. 10.1). In this chapter we briefly revisit the highlights of the evolution of adaptive management practice to date, describe the expansion of adaptive management domains of practice, introduce what we see as three main dimensions of adaptive management, present the concept of an adaptive management "definition space", examine the characteristics of good adaptive management practice as we presently understand it, and speculate about the future of adaptive management.

The Evolution of Adaptive Management: Visions and Realities

Our involvement with adaptive management began in the late 1970s and early 1980s with the graduate work of ESSA's founding and following partners at UBC under Holling and Walters. They were dizzying times with great anticipation and excitement. There was considerable hope and enthusiasm for the gains to be made in improving the scientific basis of environmental management, and the potential for reducing environmental conflicts, through increased understanding in trans-disciplinary adaptive management initiatives, built on a foundation of systems analysis and modelling. This scientific progress occurred in the context of ever increasing cultural awareness of our impacts on ecosystems and the need to bring them in balance with the biosphere's capacity to tolerate us. The early 1980s were exciting, with rapid uptake of adaptive management concepts by a number of institutions such as the Adaptive Environmental Assessment Team of the U.S. Fish and Wildlife

Service's Western Energy and Land Use Team, the Great Lakes Fishery Commission, the Bureau of Land Management, the U.S. Forest Service, the Ontario Ministry of Natural Resources, Fisheries and Ocean Canada and BC Hydro. People at ESSA, along with visionary clients within the above entities, worked to introduce the concepts broadly in a wide spectrum of contexts. In many cases we were unable to engage our clients in completing the full adaptive management cycle, but concepts of systems modelling and well-monitored management experiments were widely promoted, and significantly influenced analyses and assessments of hydrocarbon development in the Beaufort Sea, oil and gas development in the Mackenzie Valley, acid precipitation in Eastern Canada, alternative forest management policies in southeast Alaska, fisheries management in the Great Lakes, and resource development in Jackson Hole, Wyoming, to name just a few.

As the practice expanded and more experience was gained it became increasingly clear that a number of problems typically plague efforts to implement rigorous adaptive management. This led to various laments, including Kai Lee's observation that "adaptive management has been more influential, so far, as an idea than as a practical means of gaining insight into the behaviour of ecosystems utilized and inhabited by humans" (Lee 1991). Similarly, a review by Walters (1997) diagnosed multiple institutional, scientific and social causes for his assertion that "many case studies in adaptive-management planning for riparian ecosystems have failed to produce useful models for policy comparison or good experimental management plans for resolving key uncertainties." Such problems stimulated the production of a number of seminal works on adaptive management theory and direction, spanning technical excellence in analysis & approach (Walters 1986), high level policy and institutional dynamics (Lee 1993), and barriers and bridges (Gunderson et al. 1995), among others. The institutional response in a number of cases was also to develop guidance for the conduct of adaptive management, including British Columbia Forest Service training courses, the U.S. Department of Interior Adaptive Management Technical Guide (Williams et al. 2009), and a guide to Statistical Methods for AM (Sit and Taylor 1998). At the present time adaptive management continues to hold the attention of scientists and managers, and is currently undergoing both a resurgence in the environmental management domain and an expansion into new fields of practice.

Expanding Domains of Practice

Adaptive environmental assessment and management (the first explicit incarnation of adaptive management) arose in the late 1970s out of an explicit recognition of the uncertainties intrinsic in complex environmental systems, and a need for increased ecological understanding to improve environmental and resource management. Our own practice of adaptive management in the 1980s and early 1990s was focused primarily on building models and designing adaptive management experiments to better understand the effects of alternative management approaches (for forests, dams, fish harvesting, invasive species, livestock, acidic emissions, oil and gas development) on freshwater and marine ecosystems, fisheries, forest ecosystems, rangelands, tourism, and wildlife. During the 1980s these efforts were driven by the desire of resource management agencies to develop and apply the best science, predictive tools and adaptive management approaches to challenging problems, and a willingness to rigorously test traditional assumptions. Budgets were generally healthy enough to really make progress on these objectives, and this work spawned some excellent research, though only a few rigorous adaptive management experiments were implemented on the ground. But there were some notable examples of interesting management experiments, such as the creative work by Gunn and Sein (2000) in Northern Ontario demonstrating the surprising resilience of lake trout populations to spawning habitat disruption, and by contrast the vulnerability of these populations to sport fishing pressure.

During the 1990s, government budgets gradually declined. Perhaps as a result, we found that these agencies (and private companies) were less motivated purely by the desire for better management, and more driven by the necessity to comply with legal regulations, such as the Forest Practices Code in British Columbia, the National Pollution Discharge Elimination System on the east coast of the U.S., or the U.S. Endangered Species Act. Legal criteria generally outweighed a desire for learning. However we found that stakeholders struggling with adversarial conflicts and difficult decisions were often delighted to step back and ponder adaptive management experiments that could lead to new insights, rather than refighting battles over conflicting evidence (Marmorek and Peters 2001). We also learned that decision analysis created a very informative context in which to embed models and existing evidence, and to explore the benefits and costs of adaptive management experiments (Peters and Marmorek 2001, Alexander et al. 2006). In the context of decision analysis, adaptive management experiments can provide valuable information on the relative merits of alternative actions.

During the 1990s however even fewer adaptive management experiments were actually implemented than in the 1980s. Sometimes this was because government agencies in charge of large rehabilitation or restoration programs saw it as wasteful or threatening to challenge the orthodoxy of their restoration strategy (for example, in the words of a participant at one of our workshops, "why monitor control areas when you could use the money for more restoration?"). In other cases we observed that environmental groups were fearful of testing existing regulations with rigorous adaptive management experiments (for example, varying the width of buffer strips around streams to assess the actual effects of logging on fish habitat and populations) lest these experiments lead to a weakening of those regulations.

In the past decade we have witnessed further expansion of the practice of adaptive management into several new directions. One is the increasing application of adaptive management in large drainage systems in order to meet multiple objectives as social and ecological tensions rise due to the concurrent pressures of growing human water demand, decreasing water availability, and a growing list of species at risk, each dependent on various processes and attributes of aquatic ecosystems to fulfill their life history requirements. All of these interacting forces create serious social, technical and legal pressures on management agencies.

In the American Southwest and Midwest, droughts during the past decade strained the ability of major species recovery programs to implement adaptive management flow experiments. When there's less and less water each year, it's hard to create a deliberate, contrasting sequence of conditions that also includes high flows. This drying pattern was observed during the early 2000s in the Glen Canyon Dam Adaptive Management Program (Melis et al. 2005), Platte River Recovery Implementation Program (Smith 2011), and the Middle Rio Grande Adaptive Management Plan (Murray et al. 2011). With global warming, that drying trend is likely to continue in the American Southwest, which will further constrain the options of water managers (Seager et al. 2007, Bureau of Reclamation 2011). As drying trends continue, it seems likely that human water uses will increasingly trump concerns for endangered species, and adaptive management may move more into the socio-economic realm of testing which demand-side regulations and incentives are most effective.

In the dry Okanagan region of southeastern British Columbia we have developed a real time Fish\Water Management Tool that allows water managers to adaptively manage water releases from Okanagan Lake, and better meet multiple human and fisheries objectives (Alexander et al. 2008). Rigorously designed blind tests with water managers showed that using this tool could have increased past production of Okanagan sockeye smolts by an average of 55 % (median benefit of 12 %) over the period from 1974 to 2003 (Hyatt and Alexander 2005). The improved smolt production occurred in water-years with moderate inflows (neither too dry nor too wet) in which water managers had greater operational flexibility. However, we also found that under future climate regimes (downscaled climate data for the period 2041–2070 with projected future water demand increases), these fishery benefits are predicted to be greatly reduced (mean simulated sockeye egg-to-smolt survival falling by 44 %) as a higher proportion of low water supply years would leave managers with much less ability to adaptively vary flows. At the present time Okanagan sockeye are doing surprisingly well, with some of the largest returns in the last 70 years (http://www.pentictonwesternnews.com/news/161506155.html). However, in the longer term the projected large loss of early freshwater survival cohorts represents a reduction in Okanagan sockeye resilience that is expected to have negative population-level consequences in the absence of compensatory gains elsewhere in the sockeye life-history.

Many of these examples were presented at an Adaptive Management Symposium held in Seattle in September 2011 as part of the American Fisheries Society Annual General Meeting, and their geographic locations are shown in Fig. 10.2. Expansion of adaptive management to the field of water resource management during the past decade also occurred in South Africa and Australia (Stankey and Allan 2009). For example, adaptive management was used to explore variable flow releases in the Murray-Darling Basin, enabled by the launching in 2004 of a National Water Initiative that explicitly provided for adaptive management of surface and groundwater systems to meet multiple objectives (Allan et al. 2009).

While in the early years adaptive management was focused on various aspects of environmental and natural resource management, its relevance to the practice

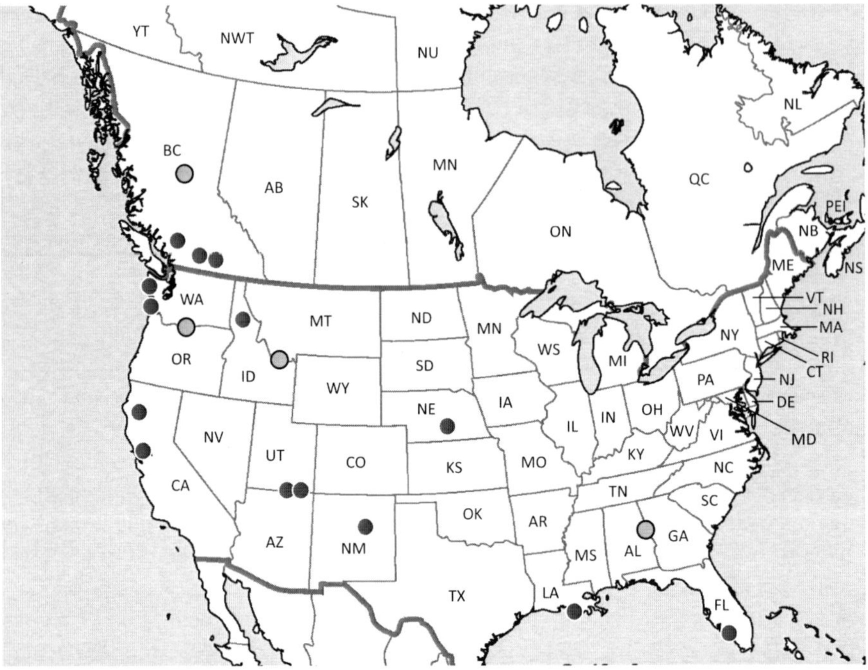

Fig. 10.2 Locations of selected adaptive management projects on aquatic ecosystem management in North America. The *lighter grey circles* indicate projects with multiple geographic locations

of environmental assessment (EA) was identified early on (Jones and Greig 1985) and the prevalence of adaptive management has grown rapidly in contemporary EA practice. One of the earliest references to adaptive management in the context of environmental assessment at the federal level in Canada is in the 1997 Panel Report of the EKATI Diamond Mine in the Northwest Territories (Olszynski 2010). Adaptive management was formally introduced into the *Canadian Environmental Assessment Act* in 2003, which focused on its use during follow-up programs for improving the quality of future assessments. Adaptive management was subsequently featured in Canadian Environmental Assessment Agency reviews and court decisions, but it was neither consistently used nor characterized, reflecting confusion about how to apply adaptive management in the context of environmental assessment, consistent with our own impressions gained through requests to critically review adaptive management plans or strategies for various development projects. The Canadian Environmental Assessment Agency released a policy statement in 2009 (CEAA 2009) which clarifies the use of adaptive management in follow-up monitoring and evaluation programs. However, EA regulators to date have had little control over what actually occurs after an EA decision has been made (a new Canadian Environmental Assessment Act (2012) includes enforcement measures for mitigations, but it is too soon to know how this will change things in practice).

The regulatory power for monitoring, evaluation and adaptation of projects after the EA decision lies within permitting agencies, and adaptive management in this context is a relatively new application that is garnering considerable interest. The Wek'èezhìi Land and Water Board in Canada's Northwest Territories is exploring how to use adaptive management in their issuance of land use permits and water licenses, towards better environmental protection, more focused requests for monitoring from proponents, and more meaningful and informative monitoring results. They drafted a response framework for aquatic effects monitoring which involves 3 action levels (low, moderate, and high) corresponding to increasing levels of environmental change, with each level having a corresponding management response (Racher et al. 2010). Adaptive management in a similar regulatory context is also being used in water use planning to iteratively revise dam operations in British Columbia (Gregory et al. 2006) and on the Columbia River (Federal Columbia River Power System 2009), and to respond to potential impacts from California's greenhouse gas cap-and-trade regulation (California EPA 2011).

Characterizing Adaptive Management: Three Main Dimensions

Just as examining a patient's history and lifestyle is a fundamental part of a medical doctor's diagnosis and treatment, we believe that characterizing the attributes of a potential adaptive management application can be helpful for pinpointing likely challenges and finding appropriate problem-solving strategies, including a decision about whether or not to apply adaptive management. At the highest conceptual level we see three dimensions to the application of adaptive management (Fig. 10.3): one pertaining to the nature of the **practice**, one pertaining to the nature of the **problem**, and one pertaining to the **social/organizational environment** in which adaptive management is being practiced. For each of these dimensions we describe the set of attributes or characteristics that would place adaptive management within the bulls-eye, "true adaptive management" centre sphere, with carefully designed and well monitored management experiments. As characteristics of an adaptive management application move away from the centre, they start to morph into what we call "pretend adaptive management", or applications that share some recognizable attributes of adaptive management but lack the rigour that gives adaptive management its learning power.

Dimension 1: Characteristics of the Practice

The term 'adaptive management' is widely used and becoming increasingly ubiquitous in environmental management spheres. Unfortunately it has been broadly misused to the point of becoming a "plastic word"—applied in so many different

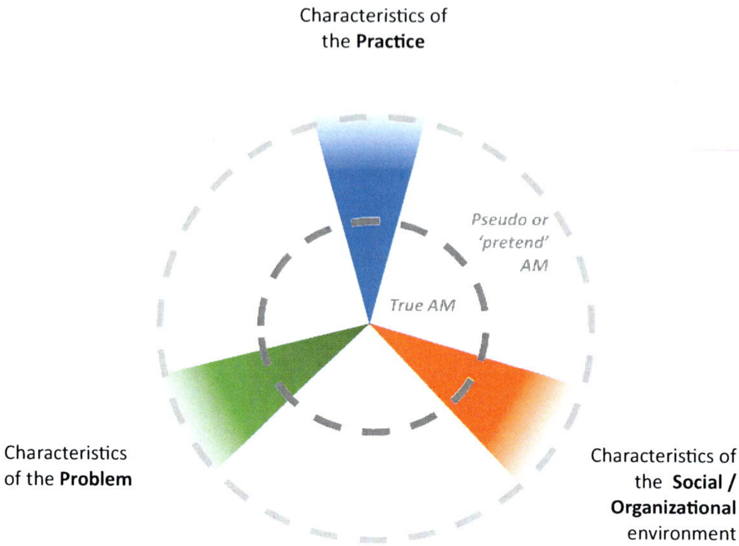

Fig. 10.3 Three conceptual dimensions of adaptive management application

circumstances that its meaning has been diluted and distorted, resulting in confusion at best, and at worst a false sense of security among those assuming that entrenching it in policy will solve their problems (Poerksen 1995). We have observed that much of the discussion of adaptive management, either by policy-makers or by purported practitioners, falls outside of what we would consider "true adaptive management." In many cases this stems from false beliefs that adaptive management is the same as adapting to serendipitous learning, the mistaken notion that adaptive management is what managers have been doing all along, false hopes that it is a magic bullet for growing uncertainty, or simply a lack of understanding about what adaptive management is. Some misunderstandings are probably facilitated by the simplicity of its name. While the idea of adaptive management is simple and elegant, the practice of adaptive management can be complex. We believe that "true adaptive management" includes all of the steps and elements shown in Table 10.1. This list of elements was initially crafted with leading practitioners in the field of forest management and honed through subsequent work on adaptive management projects in the fields of water management and ecosystem restoration. In some instances it may be difficult or unnecessary to undertake absolutely all of the listed elements, but each plays an important role; leaving any elements out should be done with full knowledge and explicit consideration of the implications.

Different practitioners, even those practicing "true" adaptive management, likely all use slightly different definitions of adaptive management. We therefore find it useful in our adaptive management discussions, practice and training to use the concept of an adaptive management "definition space". Applications of adaptive management at the most rigorous end of the definition space would ideally include all of the elements in Table 10.1. As applications of adaptive management

Table 10.1 Ideal elements of adaptive management at the most rigorous end of the adaptive management "definition space" (Murray et al. 2011)

AM steps	Ideal elements within each step
Step 1. *Assess* and define the problem	a. Clearly state management goals and objectives
	b. Review existing information to identify critical uncertainties and management questions
	c. Build conceptual models
	d. Articulate hypotheses to be tested
	e. Explore alternative management actions (experimental 'treatments')
	f. Identify measurable indicators
	g. Identify spatial and temporal bounds
	h. Explicitly state assumptions
	i. State up front how what is learned will be used
	j. Involve stakeholders, scientists, and managers
Step 2. *Design*	a. Use active adaptive management b. When and where possible, include contrasts, replications, controls c. Obtain statistical advice, building on analyses of existing data d. Predict expected outcomes and level of risk involved e. Consider next steps under alternative outcomes f. Develop a data management plan g. Develop a monitoring plan h. Develop a formal adaptive management plan for all of the remaining steps i. Peer-review (internal, external) the design j. Obtain multi-year budget commitments k. Involve stakeholders
Step 3. *Implement*	a. Implement contrasting treatments b. Implement as designed (or document unavoidable changes) c. Monitor the implementation
Step 4. *Monitor*	a. Implement the monitoring plan as it was designed b. Undertake baseline ('before') monitoring c. Undertake effectiveness and validation monitoring
Step 5. *Evaluate* results	a. Compare monitoring results against objectives b. Compare monitoring results against assumptions, critical uncertainties, and hypotheses c. Compare actual results against model predictions d. Receive statistical or analysis advice e. Have data analysis keep up with data generation from monitoring activities
Step 6. *Adjust* hypotheses, conceptual models, and management	a. Meaningful learning occurred, and was documented b. Communicate this to decision makers and others c. Actions or instruments changed based on what was learned

move further away from this they lose structure and rigor, and at some point they leave the adaptive management definition space ("true adaptive management") and move into some other learning paradigm that may bear a resemblance to adaptive management or share some of its characteristics and elements, but cannot really be considered adaptive management anymore. Rather than debate the conceptual location of the boundary between adaptive management and other approaches, we find it more productive to acknowledge that the definition space exists, that it may be larger in the minds of some practitioners than others, and that it is probably expanding as domains of practice broaden, and to try to work as close to the rigorous end of it as possible. The bottom line: rigor decreases as elements are ignored or abandoned, and at some point the application moves out of the adaptive management definition space to some other less rigorous approach to learning the effects of management actions (Table 10.1).

Dimension 2: Characteristics of the Problem

The characteristics of the problem include sub-dimensions of scale, reversibility, and the type of uncertainty being addressed. A critical requirement for adaptive management is contrast over space and time. The *spatial scale* of an adaptive management application depends on the geographic extent of the particular management uncertainties and objectives in question, and the area needed to adequately test hypotheses using experimental design elements of replicates, contrasts, and 'controls' or reference sites. Implementing adaptive management at large spatial scales increases the breadth and complexity of all of the steps: there will likely be more stakeholders, perspectives and conflicting values; broader objectives that may be difficult to quantify at larger scales; more uncertainties to examine and sequence; more management entities to align; more landowners to engage; less ability to find appropriate controls and replicates at the scale of large management actions (e.g., river basins); and more nested treatment sites to design, implement, monitor and evaluate. The *temporal scale* of an adaptive management application depends on several factors: how long it will take for the system to respond to the treatments; how long it will take to be able to *detect* this response given the monitoring resources and design; and the natural temporal variability in the system. If there is high year-to-year variability in key system drivers (e.g., precipitation, ocean conditions) or attributes (e.g., species abundance or survival), it will take longer to implement an experimental design that compares the outcomes of contrasting management actions, and controls for confounding factors, across these variable conditions. Adaptive management will also take longer if contrasting treatments cannot be done spatially and must therefore be done sequentially. For example, implementing different spring flow treatments in a single river system will need to be done across multiple years, since there is only one river and spatially concurrent contrasting treatments are not possible. If this single river system also has high natural variability, detecting the signal of management effects within the noise of natural variation may require even longer durations of monitoring and evaluation (Alexander et al. 2006).

These concepts are illustrated in Fig. 10.4. Farmers have been practicing adaptive management for years perhaps without recognizing it formally, using replicated

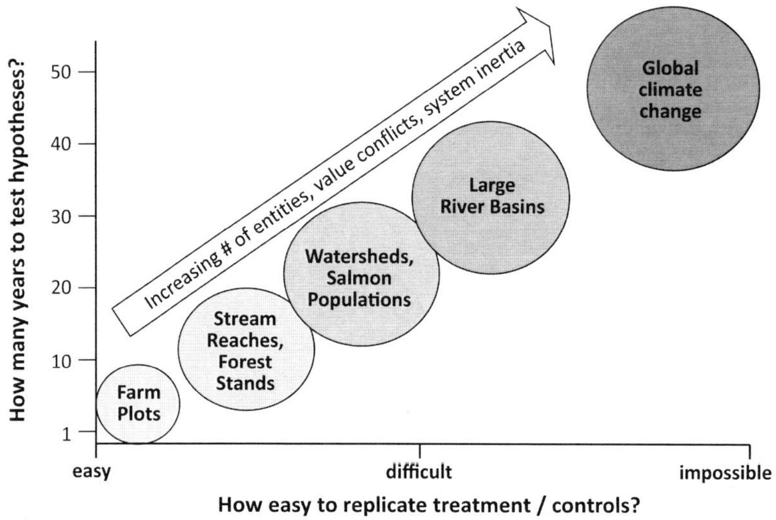

Fig. 10.4 The effects of spatial and temporal scales on the ability to implement adaptive management. Modified from a figure presented by Ray Hilborn in a course we co-taught on adaptive management at the Banff Centre in 1988

field plots receiving different treatments, and rapidly responding within 1–2 years (see Gawande 2009). Adaptive management at the scale of stream reaches and forest stands mostly involves habitat restoration and other actions to improve resource yields. Spatial replication is reasonably easy, one can control for site to site differences, and indicators of habitat quality or resource quantity generally respond within 5–10 years, within managers' attention spans and budget cycles. Salmon populations, further up in the scale progression, are inherently more variable due to variation in freshwater and particularly marine conditions, so confidently detecting population changes usually involves about a decade of monitoring before and after the restoration treatment. On the spatial axis, it's very difficult outside of public lands to have the institutional control to maintain both treatment and control watersheds for two decades. Adaptive management is possible, but difficult, and increasingly more difficult as watershed size increases. However, advances in technology can increase the ability to detect treatment effects. This has recently been found in the Columbia River where over 800,000 salmon smolts are "PIT-tagged" each year, enabling scientists to compare the survival rates of fish that are exposed to different flow and passage conditions within a year (i.e., essentially creating multiple 'treatment groups' within a year and accelerating the rate of hypothesis testing; see Haeseker et al. 2012). Adaptive management is difficult or impossible at very large scales. For global climate change (extreme right of Fig. 10.3), we believe it will take roughly 50 years to know the full ecological effects of our actions on a global scale, at which point irreversible changes are almost certain. The feedback loop to revise human actions at this scale is too slow, and there is only one Earth, rendering replicates impossible—it would be essentially impossible to do adaptive management on a planetary scale. The only viable strategy on a planetary scale is to take the most prudent actions based on modelling of human, atmospheric and ecological systems and hope that these actions are sufficient to prevent disastrous outcomes. On smaller spatial scales, it may however be possible to use an adaptive management approach (e.g., comparative case studies of the relative merits of cap-and-trade vs. carbon taxes for reducing emissions of greenhouse gases). The bottom line: adaptive management is easier and more likely to be successful at smaller spatial scales where treatments can be more readily replicated and controlled, and if the time required to test hypotheses can be measured in years rather than decades. Adaptive management gets harder at larger spatial and temporal scales, and at the extremes becomes impossible.

Reversibility is another important sub-dimension, and is related to risk. Reversibility pertains to both the action or treatment and the system responses, and different degrees of reversibility are represented by the example uncertainties in Fig. 10.4. Invasive shrub removal techniques such as cutting, pulling or prescribed burning can be stopped very quickly, with success likely to be evident either within-season or within a year. Waterfowl harvest regulations can be revised at the start of each hunting season, with population effects evident with a few years depending on the species (Johnson et al. 1993, Nichols et al. 2007). Timber harvest strategies may be defined and revised on an annual basis as well, or on longer cycles (e.g. 5-year harvest plans), with the effects lasting on the order of decades. Construction of development projects such as hydroelectric dams is much more permanent, and most adaptive management applications have focused on dam operations (e.g., Gregory et al. 2006). Dam removal is conceptually possible but rarely occurs for larger dams. A recent exception is the 2011 removal of the Elwha Dam in Washington State, which presented an opportunity to learn from this rare management action. The release of biocontrol agents is another example of an action with little or no reversibility (Shea et al. 2002). The differences in reversibility are important within a learning paradigm such as adaptive management where operational experiments are applied in real-world systems, and where surprises and 'failures' often provide the greatest insights. This also affects the risk of conducting adaptive management in rare ecosystems, where outcomes may have more dire ecological or legal implications. "Stopping rules" within an adaptive management plan can help mitigate this risk, but that implies the action can be stopped and the effects reversed. Bottom line: the appropriateness of adaptive management decreases as reversibility of the interventions decreases.

The third sub-dimension of the characteristics of the problem pertains to the *type of uncertainty* being addressed. Contrary to some beliefs, adaptive management is not a panacea for uncertainty. It is most useful when there is high management uncertainty but only a small-to-moderate degree of ecological uncertainty (upper left quadrant of Fig. 10.5). Management uncertainty refers to questions managers have about which management actions or strategies will best meet multiple management objectives, which typically span ecological, social and economic outcomes. Ecological uncertainty refers to ecosystem components, structure, function, or

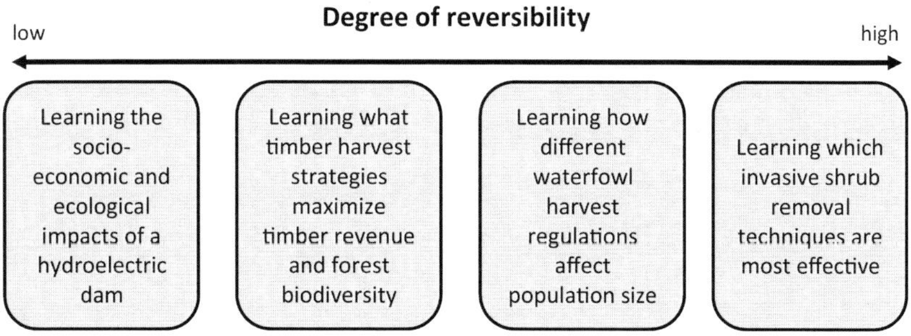

Fig. 10.5 Examples of different uncertainties with different degrees of reversibility

relationships, for example specific habitat needs of rare species and their resilience to variation in habitat characteristics over space and time. If there is a large amount of ecological uncertainty, it becomes increasingly difficult to prioritize hypotheses to be tested, to formalize management objectives, to know what indicators to monitor to assess progress towards those objectives, and to detect the management signal within the noise of natural spatial and temporal variability (common forms of ecological uncertainty).

For example, in the Middle Rio Grande, because of the short life-cycle of the Rio Grande silvery minnow and its extreme rarity, there are fundamental uncertainties about relationships between the magnitude and timing of flows, the annual recruitment of the Rio Grande silvery minnow, and the recruitment levels needed for sustainable populations (Murray et al. 2011). Indeed, the first action listed in the Recovery Plan for the minnow was to develop "a thorough knowledge of the Rio Grande silvery minnow's life history, ecology, and behavior, and the current status of its habitat"—in other words, reducing ecological uncertainty. These uncertainties make it challenging to define quantitative objectives for the silvery minnow in the Middle Rio Grande and select key management hypotheses to be tested through adaptive management. Flow-habitat and population viability models are being developed to address some of these ecological uncertainties and help managers to quantify objectives for habitat characteristics and minnow recruitment.

It is important to understand that managers and scientists may view uncertainties in very different ways, and assign different levels of importance to different questions or hypotheses. Scientists are most interested in learning more about what's not known: the uncertainties with an ecosystem. Science is conservative to prevent erroneous results from being widely applied. Managers however want to use what's known to make practical decisions as soon as possible, balancing risks and benefits. In the words of Ray Hilborn (1992), "managers must go where scientists fear to tread." Bridging these two worlds is an important aspect of adaptive management, and involves a mutual understanding of them. The bottom line: the usefulness of adaptive management increases as management uncertainty increases, but within limits of ecological uncertainty.

To summarize this dimension of adaptive management: problems are more amenable to adaptive management the less time it takes to test hypotheses and the easier it is to replicate and control treatments, the more reversible the treatments, and the greater the management uncertainty (within limits of ecological uncertainty).

Dimension 3: Characteristics of the Social and Organizational Environment

In 2001 we facilitated a workshop for the University of Washington Olympic Natural Resources Centre as part of a conference on Organizational Learning: Adaptive Management for Salmon Conservation from which we generated a list of enabling or inhibiting factors (Alverts et al. 2001). Five years later during a study into factors that enable adaptive forest management for the National Commission on Science for Sustainable Forestry we used an interview and survey process to examine 20 forest management projects that had applied the principles of adaptive management (Greig et al. 2013). Fourteen of these were considered successful in that they made it all of the way around the adaptive management cycle and adjusted management actions, such as new forest practice rules for buffer strips on fish bearing streams, measures to improve ungulate winter range, snags for cavity nesting birds and prescribed fire. The other six projects were not necessarily failures, but they didn't complete the last step. We asked each project leader to grade their project on how rigorously they applied adaptive management and also to assess the degree to which each of various listed factors inhibited or enabled adaptive management. We were particularly interested in differences between the more successful projects and the rest. Two interesting results emerged. First, everyone found that leadership was enabling, regardless of how successful the project was. Therefore strong leadership is a necessary but not sufficient factor for 'closing the loop'. Second, three factors were more enabling in the more successful projects and more inhibiting in the others: executive mandate, community involvement and adaptive management science. We reviewed the results of the study at a 2-day workshop in Portland, Oregon with some very experienced adaptive management practitioners, and by the end of the meeting had generated the hierarchy shown in Fig. 10.6. We concluded that there were five primary factors which were absolutely critical for success (the top two boxes). If those were enabling, then it was likely that the other necessary attributes for success (the bottom box) would also be established.

The first factor (context, driving problem) includes recognition that there is a problem that might be addressed through an adaptive management approach. An enabling context is one where existing legislation neither inhibits nor prevents adaptive management, and is not overly adversarial. However disagreement among stakeholders is not necessarily inhibiting, depending on the nature of the disagreement. Adaptive management cannot resolve conflicts over values—for example, if stakeholders disagree on the desired outcomes or objectives, negotiated conflict resolution may be required. AM can help find the best methods for achieving

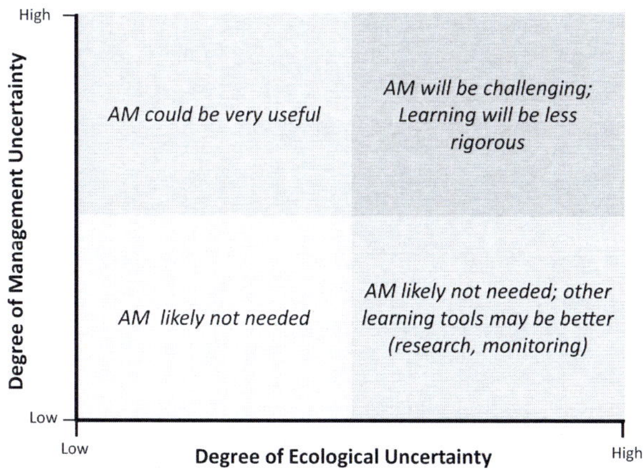

Fig. 10.6 The 'uncertainty space' where adaptive management best applies

agreed-upon outcomes, and help clarify trade-offs among objectives under different management scenarios.

The second factor, leadership, is critical for gaining the support needed to begin adaptive management and to sustain the initiative over time. This is important at the top levels of an organization where advocacy, a willingness to take risk, and strategic allocation of the necessary resources is important, as well as at lower and middle levels where much of the technical and field work occurs. With strong, creative leadership in place, the five factors in the lower box (community involvement, planning, funding, staff training, rigorous adaptive management science) are much more likely to occur.

Leadership is closely related to the third factor, executive direction. Think of this as organizational buy-in, which indicates an agreement to incorporate the results into policy and practice, and which is essential for "closing the loop". This factor is most enabling in corporate cultures that embrace management uncertainty, and actively seek ways to manage effectively despite this. In public entities, a legislative or regulatory authority to implement adaptive management is essential to provide the "hammer" behind an agency's executive mandate.

Problem definition sounds straightforward, but is tricky to get just right. Failure to be clear about what the problem really is will lead to later difficulties in maintaining an appropriate and effective focus. It is always helpful to frame a problem through the lens of management decisions, and critical uncertainties affecting the best course of action to achieve stated objectives (Peters and Marmorek 2001). This requires participation from managers in clearly defining the problem, which is wise if the expectation is that they will act on what is learned. It is best to focus on a problem that is 'durable', and not likely subject to frequent shifts in the social or political wind.

Lastly, successful adaptive management requires effective communication channels both laterally and vertically within the organization—across levels, roles, and disciplines. The field personnel need to understand the perspectives of the managers, and vice versa. Participants need a good understanding of the purpose, process, progress and outcomes of any adaptive management initiative for it to be successful. Strong leaders embrace open communication and early recognition of the need for changes in existing management strategies. The bottom line: The more these factors are enabling instead of inhibiting, the easier it will be to undertake adaptive management.

The Future for Adaptive Management

Diverse pressures are leading to an increasing need for decision-making approaches that can help chart a path through uncertainty (Fig. 10.7). These same pressures also make the application of such approaches very challenging, because as pressures on natural ecosystems and resources increase, the urgency and stakes also rise. Add global economic instability and climate change to the mix, along with a populace with unprecedented access to information and each other through electronic and social media, and managers are left with some tough choices about how to proceed—as well as some intriguing opportunities. Adaptive management is one such opportunity, under the right circumstances. What we have tried to do in this chapter is to elaborate on what those circumstances are, so that the power of the approach can be wielded most effectively *and* where it is most useful as the practice broadens into new domains.

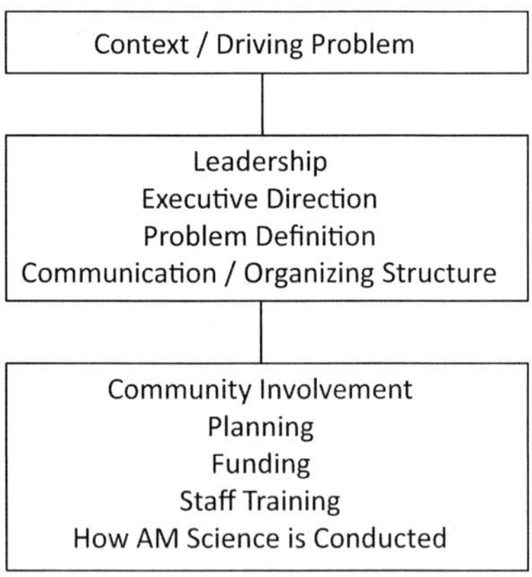

Fig. 10.7 Hierarchy of social and organizational factors that enable (or inhibit) adaptive management. (Greig et al. 2013)

References

Alexander, C. A. D., Peters, C. N., Marmorek, D. R., & Higgins, P. (2006). A decision analysis of flow management experiments for Columbia River mountain whitefish (*Prosopium williamsoni*) management. *Canadian Journal of Fisheries and Aquatic Sciences, 63,* 1142–1156.

Alexander, C. A. D., Hyatt, K. D., & Symonds, B. (Eds.). (2008). The Okanagan fish/water management tool: Guidelines for apprentice water managers. V.2.1.000. Prepared for Canadian Okanagan Basin Technical Working Group, Kamloops, Canada.

Allan, C., Watts, R. J., Commens, S., & Ryder, D. S. (2009). Using adaptive management to meet multiple goals for flows along the Mitta Mitta River in south-eastern Australia. In C. Allen & G. H. Stankey (Eds.), *Adaptive environmental management: A practitioner's guide* (pp. 59–69). Dordrecht: Springer Science+Business Media V.B., Collingwood: The Netherlands and CSIRO Publishing.

Alverts, R., Calhoun, J. M., & Lee, R. L. (2001). *Organizational learning: Adaptive management for Salmon conservation conference proceedings*. Forks: University of Washington, Olympic Natural Resources Centre.

Bureau of Reclamation. (2011). SECURE Water Act Section 9503(c)—Reclamation climate change and water, Report to Congress.

California EPA. (2011). Adaptive management plan for the cap-and-trade regulation. Prepared by the Air Resources Board. http://www.arb.ca.gov/cc/capandtrade/adaptive_management/plan.pdf. Accessed 31 Dec 2014.

Canadian Environmental Assessment Agency (CEAA). (2009). Operational policy statement: Adaptive management measures under the Canadian Environmental Assessment Act. http://www.ceaa.gc.ca/default.asp?lang=En-US&n=50139251-1. Accessed 31 Dec 2014.

Federal Columbia River Power System. (2009). FCRPS adaptive management implementation plan. Prepared under the 2008–2018 Federal Columbia River Power System Biological Opinion.

Gawande, A. (14 December 2009). Testing, testing: The health-care bill has no master plan for curbing costs. Is that a bad thing? *The New Yorker.*

Gregory, R., Failing, L., & Higgins, P. (2006). Adaptive management and environmental decision making: A case study application to water use planning. *Ecological Economics, 58*, 434–447.

Greig, L. A., Marmorek, D. R., Murray, C., & Robinson, D. C. E. (2013). Insight into enabling adaptive management. *Ecology and Society, 18*(3), 24. http://dx.doi.org/10.5751/ES-05686-180324.

Gunderson, L. H., Holling, C. S., & Light, S. S. (Eds.). (1995). *Barriers and bridges to the renewal of ecosystems and institutions*. New York: Columbia University Press.

Gunn, J. M., & Sein, R. (2000). Effects of forestry roads on reproductive habitat and exploitation of lake trout (*Salvelinus namaycush*) in three experimental lakes. *Canadian Journal of Fisheries and Aquatic Sciences, 57*(Suppl. 2), 97–104.

Haeseker, S. L., McCann, J. A., Tuomikoski, J., & Chockley, B. (2012). Assessing freshwater and marine environmental influences on life-stage-specific survival rates of Snake River spring/summer Chinook salmon and steelhead. *Transactions of the American Fisheries Society, 141,* 121–138.

Hilborn, R. (1992). Can fisheries agencies learn from experience? *Fisheries, 17,* 6–14.

Holling, C. S. (Ed.). (1978). *Adaptive environmental assessment and management.* New York: Wiley.

Hyatt, K., & Alexander, C. A. D. (2005). The Okanagan fish-water management (OKFWM) tool: Results of a 25 year retrospective analysis. Prepared for Canadian Okanagan Basin Technical Working Group, Kamloops.

Hyatt, K. D., & Bull, C. (2007). *Fish and water management tool project assessments: Record of management strategy and decisions for 2005*. Nanaimo: Fisheries and Oceans Canada, Science Branch, Pacific Region, Pacific Biological Station.

Johnson, F. A., Williams, B. K., Nichols, J. D., Hines, J. E., Kendall, W. L., Smith, G. W., & Caithamer, D. F. (1993). Developing an adaptive management strategy for harvesting waterfowl in North America. *Transactions of the North American Wildlife and Natural Resources Conference, 58,* 565–583.

Jones, M., & Greig, L. (1985). Adaptive environmental assessment and management: A new approach to environmental impact assessment. In V. W. Maclaren & B. W. Joseph (Eds.), *New directions in environmental impact assessment in Canada*. Methuen.

Lee, K. N. (1991). Appraising adaptive management. *Conservation Ecology, 3*(2), 3. http://www.consecol.org/vol3/iss2/art3.

Lee, K. N. (1993). *Compass and gyroscope: Integrating science and politics for the environment*. Washington, D.C.: Island Press.

Marmorek, D. R., & Peters, C. (2001). Finding a PATH towards scientific collaboration: Insights from the Columbia River Basin. *Conservation Ecology, 5*(2), 8. http://www.consecol.org/vol5/iss2/art8.

Melis, T., Korman, J., & Walters, C. J. (2005). Active adaptive management of the colorado river ecosystem below Glen Canyon Dam, USA: Using modeling and experimental design to resolve uncertainty in large-river management. Proceedings of the International Conference on Reservoir Operation & River Management, Guangzhou, China, September, pp. 18–23. http://www.gcmrc.gov/library/reports/adaptive_management/melis2005.pdf. Accessed 17 Oct 2011.

Murray, C., Smith, C., & Marmorek, D. (2011). Middle Rio Grande endangered species Collaborative Program Adaptive Management plan Version 1. Prepared by ESSA Technologies Ltd. (Vancouver, BC) and Headwaters Corporation (Kearney, NE) for the Middle Rio Grande Endangered Species Collaborative Program, Albuquerque. 108 p. http://www.middleriogrande.com/LinkClick.aspx?fileticket=d%2fzdKUAz9LI%3d&tabid=460&mid=1307. Accessed 31 Dec 2014.

Nichols, J. D., Runge, M. C., Johnson, F. A., & Williams, B. K. (2007). Adaptive harvest management of North American waterfowl populations: A brief history and future prospects. *Journal of Ornithology, 148*(Supplement 2), s343–s349.

Olszynski, M. Z. P. (2010). Adaptive management in Canadian environmental assessment law: Exploring uses and limitations. *Journal of Environmental Law and Practice, 21,* 1–30.

Peters, C. N., & Marmorek, D. R. (2001). Application of decision analysis to evaluate recovery actions for threatened Snake River spring and summer chinook salmon (*Oncorhynchus tshawytscha*). *Canadian Journal of Fisheries and Aquatic Sciences, 58,* 2431–2446.

Poerksen, U. (1995). *Plastic words: The tyranny of a modular language*. State College: Pennsylvania State University Press.

Racher, K., Hutchinson, N., Hart, D., Fraser, B., Clark, B., Fequet, R., Ewaschuk, P., & Cliffe-Phillipsy, M. (2010). Linking environmental assessment to environmental regulation through adaptive management. *Integrated Environmental Assessment and Management, 7,* 301–302.

Seager, R., Ting, M. F., Held, I., Kushnir, Y., Lu, J., Vecchi, G., Huang, H. P., Harnik, N., Leetmaa, A., Lau, N., Li, C. H., Velez, J., & Naik, N. (2007). Model projections of an imminent transition to a more arid climate in southwestern North America. *Science, 316,* 1181–1184.

Shea, K., Possingham, H. P., Murdoch, W. W., & Roush, R. (2002). Active adaptive management in insect pest and weed control: Intervention with a plan for learning. *Ecological Applications, 12,* 927–936.

Sit, V., & Taylor, B. (Eds.). (1998). *Statistical methods for adaptive management studies*. Victoria: British Columbia Ministry of Forests.

Smith, C. B. (2011). Adaptive management on the central Platte River—Science, engineering, and decision analysis to assist in recovery of four species. *Journal of Environmental Management, 92,* 1414–1419.

Stankey, G., & Allan, C. (2009). Introduction. In C. Allen & G. H. Stankey (Eds.). *Adaptive environmental management: A practitioner's guide* (pp. 3–8). Dordrecht: Springer Science+Business Media V.B., Collingwood: The Netherlands and CSIRO Publishing.

United States Fish and Wildlife Service (USFWS). (2010). Rio Grande Silvery Minnow (Hybognathus amarus) Recovery Plan, First Revision. Albuquerque. viii + 210 p. http://www.fws.gov/southwest/es/Documents/R2ES/Rio_Grande_Silvery_Minnow_Recovery_Plan_First_Revision.pdf.

Walters, C. (1986). *Adaptive management of renewable resources*. New York: MacMillan Publishing Company.

Walters, C. (1997). Challenges in adaptive management of riparian and coastal ecosystems. *Conservation Ecology, 1*(2), 1. http://www.consecol.org/vol1/iss2/art1/.

Williams, B. K., Szaro, R. C., & Shapiro, C. D. (2009). Adaptive management: The U.S. Department of the interior technical guide. Adaptive Management Working Group, U.S. Department of the Interior, Washington, D.C.

Chapter 11
Practical Resilience: Building Networks of Adaptive Management

Jim Berkley and Lance Gunderson

Keywords Resilience · Adaptive management · Uncertainty · Organizations · Environmental management

Introduction

Holling (1973) proposed the word resilience as an alternative paradigm to explain abrupt changes in ecosystem structure and function. In his seminal paper, he questioned deeply held assumptions that relationships among key variables in an ecosystem were not only stationary and persistent, but that the resulting dynamics were predictable. He also suggested that ecosystems were adaptable in the face of changing external influences, due to shifting relationships among state variables and parameters in ways that allowed for persistence over time. The key conceptual breakthrough of this work, however, was the proposition that even though many systems could be defined in terms of stable configurations (both structure and functions) that absorb external disturbances, a wide range of systems could flip or rapidly transition into alternative configurations. Hence ecological resilience was used to describe 'far from equilibrium behavior' and the processes that mediate transitions among multiple configurations or stable states (Holling 1973). The past forty years has resulted in numerous studies that, have documented alternative states and transitional dynamics in a wide range of ecological systems (Gunderson and Pritchard 2002, Folke et al. 2004). Additionally, the practical implications of such non-linear dynamics in ecosystems for ecosystem management have been realized (Gunderson 1999, Chapin et al. 2009), especially for adaptive management.

The processes of adaptive management as described by Holling (1978), Walters (1986), and others, including chapters of this book, are conceptually rooted in theories of ecological resilience (Holling 1973). The most overt linkage is the contribu-

J. Berkley (✉)
U.S. Environmental Protection Agency, 1595 Wynkoop Street, EPR-EP, 80203 Denver, CO, USA
e-mail: berkley.jim@epa.gov

L. Gunderson
Department of Environmental Sciences, Emory University, 524 Mathematics and Science Center, 400 Dowman Drive, 30322 Atlanta, GA, USA
e-mail: lgunder@emory.edu

C. R. Allen, A. S. Garmestani (eds.), *Adaptive Management of Social-Ecological Systems*,
DOI 10.1007/978-94-017-9682-8_11

tion of C.S. Holling to both the concepts of resilience and the development of adaptive management. At least three other manifestations of how the theory influences practice include: (a) the uncertain and surprising nature of ecosystem dynamics, (b) difficulties in environmental assessment as prelude to adaptive management, and (c) the institutional complexities of managing such non-linear ecological systems. Each of these is briefly described below.

The presence of alternative ecosystem configurations creates many difficulties for managers, one of which is the level of uncertainty associated with regime shifts. Simplistically managers attempt to either (a) maintain ecosystems in a particular state (such as a park or preserve), or (b) attempt to transition an ecosystem from one state to another. Although resilience is a key property that mediates transition among states, it remains an elusive target for analysis and prescription. Recent work on quantifying system states, thresholds and tipping points (Scheffer et al. 2001, Brock and Carpenter 2010) indicates that no single indicator or surrogate can be found, and that *a priori* signals of regime shifts require immense amounts of data that are generally lacking for most managed resource systems. Moreover, while many regime shifts have been observed and studied (Gunderson and Pritchard 2002, Folke et al. 2004), the mechanisms of moving systems among regimes cannot be determined through planning and design, but must be learned through practice. This key uncertainty, the inability to bridge the knowledge to action gap which resilience theory posits, was one of the key reasons for developing adaptive management (Holling 1978).

Adaptive assessments were developed to help understand processes that underpin ecosystem regime shifts and how to design management actions to navigate such transitions. The process was developed in part to attempt to bring together understandings, models and theories held by scientists and practitioners with different disciplinary backgrounds and training (Holling 1978). As such, the models have been used to integrate understanding, highlight uncertainties, and develop imaginative solutions (Walters 1986). Indeed some of the failures of the adaptive assessment process have been attributed to the difficulties of modeling cross-scale dynamics that generate certain types of regime changes (Walters 1997). However, the assessment process has been critical in generating robust management actions that have led to learning how to navigate regime transitions, and moreover the importance of experimentation through adaptive management. One example is the environmental assessment workshops that contributed to the large-scale ecosystem restoration plans in the Everglades (Walters et al. 1992), with the restoration objective being to flip the current degraded system into a more desired ecological state.

The implementation of adaptive management by resource managers in the United States over the past three decades has been problematic (Allen and Gunderson 2011). While hundreds of adaptive assessments have led to successfully redefining or creating new policies and programs (Walters 1997), adaptive management programs have had mixed results (Johnson 1999). Lee (1993) chronicled the lessons from the adaptive management program in the Columbia River and was among the first to distinguish between the scientific/technical aspects of resource management and the social/political dimensions. Gunderson (1999) suggested that resource systems had to have two key properties to flourish: ecological resilience, which provides the insurance against failure of management actions, and institutional flexibility, which allows for expansion of narrowly defined missions to ones of learning and adaptation. Peterson et al. (2003) argued that adaptive management was appropriate for resource settings characterized by high uncertainty and high degrees of resource controllability for experimentation. Walters (2007) painted a gloomy picture from his experiences; hundreds of programs have failed to implement experimental management due to a lack of resources to monitor and learn from actions, an unwillingness to admit uncertainty, and from failures of leadership. Recent work (Benson and Stone 2013) found that legal and institutional constraints hinder the implementation of adaptive management. Yet, in spite of these failures and obstacles, adaptive management has continued to be applied to federal level resource management agencies including many in the U.S. Department of Interior, U.S. Forest Service, and U.S. Environmental Protection Agency (Williams et al. 2009, Benson and Stone 2013, Scarlett 2013).

One response to the institutional and bureaucratic barriers faced by practitioners of adaptive management has been the advent of adaptive governance (Brunner et al. 2006). Adaptive governance is an emergent framework that unites formal and informal institutions to manage complex, dynamic and turbulent resource systems (Folke et al. 2005). Adaptive governance arose in response to the failures of implementing singular solutions through a top down, technocratic system that failed to adequately navigate the social and political realms of resource issues (Brunner et al. 2006). As such it has developed into a collaborative framework of collective decision-making (Benson and Stone 2013, Scarlett 2013) on how to implement, monitor and learn from management actions. The lack of such governance can lead to not only failures of adaptive management, but can be indicative of managed resource systems being very resilient or trapped by extant policies and politics (Gunderson and Light 2006). As adaptive governance continues to evolve through the practice of adaptive management, it will change as will our understanding of how policy and governance best interfaces with management. But such changes are based upon agency or organizational cultures, resolving multiple stakeholder values, ecological understanding, and legal mandates, which are also dynamic entities (Westley 2002). Moreover, as other chapters in this book indicate, the concept and theories of resilience are just beginning to be applied in context of the many dimensions of adaptive governance.

Even though there have been conceptual linkages between adaptive managers and theories of ecological resilience, few cases exist of explicit linkage of adaptive management to ecological resilience. In order to explore the opportunities for linkages among adaptive management, adaptive governance and resilience, the senior author interviewed ten key decision makers in natural resource management. The participants are part of a formal network, the Collaborative Adaptive Management Network (http://www.adaptivemanagement.net/). The practitioners who participated work, or have worked, in U.S. federal agencies (U.S. Army Corps of Engineers, Bureau of Reclamation, U.S. Institute for Environmental Conflict Resolution, U.S. Geological Survey, Bureau of Land Management) or non-governmental organizations (the Meridian Institute, Foundations of Success and other private science/engineering/restoration consulting firms), and are experienced in the application of adaptive management. The interviewees were selected from a cross-sectional sample of the two organizational types. The discussions were driven by a series of queries, such as do adaptive management organizations engage in managing for resilience? If so, how does that manifest in agency statements and/or missions? What roles do organizations (both agencies and networks) play in resilience management? What, if anything, could be monitored to independently test a system's resilience? Their responses tended to fall into three categories: managing ecological resilience, the development of networks to facilitate adaptive management and resilience-based management, and preliminary reflections on networks of practice and resilience. Each of these is discussed in the following sections.

Managing Ecological Resilience

Managing ecological resilience has proven enigmatic. Part of the puzzle is in understanding the relationship between concepts of ecological resilience and application of the concepts by land managers in the United States (and elsewhere). A number of facets of this relationship can be identified. One has to do with the definition(s) of resilience and implications for resource management. Another complication has to do with whether one assumes resilience to be a normative property of managed ecosystems. Another complication arises from the relationship between resilience and the organizational, institutional or agency mission and directives. Yet another problematic area is in the operationalizing of resilience.

Resilience remains an ambiguous term for both scholars and managers. While most of the applied definitions recognize the presence of alternative states (Folke et al. 2004), many focus on protecting and maintaining a particular ecosystem state (Levin and Lubchenco 2008). Walker et al. (2002) suggest that resilience management is a goal to prevent a system from "moving into undesirable configurations." Certainly, this is appropriate and consistent with many ecosystem settings, such as protecting endangered species from extinction (an irreversible regime shift). However it does not deal with situations in which a managed resource is in a pathologically stable state and restoration is a goal (Zellmer and Gunderson 2009). The managers we interviewed suggest that resilience is defined broadly, which makes it appropriate for talking about general goals, such as ecosystem restoration. This is consistent with Brand and Jax (2007) who define the use of the term as a 'boundary object', or useful term for bridging different perspectives. However, some managers felt that resilience still remains too abstract a concept to be used in operations, plans or management actions. Many of the interviewed managers understood various definitions of resilience, but had different views on resilience as a management goal.

Resilience according to some managers is a normative concept; in other words the ecological systems that they manage should be or ought to be resilient. Those managers who focus on conservation, tended to view systems in this way. That is, systems ought to be resilient to broad external forces, such as human development or climate change. Other managers didn't use the word resilience in this context, but rather used the word health or healthy as a synonym of resilience to describe the way in which a particular area or unit should be managed. Most of the managers indicated that resilience was a positive property of systems. In other words, many thought that resilience should be considered as one of many general objectives of management, but difficulties in defining and making it operational would preclude it as a specific goal.

Many managers discussed the relationship between managing for resilience and objectives or culture of their organization. In this context, resilience is viewed as a description of ecosystem properties (Brand and Jax 2007). One interviewee stated that their organization still fundamentally views nature as a static entity and doesn't recognize the need for either resilience concepts or adaptive management. For some of the interviewees, resilience was either implicit in or unnecessary to their implementation of adaptive management. A few interviewees stated that resilience has not been considered because it is not mentioned in laws that guide and structure agency or organizational missions. Other scholars, however, have argued that existing laws and regulations, which drive many agency and organizational actions, can be interpreted through a resilience lens. Scholars are now suggesting that laws such as the National Environmental Protection Act (Benson and Garmestani 2011) or the Endangered Species Act (Benson 2012) could be administered in such a way to reflect resilience concepts of multiple states or regimes. For those interviewees who thought that resilience ideas could be addressed with existing organizational missions, they pointed to issues of implementation.

For those interviewed managers who understand ecological resilience, they listed a number of key challenges that preclude managing for resilience. These challenges are related to how individuals and agencies deal with uncertainty and risk. Many of the managers state that the lack of knowledge about linking a particular action with a particular magnitude of response will preclude any action. Others recognize that the uncertainties of regime shift dynamics will generally stymie management actions as well, and bluntly state that they do not have sufficient information to be able to manage for resilience. Other managers relayed difficulties in establishing ecological attributes and indicators that reveal information about ecological resilience because most of the variables are monitored for other reasons. Others suggest that the ecosystem indicators were established to assess progress towards management goals, or performance measures. That is, these measures allow for evaluation of success of strategies and actions. Managers suggest that quantification of actions and consequences are needed to justify actions and to report organizational progress to funding sources.

Creating Networks for Adaptive Management Implementation

In recent history in the United States, decision makers and scientists struggled with making progress toward their objectives when attempting species restoration and recovery. There were significant barriers to progress. Brunner (2006) discusses these and other factors in his discussion about the genesis of adaptive governance. Most of the work related to managing these efforts was directed by federal agencies which previously had different missions and roles than that of restoration and recovery. Much of this work was completed during the era when it was thought that science and decision making could rise above politics and could be executed by a single bureaucratic organization (Brunner 2006). Ironically, organizations such as the U.S. Army Corps of Engineers and the Bureau of Reclamation were asked to undo or ameliorate the unintended consequences of decades of their own work; work that successfully addressed the pressing problems of the time such as the ability to navigate waters for purposes of commerce and national defense, the provision of irrigation for an adequate food supply, flood control, and the provision of hydroelectric power. The task of shifting long established governmental organizations from their existing culture and mission to a new one was formidable. This was a factor along with others (Brunner 2006) that gave rise to groups of stakeholders creating informal efforts to network outside of the normal federal process; early examples of adaptive governance.

The restoration of the Kissimmee River, Florida, U.S.A, is a specific example of the early genesis of adaptive governance. The river is located in south-central Florida and forms the headwaters of the Kissimmee River-Lake Okeechobee-Everglades ecosystem. Beginning in the 1960s, the river was channelized and six locks and dams were built by the Army Corps for flood control. In 1992, Congress authorized efforts to restore the Kissimmee River because of negative and unintended environmental effects of the flood control projects. The Army Corps was charged with the task of restoring the Kissimmee. Initially, progress towards restoration was slow. Shifting the focus of the Army Corps' mission from straightening and maintaining the river and controlling flow, to filling some of the channelized portions and restoring river sinuosity required a significant organizational shift. Progress went slowly from some stakeholders' perspectives. This led to the creation of informal groups that convened outside the formal organizational structure of the Army Corps. These informal groups included scientists and non-governmental organizations that worked on problems such as physical models that stakeholders could agree were representative of the natural system. The ability to have the discussions and create such models had been hamstrung by litigation, litigation fears, and reticence to share information among parties with diverse interests. The ability of these informal groups to meet outside of the formal organizational structure led to breakthroughs as stakeholders worked together and moved forward with restoration instead of gridlock.

The Kissimmee provides one of the first examples of the potential for informal networks and innovative governance (Brunner 2006) providing a pathway to progress that may not be possible by working within formal organizational structures alone. Below we provide a description of the roles and missions of agencies and non-governmental organizations, and how they formed networks of adaptive management practice, evolved and matured.

Federal Agencies

Federal agencies' formal role in adaptive management is implementation. Some of the main agencies responsible for this activity in the United States are the U.S. Army Corps of Engineers, the Bureau of Reclamation, the U.S. Forest Service and the Bureau of Land Management. Currently there are several federally managed adaptive management efforts across a range of ecosystem types. Federal agencies are organizations created by Congress with specific goals. Inherent in each agency's enabling legislation is a unique mission. The federal system includes a multilayered system of accountability which overlays the mission and any process pursued by an agency. This system of accountability heavily influences the ability of agencies' flexibility in the process. The system is a hierarchical one of checks and balances via the judicial and legislative branches of government. Within each agency is its own system of accountability, including an office of the Inspector General, statutes, regulations and agency cultural norms and culturally related decision-maker performance measures.

This structure influences federal agencies' adaptive management implementation and flexibility. The current system rewards rule following, a focus on short term deadlines and incremental project component completion. Thus the guide for management action can be characterized by time scales that are shorter than appropriate for natural system restoration, i.e., the need to produce results on a time frame that may more nearly match that of an election cycle and is based on congressional expectations of high levels of certainty in results. This position is supported by statements by two practitioners: "Congress wants a lot of certainty that we're going to get certain benefits and costs and not affect these constraints [referring to limiting impacts to current interests] before they approve something,…which takes longer," and "No one is getting held responsible for the long term condition of the outcome of the project. The accountability system is based on a quarter or a year, not the time frame for adaptive management projects. No incentive to be accountable for the long term."

One of the interviewee's statements characterizes one of the problems related to creating an agency environment for innovation, which is necessary to manage for resilience: "Change or innovation is not rewarded because so many interest groups are challenging them. They are challenged by stakeholders using the same accountability system by which the agencies are guided." Additional interviewees characterize the influence of the accountability system on the ability to shift to a new mission oriented focus: "If management doesn't have something as a measure in their incentives, it isn't going to happen", "[a]gencies rarely measure the health of the ecosystem", "your measures should match up with what you're trying to achieve in the ecosystem", and "if you have some other measure for Congress, you're not going to achieve your ecosystem objectives."

The challenge in the past was for the agencies to accept that adaptive management was a form of management that is useful and could meet their needs and replace their existing approaches. Non-governmental organizations initially spent considerable time trying to convince agencies that adaptive management should be employed on their restoration projects. Since those initial efforts, agencies have moved toward adopting adaptive management and have officially incorporated it into the accountability system, e.g., the Department of Interior Technical Guide (Williams et al. 2009). They have also developed training to roll out their approach to adaptive management implementation. Implementation is still in the development stage, as indicated by the quote from one of the practitioners interviewed: "The Feds have to figure out systemically, administratively and legally how to do this more effectively if they're committed to it." Once a new federal policy is adopted it can bring large scale change in thinking and policy quite quickly across the country. The publication of the Department of Interior Adaptive Management Technical Guide (Williams et al. 2009) led to almost immediate adoption of adaptive management approaches to resource management across agencies within the Department of Interior.

Non-Governmental Organizations

Non-governmental organizations have been involved in adaptive management efforts for as long as agencies have. They cover a wide range of adaptive management interests, organizational size and complexity, and sophistication levels and types of knowledge. They generally view an adaptive management project from a social-ecological orientation and thus not only address natural systems problems but also include social aspects. Non-governmental organizations originally began their participation in adaptive management processes because federal and state agencies left critical gaps in their approach and non-governmental organizations had the interest, knowledge, expertise and experience to fill those gaps. Specifically, non-governmental organizations stepped in to fill these gaps by supplementing natural and social science expertise not found in the agencies. Example areas included collaborative processes experience, systematic approaches to ecosystem and watershed scale assessment, and management and species specific knowledge and experience. Their ability to fill these gaps was enhanced by their relative organizational flexibility and nimbleness as compared to their agency counterparts.

The Las Cienegas National Conservation Area is representative of this type of effort. The Nature Conservancy, individual ranchers and the Bureau of Land Management have worked together to restore and manage this area using adaptive management. The types of knowledge and thinking that The Nature Conservancy has brought to the effort are a deep and sophisticated technical understanding of the systems, and systematic approaches to management, while individual ranchers have contributed sophisticated thinking and knowledge about local, social and agricultural systems.

The agency-non-governmental organization linkage also provides an opportunity to fill additional gaps, especially ones that agency personnel cannot easily fill. Non-governmental organizations can supplement expertise that is not found within the agencies, can provide a safe-alternative place to talk about problems, can be a source of adaptive management education, can provide a nexus for stakeholder collaboration, and can provide independent science advice and review. Additionally, non-governmental organizations can act as a team builder, trusted convener and honest broker to bring practitioner expertise and knowledge together. Non-governmental organizations can help bring credibility to the process because, as one practitioner pointed out: "They also can be important in achieving credibility for the process with the public, because they do not suffer the stigma associated with government-sponsored organizations." They bring fresh thinking to the process and have a unique ability to push on issues at higher decision-making levels because they are not restricted, as agencies are, by rules prohibiting political activity and lobbying. On the other hand, agencies have broad influence at a large scale, as indicated by this interviewee: "DOI guidebook had a lot of impact…when it comes from the Feds it has a lot of impact. It codifies and normalizes this kind of approach in a way that [is] exemplary to the rest of the practitioner community."

The Network Approach to Adaptive Management

Networks provide an avenue for bringing together entities that are influential over the components of adaptive governance, science, policy and decision making; helping bureaucracies shift from a scientific management orientation to adaptive governance. The Kissimmee River successes indirectly resulted in others' recognizing the utility of the approach and the usefulness of forming informal networks to solve problems. Based on that model, organizations such as the Collaborative Adaptive Management Network and the Adaptive Management Conference Series arose. The membership of these groups consists of agency and non-governmental organization personnel. Initially, organizations like the Collaborative Adaptive Management Network were focused on encouraging agencies to officially adopt adaptive management as their approach to the management of complex problems.

As agencies shifted their approach to adaptive management, networks have shifted their focus to other aspects of practice, such as implementation. Over time agencies and non-governmental organizations have developed a symbiotic working relationship through networks. In this context, networks provide a forum by which agency and non-agency adaptive management practitioners can exchange information about such topics as current agency thinking and practice which allows non-governmental organizations to provide information about appropriate indicators, measures, incentives and goals for agencies. Non-governmental organizations also act as team builders and conveners to bring practitioner expertise and knowledge together.

There is an opportunity to make more use of the agency-non-governmental organization linkage to examine additional possibilities for taking advantage of each organizational type's strengths for adaptive management problem solving and advancement. In other words, organizations like the Collaborative Adaptive Management Network should develop strategies within its membership to facilitate positive influences toward further development of adaptive management in agency culture. Meanwhile non-governmental organizations should work toward developing their capacity to network with and influence high level decision makers with the goal of accelerating the education, evolution and acceptance of adaptive management implementation in agencies. This includes the further development and understanding within agencies about what it means to manage for resilience.

There are tools and levers available that can be used to further help networks facilitate a shift towards change and managing for resilience. For example, there is a U.S. Office of Management and Budget initiative focusing on evidence-based decision making, which suggests that agencies have incentive and will be rewarded for using an evidence-based approach to management and decision making. This supports agencies' use of adaptive management experimentation in their management approach. For non-governmental organizations to help accelerate the acceptance of managing for resilience, they will have to further develop their capacity to influence high-level decision makers who then influence agencies towards more complete adaptive management development and implementation. Networks also need to develop capacity to provide education and training beyond what the federal government already provides to educate current and future practitioners, as one practitioner notes: "education is key for getting the lessons from implementing adaptive management by others."

Networks of Practice: What Has Collaborative Adaptive Management Network Learned?

The Collaborative Adaptive Management Network acts as a hub for Agency and non-governmental organization learning and information exchange through meetings in which participants visit selected Collaborative Adaptive Management Network projects. In addition, the Collaborative Adaptive Management Network provides a web presence, disseminating summaries of lessons learned from its gatherings and acts as a training and advising resource on an ad hoc basis for practitioners and projects. As one practitioner notes: "...integration of new ideas and adoption of new processes likely will be better served by encouraging the informal organizations [such as Collaborative Adaptive Management Network] to serve as hubs of idea development and communication of those ideas to the formal organizations [e.g., Army Corps, Bureau of Reclamation]." Collaborative Adaptive Management Network participants include a range of stakeholders, such as federal agency personnel, non-governmental organizations with collaboration and natural resource expertise, state conservation agencies and universities. Participation includes personnel from a broad range of Collaborative Adaptive Management Network efforts including the Glen Canyon Dam Adaptive Management Program, the Comprehensive Everglades Restoration Program, the Missouri River Recovery Program, the Malpai Borderlands Restoration Project, the Las Cienegas Landscape Conservation Area, and others. The broad range of agency and non-agency representation provides a similarly broad range in experiential knowledge and expertise.

Role in Resilience of Agencies

Similar to ecological systems, most organizational systems are dynamic, i.e., they need to adapt to changing contextual conditions to function well. For agencies to shift from managing nature-as-static to managing under uncertainty and for resilience, one first needs to understand what it means to do so. None of the interviewees claimed that their organization or the adaptive management effort on which they were working was managing for resilience. Additionally, there was no consistent definition of resilience among the interviewees.

Resilience is a measure of the amount of change a system can absorb before the system reorganizes into an alternate regime characterized by a different set of processes and structures. Agencies attempted to make a shift to adaptive management, but without explicitly taking resilience into account. For example, many federal agencies shifted from management of natural resources based upon nature-as-static, to using adaptive management. Lee's (1999) statement that "adaptive management has been much more influential as an idea than as a way of doing conservation so far," still holds. Education and motivation are needed to spur agency changes to manage for resilience. Agency personnel need to be exposed to education from other practitioners about what can and has been done to manage for resilience. In addition, they need to have information about how to recognize, prepare for and develop potential responses to large magnitude social-ecological and organizational disturbances.

The Collaborative Adaptive Management Network provides this education to agency personnel and non-governmental organizations. Its activities also facilitate the infusion of resilience thinking into agencies through a number of avenues. One is by providing a safe place for agency personnel to be exposed to, learn about and exchange fresh ideas with agency and non-agency experts, and practitioners to which they might not normally have access. This interaction allows agency practitioners to bring new ideas to their agencies that influence and potentially advance practice from within, and vice versa.

Learning While Doing

The Collaborative Adaptive Management Network has learned that treating policies as experiments is not rewarded at federal agencies. If learning by doing is to be embraced, managers will need to have the practice incorporated as a performance measure, as indicated by this practitioner: "If management doesn't have something as a measure in their incentives, it isn't going to happen." Another difficulty alluded to previously is that learning while doing requires innovation and change and there are too many interests that do not trust government agencies to allow them to make changes to policy very easily: "Change or innovation is not rewarded in Fed agencies because so many interest groups are challenging them." There are legal barriers, such as lawsuits, that prevent policy change. The good news is that networks such as the Collaborative Adaptive Management Network provide an opportunity for agencies and non-governmental organizations to work together to strategize for change. This approach can provide an avenue around particular barriers for advancing adaptive management and managing for resilience.

Because Collaborative Adaptive Management Network participants are practitioners from a wide variety of projects, it provides an opportunity to learn while working. Agency participants are learning while practicing adaptive management on their respective projects. As one practitioner notes: "…[I]ntegration of new ideas and adoption of new processes likely will be better served by encouraging…[non-governmental organizations]… serv[ing] as hubs of idea development and [then agency participants] communicati[ng]…those ideas to the formal organizations, e.g., Army Corps, etc." The network also provides other avenues of sharing and learning through electronic communication venues, exposure to new thinking from academics through educational training at conferences like the Conference on Ecological and Ecosystem Restoration, sharing of experiences and providing a source of experience.

Sharing Experience

Adaptive management and adaptive governance are evolving. They are endeavors that evolve based on practitioner experience. Coupling experience with theory is important to keep the development of practice moving forward. The Collaborative Adaptive Management Network has developed and is in the process of developing additional methods of sharing practitioner experience. The three main modes of experiential sharing employed to date have been: (1) convening practitioner network gatherings; (2) conducting ad hoc consultations for particular adaptive management projects; and (3) passing on experience through the publication.

Practitioner gatherings through the Collaborative Adaptive Management Network have occurred at the Cal-Fed Bay Delta project, the Las Cienegas Landscape Management Area, the National Elk Wildlife Refuge, the Upper Mississippi River,

Rocky Mountain National Park, Puget Sound, the South Platte River Restoration Project and others. These experiences allow participants to learn about alternative approaches to solving complex problems, and also allows practitioners from other projects to learn from practice approaches used at these project locations. The settings provide an opportunity for agency practitioners to discuss implementation difficulties for their project safely outside of their organizations. It also provides non-agency participants an opportunity to learn about current internal agency practice. Ideally this provides an opportunity to gather information that can be used to push for change at higher agency levels.

Conclusion

A theme of this chapter is defined by the question: What is necessary for adaptive management decision makers to manage for resilience? In response, this chapter is focused on adaptive management organizations. The organizations were parsed into two general categories, federal agencies and non-governmental organizations. Non-governmental organizations involvement in adaptive management arose because they filled a gap in adaptive management education and implementation. As time has passed and adaptive management has developed, practitioners have come to view the two types of organizations as necessary and having a symbiotic relationship, each playing important roles in the development and implementation of adaptive management. Interviews of a cross-section of individuals from the two types of organizations did not yield a consensus about a definition for resilience and what it meant to manage for it. Moreover, none of the organizations involved claimed that they were managing for resilience.

Key to understanding what is needed for decision makers to manage for resilience in the future is to first comprehend the current state of practice and where gaps exist. Currently, federal agencies manage from a static systems model where the concept of stationarity holds; nature is static and historical data can be used as a predictor of the future. Federal agency decision making, accountability and incentives are not geared toward rewarding innovation and dynamic change in processes and policies. The current system rewards meeting short-term deadlines and incremental project component completion. External political forces and perceived legal risks associated with expedited policy change act as barriers to timely federal action.

To manage for resilience is to accept that the concept of stationarity does not hold and one must use policies and management approaches that can accommodate uncertainty. Movement toward acceptance of, and methods to, address uncertainty as part of the natural resources management landscape is surfacing in federal discussions regarding climate change. Thus, there is a need to have this dialogue broadened and accelerated. Non-governmental organizations can help influence federal agencies to introduce and accept new concepts and practices.

The gaps which the non-governmental organizations have filled have shifted over time as adaptive management acceptance and implementation has shifted. As adaptive management has become more common, federal agencies have begun to codify established adaptive management approaches. As these changes have taken place, non-governmental organizations have shifted their focus to other aspects of adaptive management, such as implementation. Although federal agencies have a difficult time moving at speed in implementing new policies and approaches, they obviously have a strong ability to influence policy.

Non-governmental organizations are more nimble and can work on aspects of adaptive management that federal agencies cannot. Non-governmental organizations are able to push for change without as many constraints as federal agencies. Non-governmental organizations such as the Collaborative Adaptive Management Network can also provide a safe forum for practitioners (both federal and non-federal) to explore new adaptive management ideas and policy adjustments (Pratt-Miles 2013), and can help bring credibility to adaptive management and fresh thinking to the process (Smedstad and Gosnell 2013). Organizations like the Collaborative Adaptive Management Network act as a hub of information where agency stakeholders can gain new knowledge.

Much progress has been made toward adaptive management implementation by both federal agencies and non-governmental organizations. For federal agencies to manage for resilience effectively, resilience will need to be defined as a management objective. Agencies will also need to determine how to systematically, administratively and legally implement adaptive management more effectively. Non-governmental organizations can advocate for adaptive management, but need to focus on increasing their capacity for influence. More education is needed at all levels of government. Governments expect certainty about what benefits and costs are associated with employing adaptive management and managing for resilience, while non-governmental organizations need to continue to press the case for using adaptive management methods to manage in an increasingly uncertain world.

References

Allen, C. R., & Gunderson, L. H. (2011). Pathology and failure in the design and implementation of adaptive management. *Journal of Environmental Management, 92,* 1379–1384.

Benson, M. H. (2012). Intelligent tinkering: The endangered species act and resilience. *Ecology and Society, 17*(4), 28. http://dx.doi.org/10.5751/ES-05116-170428.

Benson, M. H., & Garmestani, A. S. (2011). Embracing panarchy, building resilience and integrating adaptive management through a rebirth of the National Environmental Policy Act. *Journal of Environmental Management, 92,* 1420–1427.

Benson, M. H., & Stone, A. B. (2013). Practitioner perceptions of adaptive management implementation in the United States. *Ecology and Society, 18*(3), 32. http://dx.doi.org/10.5751/ES-05613-180332.

Brand, F. S., & Jax, K. (2007). Focusing the meaning(s) of resilience: Resilience as a descriptive concept and a boundary object. *Ecology and Society, 12*(1), 23. http://www.ecologyandsociety.org/vol12/iss1/art23/.

Brock, W. A., & Carpenter, S. R. (2010). Interacting regime shifts in ecosystems: Implication for early warnings. *Ecological Monographs, 80,* 353–367.

Brunner, R. D., Steelman, T. D., Coe-Juell, L., Cromley, C. M., Edwards, C. M., & Tucker, D. W. (2006). *Adaptive governance: Integrating science policy and decision making*. New York: Columbia University Press.

Chapin, F. S., Kofinas, G., & Folke, C. (2009). *Principles of ecosystem stewardship*. New York: Springer.

Folke, C., Carpenter, S., Walker, B., Scheffer, M., Elmqvist, T., Gunderson, L., & Holling, C. S. (2004). Regime shifts, resilience, and biodiversity in ecosystem management. *Annual Review of Ecology Evolution and Systematics, 35,* 557–581.

Folke, C., Hahn, T., Olsson, P., & Norberg, J. (2005). Adaptive governance of social-ecological systems. *Annual Review of Environment and Resources, 30,* 441–473.

Gunderson, L. (1999). Resilience, flexibility and adaptive management—antidotes for spurious certitude? *Conservation Ecology, 3*(1), 7. http://www.consecol.org/vol3/iss1/art7/.

Gunderson, L., & Light, S. S. (2006). Adaptive management and adaptive governance in the everglades ecosystem. *Policy Sciences, 39,* 323–334.

Gunderson, L., & Pritchard, R. (2002). *Resilience and the behavior of large scale systems*. Washington, D.C.: Island Press.

Holling, C. S. (1973). Resilience and stability of ecological systems. *Annual Review of Ecology and Systematics, 4,* 1–23.

Holling, C. S. (1978). *Adaptive environment assessment and management*. London: Wiley.

Johnson, B. L. (1999). The role of adaptive management as an operational approach for resource management agencies. *Conservation Ecology, 3*(2), 8. http://www.consecol.org/vol3/iss2/art8/.

Lee, K. N. (1993). *Compass and gyroscope*. Washington, D.C.: Island Press.

Lee, K. N. (1999). Appraising adaptive management. *Conservation Ecology, 3*(2), 3. http://www.consecol.org/vol3/iss2/art3/.

Levin, S. A., & Lubchenco, J. (2008). Resilience, robustness, and marine ecosystem-based management. *BioScience, 58,* 27–32.

Peterson, G. D., Cumming, G. S., & Carpenter, S. R. (2003). Scenario planning: A tool for conservation in an uncertain world. *Conservation Biology, 17,* 358–366.

Pratt-Miles, J. D. (2013). Designing collaborative processes for adaptive management: Four structures for multistakeholder collaboration. *Ecology and Society, 18*(4), 5. http://dx.doi.org/10.5751/ES-05709-180405.

Scarlett, L. (2013). Collaborative adaptive management: challenges and opportunities. *Ecology and Society, 18*(3), 26. http://dx.doi.org/10.5751/ES-05762-180326.

Scheffer, M., Carpenter, S., Foley, J. A., Folke, C., & Walker, B. (2001). Catastrophic shifts in ecosystems. *Nature, 413,* 591–596.

Smedstad, J. A., & Gosnell, H. (2013). Do adaptive comanagement processes lead to adaptive comanagement outcomes? A multicase study of long-term outcomes associated with the National Riparian Service Team's place-based riparian assistance. *Ecology and Society, 18*(4), 8. http://dx.doi.org/10.5751/ES-05793-180408.

Walker, B., Carpenter, S., Anderies, J., Abel, N., Cumming, G. S., Janssen, M., Lebel, L., Norberg, J., Peterson, G. D., & Pritchard, R. (2002). Resilience management in social-ecological systems: A working hypothesis for a participatory approach. *Conservation Ecology, 6*(1), 14. http://www.consecol.org/vol6/iss1/art14/.

Walters, C. J. (1986). *Adaptive management of renewable resources*. New York: McGraw Hill.

Walters, C. (1997). Challenges in adaptive management of riparian and coastal ecosystems. *Conservation Ecology, 1*(2), 1. http://www.consecol.org/vol1/iss2/art1/.

Walters, C. J. (2007). Is adaptive management helping to solve fisheries problems? *Ambio, 36,* 307–307.

Walters, C., Gunderson, L., & Holling, C. S. (1992). Experimental policies for water management in the Everglades. *Ecological Applications, 2,* 189–202.

Westley, F. (2002). The devil is in the dynamic. In L. H. Gunderson & C. S. Holling (Eds.), *Panarchy* (pp. 333–360). Washington, D.C.: Island Press.

Williams, B. K., Szaro, R. C., & Shapiro, C. D. (2009). *Adaptive management: The U.S. department of the interior technical guide*. Washington, D.C.: Adaptive Management Working Group, U.S. Department of the Interior.

Zellmer, S., & Gunderson, L. (2009). Why resilience may not always be a good thing: Lessons in ecosystem restoration from Glen Canyon and the Everglades. *Nebraska Law Review, 87*, 894–947.

Chapter 12
Optimization and Resilience in Natural Resources Management

Byron K. Williams and Fred A. Johnson

Keywords Resilience · Optimization · Adaptive management · Uncertainty · Natural resources management

Introduction

Over the last several decades optimization methods for system control and management have been applied with increasing frequency to the assessment and management of natural resources (Shenk and Franklin 2001, Williams et al. 2002, Simonovic 2009, Kangas et al. 2010). The generic problem motivating these methods involves objective-based decision making, in which performance objectives are used to compare and contrast the consequences of alternative actions so that optimal actions can be identified. Under very general conditions on the structure of the resource system and objective function (Sniedovich 2011), optimization can be usefully applied to the management of dynamic natural resources. Optimization theory and methodology continue to be rapidly growing areas of research and practice in natural resources, especially as concerns the incorporation of uncertainty in optimal decision making (Williams et al. 2002).

Yet despite the sophistication of optimization theory and the extraordinary range of its applications (Stengel 1994, Hull 2003, Bertsekas 2007), there is a growing sentiment among some ecologists that optimization in natural resources and ecology is fundamentally detrimental to long-term sustainable management of ecosystems (Walker and Salt 2006). The idea seems to be that the very principle of optimization is antithetical to resilient and sustainable ecosystems. The argument has been made repeatedly that optimization practiced over an extended time can reduce resilience

B. K. Williams (✉)
The Wildlife Society, 5410 Grosvenor Lane,
Suite 200, Bethesda, MD 20814-2144, USA
e-mail: ken.williams@wildlife.org

F. A. Johnson
U.S. Geological Survey, Southeast Ecological Science Center,
7920 NW 71 Street, Gainesville, FL 32653, USA
e-mail: fjohnson@usgs.gov

C. R. Allen, A. S. Garmestani (eds.), *Adaptive Management of Social-Ecological Systems*,
DOI 10.1007/978-94-017-9682-8_12

and thereby set an ecosystem up for crisis, collapse and reconfiguration (e.g., Holling and Meffe 1996, Carpenter et al. 2002, Peterson et al. 2003).

The difficulty with this argument is that it fails to account for the structure of systems subjected to human interventions and the imperative to manage them. It seems obvious that with managed natural systems there must be a principle by which to guide decision making, which at a minimum allows for a comparison among decision alternatives. If decision making and management are to be based on more than intuition or chance, the decision process must include some criterion for measuring the relative value of management alternatives, and a mechanism for selecting among them. This is true even if the primary concern of decision making is the preservation of system resilience.

In fact, we claim that it is neither optimization nor optimal management per se that is the enemy of system resilience and sustainability; rather, it is the focus on equilibrium conditions, as is often associated with familiar strategies such as maximum sustainable exploitation (Walker and Salt 2006). Indeed, optimization can be applied as effectively to sustain ecosystems as it can be to maximally exploit them. We argue here that in the face of deep uncertainties about resource states and functions, it may well be appropriate to guide decision making so as to minimize the risk of long-term damage to ecological processes. Such a management approach, which potentially sacrifices short-term gains that could otherwise be extracted, is nonetheless "optimal," albeit with a different objective.

Optimization and Resilience Frameworks

It is useful to consider commonalities between the frameworks for optimization and resilience. Optimization focuses on approaches to decision making, i.e., the choice of an action (or strategy) from some set of feasible alternatives in order to maximize (or minimize) an objective function. Decisions typically account for potential resource consequences and constraints on the alternatives and their consequences. In conservation of natural resources, optimization is often thought of as a "tool" for making defensible decisions for a well-defined decision problem. Elements include (*i*) objective function(s) that incorporate stakeholder values and desires, (*ii*) decision alternatives, (*iii*) response model(s), (*iv*) constraints on alternatives and allowable consequences, (*v*) uncertainties, and (*vi*) stakeholder engagement (Williams 2001, Williams et al. 2002).

On the other hand, resilience focuses on the capacity of a system to absorb disturbance and retain essentially the same structure, function, and identity (Walker et al. 2004). It emphasizes the interplay of gradual change and sudden disturbance, feedbacks, nonlinear behaviors, and adaptive cycles (Fischer et al. 2009). Resilience thinking promotes analysis of social-ecological systems, with key elements that include: (*i*) coupled social-ecological systems, (*ii*) cross-scale relationships, (*iii*) non-linearities (including multiple stable states, thresholds, hysteresis, adaptive cycles), (*iv*) uncertainties, and (*v*) stakeholder engagement.

There are recognizable complementarities between these two frameworks. For example, both emphasize stakeholder involvement and the importance of uncertainties in natural resource conservation. Cross-scale relationships and nonlinearities in resilience can factor into optimization in the articulation of a response model. Social-ecological factors in resilience thinking can be incorporated into the response model, or included as constraints in an optimization framework. With one exception, the features of resilience thinking can be seen to have a natural place in the optimization framework, and features of optimization have a natural place in the resilience framework.

The one feature for which there appears not to be a complementary analogue is the objective function of optimization. On reflection this makes sense: the heart of optimization is an expression of value that attends decision making, whereas the emphasis of resilience is on the analysis of patterns of change in coupled systems. While we recognize that resilience thinking is a useful perspective for analyzing complex systems (Fischer et al. 2009), it is important to stress that it does not obviate the need for managers to compare alternatives and make decisions.

In what follows, we employ a widely used framework for decision making (Stengel 1994, Williams et al. 2002), in which a resource system is characterized by a set of attributes (system parameters or states, rates of change, process functions, aggregations of these variables, other features) at one or more times. An action taken at a particular time generates an immediate response (e.g., costs and returns of some kind) and (possibly) a longer-term response through its influence on future system behavior. Returns and/or costs are generalized into utility to capture differential preferences of the decision maker, and utilities (or utility moments) in turn are aggregated across time into a function that guides decision making. Here utility is held to measure preference for decision-specific outcomes and value aggregates those (possibly uncertain) utilities for a strategy composed of one or more decisions. The function expressing the influence of actions on aggregate value is referred to as a value or objective function.

We use robust decision making to consider the linkage of optimization and resilience, recognizing that robustness and resilience are not identical concepts. However, we argue that they are closely related, especially in the context of dynamic, long-term decision making. Robust decision making is taken here to focus on sustaining a minimal level of value (i.e., accumulated utility). As above, resilience measures the ability of a system to sustain shocks and retain structure, function, and identity. The basic idea of resilience thinking is that decisions failing to account for key structural features can reduce resilience, and thereby lead to system changes beyond a critical threshold that can fundamentally alter the system and its productive capacity. In that sense, decision making that neglects resilience is not robust, since it fails to sustain a system and its productivity through time. Furthermore, decision making that is robust in sustaining productivity through time must retain the resilience to do so. In the context of dynamic decision making, it is hard to think about either attribute in the absence of the other. A management strategy that sustains an acceptable level of productivity over time must be resilient in order to do so; and resilient systems, by sustaining structure, functions, and identity, retain the productive potential of that system. We build on these associations to argue that robust decision making, and thus optimization, is not automatically at odds with resilience.

Decision Making and Uncertainty

Here we consider decisions with potentially uncertain consequences, building on the framework of structured decision making (Gregory and Keeney 2002, Runge 2011) for one-time decisions, and adaptive management (Walters 1986, Williams et al. 2002) for dynamic decision making in the face of uncertainty. We also consider robust decision making under deep uncertainties, utilizing the work of Regan et al. (2005), Ben-Haim (2006) and Lempert and Collins (2007). We focus on two types of uncertainty, namely partial observability and structural uncertainty. Partial observability refers to an inability to fully observe a system, so that state-dependent decisions must be made under uncertainty as to the actual state of the system at the time the decision is made. Uncertainty about system state is represented by a belief state that describes the strength of belief one has that the system is in each of its possible states. The belief state is essentially a distribution of state-specific probabilities, and it may or may not be recognized.

The second form of uncertainty that is considered here is structural uncertainty, which refers to a situation in which the processes that determine resource dynamics and the production of utilities are not fully understood. Uncertainty about process structure is represented by a set of process models and a model state that expresses the confidence one has in each of them. The model state is a time-dependent distribution of model-specific probabilities that may or may not be known.

In what follows we describe in detail a family of decision problems with varying levels of uncertainty, and show that different formulations of optimal decision making are appropriate within this family. We begin with the relatively simple problem of making one-time decisions, and consider variations of the kinds and amounts of problem uncertainty. We then expand the decision context to the more complex problem of iterative decision making. It will be seen that the structure of the decision process varies considerably depending on the timeframe being considered, and the amount and quality of information that is available to guide decisions. Importantly, in the absence of any information to characterize key system features (even probabilistically) and to project the consequences of potential decisions, the decision making process can differ fundamentally in its objectives and decision approaches.

One-time Decision Making

We initially consider a problem involving one-time decision making, i.e., a single decision is to be made and a single utility is produced that is tied to the decision. Let x represent the state of the system, and a represent the action taken. Assume for now that the set X of possible states is finite, as is the set A of potential actions. A utility $U(a \mid x)$ is produced with action a when the system is in state x. The expression for utility can take many forms. Examples include a function of system state $\left(\text{e.g., } U(a \mid x) = f(x)\right)$, a product of the state and action $\left(\text{e.g., } U(a \mid x) = g(ax)\right)$, or a measure of deviation from a critical state value $\left(\text{e.g., } U(a \mid x) = r(a)h[x - x_0]^2\right)$. In particular, the utility function $U(a \mid x)$ can readily incorporate system stability domains and thresholds that are the focus of ecological resilience. Note that this framework can readily absorb multi-criteria decision making by allowing the utility function to be a single expression of weighted outcomes across multiple objectives.

Known Process, Observable State

If x is fully observable and the utility model is known with certainty, the value function $V(a \mid x) = U(a \mid x)$ can be used to compare and contrast actions and guide decision making. The identification of an optimal action is a simple matter of comparison among options: Choose a to maximize $V(a \mid x)$. Because the number of options is finite, this amounts to no more than selecting the largest utility from a finite list of action-specific utilities. Since the state space is finite, optimal actions can be identified for all states in finite time.

Information requirements for this problem are not extensive, but nevertheless can be problematic. One must be able to specify the utility function $U(a \mid x)$, since it is on the basis of utilities that actions are to be compared and prioritized. In addition, one must be able to identify a set of feasible and acceptable alternative actions, and provide an accurate determination of system state. Each of these requirements can pose a serious challenge to decision making. The identification of system processes and states can be especially challenging for ecosystem management, given that ecological systems are almost never fully understood and observed (see below).

Known Process, Unobservable State

Partially Observable State. Often the state of a resource system is not known, but a belief state b is. Such may be the case when, e.g., probability-based monitoring of system state produces estimates of moments that can be used in constructing a distribution of state values. Under these circumstances a useful form of the value function averages utilities using the distribution probabilities $b(x)$ for each of the possible states:

$$\bar{V}(a \mid b) = \sum_x b(x)V(a \mid x).$$

Because specification of a belief state is required for the value function, this variable appears in the expression for expected value. Assuming a known belief state

b, an appropriate strategy is to maximize expected utility. The identification of an optimal action is a simple matter of comparison among options:

$$\text{Choose } a \text{ to maximize } \bar{V}(a \mid b) = \sum_x b(x)V(a \mid x).$$

Again, this amounts to the selection of the largest value from a finite list of action-specific expected utilities. The problem, of course, is that the space of all possible belief states is a continuous simplex, so it is not possible to enumerate such a list of values for every belief state over the entire belief space. A good deal of research has been done on ways to identify optimal or near-optimal strategies over the whole belief space, under the rubric of partially observable Markov decision processes or POMDPs (see Williams 2011b).

Deeply Uncertain State. Now assume that neither the system state nor the belief state is known. Such a situation might correspond to the lack of any observation data, or a monitoring protocol that is flawed in some unrecognizable and/or uncorrectable way. Then it no longer is meaningful to maximize an average of utilities, because there is no known distribution on which to base the averaging. A different criterion is needed to guide decision making.

One such candidate is "good enough" or robust decision making. Here the idea is not to maximize a measure of utility, but rather to produce values exceeding some specified lower limit V_c over as large a range of belief states as possible. Said differently, the intent of robust decision making is to choose an action that will maximize the range of belief states for which the expected utility for every belief state in that range will be "good enough." This shifts the focus from maximizing expected utility to maximizing coverage of "good enough" utility. One seeks the greatest extent of system states for which a minimal performance requirement is met, by employing a two-step process: (i) for each action, a region is sought in a parameter or state space over which some minimal value is sustained, and then (ii) the action maximizing the regional coverage is selected.

More formally, robust decision making is defined in terms of a range of belief states, which in turn is specified for a belief state $\tilde{b}$ in terms of a parameter α, called the uncertainty horizon:

$$\begin{aligned} &R(\alpha, \tilde{b}) \\ &= \left\{ b : \sum_x b(x) = 1 \text{ and } \max\left[0, (1-\alpha)\tilde{b}(x)\right] \le b(x) \le \min\left[1, (1+\alpha)\tilde{b}(x)\right] \forall x \in X \right\}. \end{aligned}$$

The belief state $\tilde{b}$ roughly plays the role of a location parameter, and α plays the role of a shape or spread parameter. The range essentially specifies a set of belief states located around $\tilde{b}$, with an extent given by α. It is clear that more belief states are included in a range corresponding to a larger uncertainty horizon α. A key question is how large α should be.

Robust decision making is framed in terms of an action-specific "robustness function" that incorporates a range of belief states and a performance measure V_c. Thus, for a given action one seeks the largest uncertainty horizon α for which $\bar{V}(a \mid b)$ is greater than the critical value V_c for every belief state in $R(\alpha, \tilde{b})$. The robustness function $\hat{\alpha}(a \mid V_c, \tilde{b})$ gives the uncertainty horizon identified by this maximization:

$$\hat{\alpha}(a \mid V_c, \tilde{b}) = \arg\max_{\alpha} \left[\min_{b \in R(\alpha, \tilde{b})} \bar{V}(a \mid b) \geq V_c \right].$$

Robust decision making is then defined for a given critical value V_c and guesstimate $\tilde{b}$ by the selection of the action with the largest uncertainty horizon given by the robustness function: Choose a to maximize $\hat{\alpha}(a \mid V_c, \tilde{b})$.

The form of this maximization criterion makes it clear that robust decision making focuses on maximizing the reliability or "robustness" of expected utility, rather than the expected utility itself. The range of reliability declines as the performance requirement increases; conversely, the performance requirement must decline to obtain an expanded range of reliability. At one extreme the range of reliability shrinks to a single belief state $\tilde{b}$ as the performance criterion converges to the optimal expected utility $V(a^* \mid \tilde{b})$. At the other extreme, the range of reliability expands to include the entire belief space as the performance criterion shrinks to 0.

Because belief space is continuous, it is not possible to conduct such an assessment for every belief state $\tilde{b}$ individually. One way to avoid this problem is to seek a robust decision at only one belief state $\tilde{b}$ that is identified *a priori*. Another is to select a finite set of belief states, identify a robustness function for each, and use the results to make inferences to the remainder of the belief space.

Unknown Process, Observable State

Structurally Uncertain Process. In a context of one-time decision making, "structural uncertainty" is expressed as a lack of certainty about the processes that lead to production of utilities. One way to represent structural uncertainty with finite-state, finite-action decision processes is to recognize variation in the utilities $U(a \mid x)$ within the actions (Regan et al. 2005). Another is to assume a set of K process models that produce the utilities by $U_k(a \mid x)$. We use the latter approach here.

For now, system state is assumed to be known but the process that produces utilities is not. Structural uncertainty is expressed in terms of K process models and a model state q that assigns probability $q(k)$ to model k. The value function averages model-specific utilities $U_k(a \mid x)$ over the model state to obtain an expected value function:

$$\bar{V}(a \mid x, q) = \sum_k q(k) U_k(a \mid x).$$

Because specification of both a system state and model state are required for the value function, both conditioning variables appear in the expression for expected value.

Assuming system observability and a known model state q, the identification of an optimal action is a simple matter of comparison among options: Choose a to maximize $\bar{V}(a \mid x,q) = \sum_k q(k)V_k(a \mid x)$. Again, this amounts to the selection of the largest value from a finite list of action-specific expected utilities. The problem, of course, is that the space of all possible model states is a continuous simplex, so it is not possible to enumerate such a list of values for every model state over the entire model space. Williams (2011b) has shown how the approaches developed for POMDPs can be applied to the problem of structural uncertainty.

Deeply Uncertain Process. Just as it is possible with partially observable systems to shift the focus from maximizing expected utility to robust decision making, so is it possible to make such a shift with structurally uncertain systems. Assuming system state is observable but neither the appropriate model nor the likelihoods of the alternative models are known, it no longer is meaningful to maximize expected utilities over a model state. A reasonable alternative is robust decision making. In this case an uncertainty horizon and range of model states can be identified, following the same argument as above for belief states. Thus, the range of model states for an uncertainty horizon α and guesstimate $\tilde{q}$ of the model state is given by

$$R(\alpha,\tilde{q}) = \left\{q : \sum_k q(k) = 1 \text{ and } \max\left[0,(1-\alpha)\tilde{q}(k)\right] \le q(k) \le \min\left[1,(1+\alpha)\tilde{q}(k)\right], k = 1,\ldots,K\right\}.$$

Here one seeks the largest uncertainty horizon such that expected values for all model states in the associated range $R(\alpha,\tilde{q})$ exceed a minimum value V_c. As above, this condition can be specified by

$$\max_{\alpha}\left[\min_{q\in R(\alpha,\tilde{q})} \bar{V}(a \mid x,q) \ge V_c\right].$$

A robustness function $\hat{\alpha}(a \mid V_c,\tilde{q},x)$ identifies the uncertainty horizon identified by this maximization. Robust decision making then consists of choosing the action that maximizes this function.

Again, this form of decision making involves the replacement of a selection criterion based on maximizing expected utility with one based on maximizing the broadest possible range of minimally acceptable values of expected utility. Of course, the challenge of finding an optimal strategy over the whole model space is even greater than with partially observable systems, because a different strategy is required for every system state x.

Deeply Uncertain Process, Deeply Uncertain State

We also can consider a situation in which both forms of deep uncertainty are present. Williams (2009) discussed parameterizations and computing forms for iterative decision making with models that include structural uncertainty and partial observability. Here we consider robust decision making in the face of deep uncertainty about both factors.

In principle the approach is straightforward, in that it builds on the development above for each uncertainty factor. One can define a value function in terms of both structural uncertainty and partial observability, as

$$\bar{V}(a \mid b,q) = \sum_k \sum_x q(k)b(x)U_k(a \mid x).$$

Similarly, a range of belief and model states can be defined by

$$\begin{aligned} R(\alpha,\tilde{b},\tilde{q}) = \Big\{ & (b,q) : \sum_x b(x) = 1,\ \sum_k q(k) = 1, \\ & \max\left[0,(1-\alpha)\tilde{b}(x)\right] \le b(x) \le \min\left[1,(1+\alpha)\tilde{b}(x)\right] \forall x \in X, \\ & \max\left[0,(1-\alpha)\tilde{q}(k)\right] \le q(k) \le \min\left[1,(1+\alpha)\tilde{q}(k)\right], k = 1,\ldots,K \Big\}. \end{aligned}$$

The function $R(\alpha,\tilde{b},\tilde{q})$ gives a range for the joint occurrence of model and belief states around $(\tilde{b},\tilde{q})$, with an extent given by α. For a given action a the largest range of combined model and belief states with expected values greater than V_c is given by

$$\max_{\alpha}\left[\min_{(b,q)\in R(\alpha,\tilde{b},\tilde{q})} \bar{V}(a \mid b,q) \ge V_c\right].$$

Robust decision making consists of choosing the action that maximizes this function.

The requirement to incorporate both belief and model states in identifying the range of reliable coverage limits the extent of that coverage. Thus, the joint inclusion of deep uncertainties about system and model states effectively limits the range of reliable values for either uncertainty factor to be smaller than the range that would be identified for the factor considered alone.

It is clear from the above that structural uncertainties and partial observability impose additional challenges on the problem of optimal decision making, and the presence of deep uncertainties about either system state or system processes requires a fundamental restructuring of the decision problem. In particular, robust decision making in the face of deep uncertainty focuses on maximizing the reliability or "robustness" of expected utility, rather than the expected utility itself. As seen in the next section, the shift in emphasis away from maximizing a measure of expected utility occurs as well in a dynamic decision framework.

Dynamic Decision Making

In the development above decision making was restricted to a single decision, with a single utility tied to that decision. However, many natural resource problems involve iterative decision making, with decisions made at one time influencing system behavior (and therefore future decisions) from that time forward. In what follows we consider decision making over a timeframe of equal intervals, from some initial time 0 to some terminal time T. Because system states and actions are time-specific, we use x_t and a_t to denote these variables. Here we assume a Markov decision process (MDP), with time-specific utilities $U(a_t \mid x_t)$ and Markovian transition probabilities $P(x_{t+1} \mid x_t, a_t)$.

Known Process, Observable State

In the absence of partial observability and structural uncertainty, dynamic optimization typically seeks to maximize the expected sum of current utilities and future values. The value function in this case is the expected accumulation of utilities over the remainder of the timeframe:

$$V(A_t \mid x_t) = E\left[\sum_{\tau=t}^{T} U(a_\tau \mid x_\tau) \mid x_t\right]$$

where A_t is a strategy that specifies actions for each state at each time over the remainder of the timeframe. The challenge is to choose a strategy A_t^* that maximizes $V(A_t \mid x_t)$ for any initial state x_t:

$$\begin{aligned} V\left[x_t\right] &= \max_{A_t} V(A_t \mid x_t) \\ &= V(A_t^* \mid x_t). \end{aligned}$$

The value function can be expressed in iterative form as

$$\begin{aligned} V(A_t \mid x_t) &= U(a_t \mid x_t) + E\left[\sum_{\tau=t+1}^{T} U(a_\tau \mid x_\tau) \mid x_{t+1}\right] \\ &= U(a_t \mid x_t) + \sum\nolimits_{x_t+1} P(x_{t+1} \mid x_t, a_t) V(A_{t+1} \mid x_{t+1}), \end{aligned}$$

with an optimal value function

$$V\left[x_t\right] = \max_{a_t}\left\{U(a_t \mid x_t) + \sum\nolimits_{x_t} P(x_{t+1} \mid x_t, a_t) V\left[x_{t+1}\right]\right\}.$$

Dynamic programming (Bellman 1957, Bellman and Dreyfus 1962) offers a solution for this problem. Dynamic programming uses backward iteration starting at T to choose actions at each time that maximize accumulated utilities, based on the assumption that future actions are optimal.

Known Process, Unobservable State

Partially Observable State. We assume here that the structure of the process driving system dynamics is known but the system state is not observable. If x_t is not known but a distribution b_t for the system states is, a useful form of the value function averages utilities using the distribution probabilities $b(x_t)$ for each of the possible states:

$$\bar{V}(A_t \mid b_t) = \sum_x b(x_t) V(A_t \mid x_t).$$

The distribution evolves based on observations that are collected through time. Here we assume that observations are associated with, but not the same as, system states, with a conditional probability distribution

$$f(y \mid x, a)$$

that links observation y to system state x. As observations accumulate, the belief state evolves through time according to Bayes' theorem (Kaelbling et al. 1998, Williams 2009). The evolving belief states are used in computing the expected sum of current and future utilities

$$\bar{V}(A_t \mid b_t) \; = \; E\left[\sum_{\tau=t}^{T} U(a_\tau \mid b_\tau) \middle| b_t \right],$$

which can be expressed recursively in terms of current and future utilities by

$$\bar{V}(A_t \mid b_t) = \; U(a_t \mid b_t) + \sum_{b_{t+1}} \Pr(b_{t+1} \mid b_t, a_t) \bar{V}(A_{t+1} \mid b_{t+1}).$$

Optimal decision making is given by Bellman's equation for this expression, by

$$\bar{V}[b_t] = \max_{a_t} \left\{ U(a_t \mid b_t) + \sum_{b_{t+1}} \Pr(b_{t+1} \mid b_t, a_t) \bar{V}[b_{t+1}] \right\}$$

(Williams 2009).

As above, the resulting control and system trajectories are determined by the value function, the structure of the system being managed, and the degree and nature of the uncertainties to which it is subjected. In this context the idea of an unchanging "steady state" is not particularly useful or even meaningful. Nor is it reasonable to assume that optimal control would automatically engender the elimination of resilience. Limits on controllability and the presence of environmental variation act to keep a system out of equilibrium, and a closed-loop policy derived as above involves policy adaptation when the future unfolds differently than expected. This is very different from approaches that identify a trajectory of decisions to be applied irrespective of system behavior (e.g., stationary decision making).

Deeply Uncertain State. Consider the situation in which the system process is understood but neither the system state nor the belief state is known. A range $R(\alpha,\tilde{b}_t)$ can be defined that is centered on a guesstimate $\tilde{b}_t$ with extent given by an uncertainty horizon α, with the idea of identifying a value for α such that every belief state in $R(\alpha,\tilde{b}_t)$ will produce a value that exceeds some critical value V_c. This condition is specified by

$$\min_{b_t \in R(\alpha,\tilde{b}_t)} \left\{ U(a_t \mid b_t) + \sum_{b_{t+1}} \Pr(b_{t+1} \mid b_t, a_t) \bar{V}\left[b_{t+1}\right] \right\} \geq V_c.$$

The largest possible range that satisfies the condition is found by maximizing over the choice of α:

$$\hat{\alpha}(a_t \mid V_c, \tilde{b}_t) = \arg\max_{\alpha} \left[\min_{b_t \in R(\alpha,\tilde{b}_t)} \left\{ U(a_t \mid b_t) + \sum_{b_{t+1}} \Pr(b_{t+1} \mid b_t, a_t) \bar{V}\left[b_{t+1}\right] \right\} \geq V_c \right].$$

An action-specific robustness function $\hat{\alpha}(a \mid V_c, \tilde{b})$ gives the uncertainty horizon identified by this maximization. Robust decision making with uncertain belief state then is defined for a given critical value V_c and guesstimate $\tilde{b}_t$ by the selection of the action a_t with the largest uncertainty horizon produced by the robustness function:

Choose a_t to maximize $\hat{\alpha}(a_t \mid V_c, \tilde{b}_t)$.

Unknown Process, Observable State

Structurally Uncertain Process. We assume here that the system state is observable over the timeframe, but there is uncertainty about the structure of the process driving system dynamics. This process uncertainty is captured by a set of K process models $P_k(x_{t+1} \mid x_t, a_t)$ and an observable model state q_t that evolves through time according to Bayes' theorem (McCarthy 2007). For each model there is a value function

$$V_k(A_t \mid x_t) = U(a_t \mid x_t) + \sum\nolimits_{x_t} P_k(x_{t+1} \mid x_t, a_t) V_k(A_{t+1} \mid x_{t+1}),$$

and an overall process function averages these model-specific functions over the model state:

$$\begin{aligned} \bar{V}(A_t \mid x_t, q_t) &= \sum\nolimits_k q_t(k) V_k(A_t \mid x_t) \\ &= U(a_t \mid x_t) + \sum\nolimits_{x_t} \bar{P}(x_{t+1} \mid x_t, a_t, q_t) \bar{V}(A_{t+1} \mid x_{t+1}, q_{t+1}) \end{aligned}$$

(Williams 2009). Optimal decision making for this situation is given by

$$\bar{V}[x_t, q_t] = \max_{a_t} \left\{ U(a_t \mid x_t) + \sum\nolimits_{x_t} \bar{P}(x_{t+1} \mid x_t, a_t, q_t) \bar{V}[x_{t+1}, q_{t+1}] \right\}.$$

This is the computing algorithm for adaptive optimization (Williams 2009, 2011a). In this equation a single utility model $U(a_t \mid x_t)$ is assumed to apply for every process model $P_k(x_{t+1} \mid x_t, a_t)$. Alternatively, utility can be expressed as $U_k(a_t \mid x_t)$ and an expected utility

$$\bar{U}(a_t \mid x_t, q_t) = \sum\nolimits_k q_t(k) U_k(a_t \mid x_t)$$

can be used. Reframing a discrete decision process under structural uncertainty into a continuous process in model state allows one to recognize the process as a belief Markov decision process (Kaelbling 1998, Williams 2011b).

In this framing of the control problem, optimal decision making responds to an evolving model state that itself is responsive to stochastic system dynamics. The resulting control and system trajectories are determined by the value function, the structure of the system being managed, and the degree and nature of the structural uncertainties, environmental variation, and partial controllability to which it is subjected (Williams 2001, Williams et al. 2002). Again, it is the influence of structural uncertainty on the process of decision making that defines adaptive management.

Given the fundamentally stochastic and dynamic nature of the decision making environment, it is difficult to ascribe meaning to the idea of a "steady state." And unless the objective of management involves an aggressive pursuit of an equilibrium target for a system or control variable, it is even more difficult to see how optimal control under these circumstances would lead naturally to the elimination of resilience. Quite the contrary; the responsive nature of decision making means that any unanticipated shock to the system typically would lead to a change in management strategy as the system and model states respond to it. Such is the case for stationary as well as non-stationary processes, with the absence of any notion of equilibrium.

Deeply Uncertain Process. Robust decision making can be used with a dynamic system for which there is an absence of information about the model state. In this case a range can be defined that is centered on a guesstimate $\tilde{q}_t$ with extent again given by an uncertainty horizon α, and an action-specific robustness function $\hat{\alpha}(a \mid V_c, \tilde{q}, x)$ that gives the uncertainty horizon identified by

$$\hat{\alpha}(a_t \mid V_c, \tilde{q}_t, x_t)$$
$$= \arg\max_{\alpha} \left[\min_{q_t \in R(\alpha, \tilde{q}_t)} \left\{ U(a_t \mid x_t) + \sum\nolimits_{x_t} \bar{P}(x_{t+1} \mid x_t, a_t, q_t) \bar{V}[x_{t+1}, q_{t+1}] \right\} \geq V_c \right].$$

Robust decision making with uncertain model state is then defined for a given critical value V_c and guesstimate $\tilde{q}_t$ by the selection of the action a_t with the largest uncertainty horizon produced by the robustness function:

Choose a_t to maximize $\hat{\alpha}(a_t \mid V_c, \tilde{q}_t, x_t)$.

Deeply Uncertain Process, Deeply Uncertain State

Finally, robust decision making can accommodate uncertainty in both belief and model states, by defining a range $R(\alpha, \tilde{b}, \tilde{q})$ around $(\tilde{b}, \tilde{q})$ with

$$\min_{(b_t, q_t) \in R(\alpha, \tilde{b}_t, \tilde{q}_t)} \left\{ U(a_t \mid b_t) + \sum_k \sum_{b_{t+1}} q_t(k) \Pr_k(b_{t+1} \mid b_t, a_t) \bar{V}_k\left[b_{t+1}\right] \geq V_c \right\}$$

(Williams 2009) and maximizing the size of this range for a given critical value V_c:

$$\hat{\alpha}(a_t \mid V_c, \tilde{b}_t, \tilde{q}_t)$$

$$= \arg\max_{\alpha} \left[\min_{(b_t, q_t) \in R(\alpha, \tilde{b}_t, \tilde{q}_t)} \left\{ U(a_t \mid b_t) + \sum_k \sum_{b_{t+1}} q_t(k) \Pr_k(b_{t+1} \mid b_t, a_t) \bar{V}_k\left[b_{t+1}\right] \geq V_c \right\} \right]$$

to produce a robustness function $\hat{\alpha}(a_t \mid V_c, \tilde{b}_t, \tilde{q}_t)$ that yields the maximizing value of α. Robust decision making with uncertain belief and model states then is defined for a given value V_c and guesstimate $(\tilde{b}_t, \tilde{q}_t)$ by the selection of the action with the largest uncertainty horizon given by the robustness function:

Choose a_t to maximize $\hat{\alpha}(a_t \mid V_c, \tilde{b}_t, \tilde{q}_t)$.

Conclusion

In the foregoing, natural resources management is framed in terms of objective-driven decision making that accounts for uncertainty about the consequences of management actions. The approach relies on a specification of objectives and management options, which fold into a decision apparatus based on the comparative assessment of options. This structure is distinguished from a trial-and-error approach, in which one selects an action absent the deliberate assessment of alternatives, then tracks resource responses to the action and (possibly) selects another action if the results are not as anticipated or desired.

It seems obvious that there should be some principle by which to guide decision making that is grounded in a comparison among decision alternatives. In each of the cases above, a generic value function represents accumulated utilities, with a decision criterion that utilizes the value function in a comparative assessment of management options. By expressing the relative values of decision alternatives, the value function plays a predominant role in this framework. The technical requirements for the value function are few and quite general (Sniedovich 2011). In particular, a wide range of functions can be accommodated that are relevant to the issue of resilience. For example, these functions might describe state-specific values for ecological services (or processes) as well as goods, with the shape of the function determined by the managers' attitude toward the risk of unlikely, but highly undesirable, outcomes (Burgman 2005).

Above and beyond the form of the value function, the nature of the decision making apparatus, especially the approach used for prioritizing and selecting actions, will differ in key ways depending on the time frame for decision making and the amount of information that is available to inform the process. In particular, it is clear from the above that a different form of optimal decision making is required in the absence of a stochastic structure for system features (Polasky et al. 2011). It then is not possible to identify optimal strategy (whether exploitative or otherwise) based on an average of value functions, since the probabilities on which such an average depends are not available. Under these circumstances it is necessary to reorient one's decision making framework so that it does not depend on such a probability structure. The decision-analytic framework described here is seen to encompass this as well as other resource situations that occur in natural resource management.

The presumed failure of optimization is that by not accounting for key features of an ecosystem and locking in on only a part of the ecosystem structure and functions, it forces the system into patterns of behavior that ultimately reduce resilience. But this only holds under certain circumstances, for example when management seeks to hold actions and/or system conditions constant over time in some state. Other tendencies inhibiting resilience include the lack of an appropriate framing of the decision problem, resulting from an over-focused effort to find solutions that restricts one's perception of the problem and possible solutions. The solution of such a narrowly defined problem can be at variance with resilience-inducing solutions that could be identified by a broader and more inclusive decision making context. We fully appreciate that key issues raised in the context of resilience thinking, including nonlinear system dynamics, the limits to controllability, and the need to value ecosystem processes in addition to products, greatly complicate the analysis of decision problems. But when the system is appropriately represented, its stochastic behavior is incorporated, and uncertainty elements like partial observability and structural uncertainty are accounted for in decision making, there is no reason to believe that optimization will automatically lead to the loss of resilience. Indeed, long-term sustainability can itself be the focus of management, with the idea of re-orienting management to ensure that decision making sustains productive capacity.

Robust decision making as described above provides such a re-orientation of optimal control, by focusing not on maximizing an average of values but rather on maximizing the range of "good-enough" decisions over system states. This shift in focus induces potentially profound changes in management strategy, and alters the exposure of the system to the loss of resilience that typically attends its systematic maximization for sustained yield. By shifting the focus of decisions away from maximizing average value while emphasizing the future consequences of present actions in an iterative decision context, robust decision making can contribute to long-term sustainability. Optimal decision making thus can become an ally rather than an adversary of resilience.

References

Bellman, R. (1957). *Dynamic programming*. Princeton: Princeton University Press.

Bellman, R., & Dreyfus, S. (1962). *Applied dynamic programming*. Princeton: Princeton University Press.

Ben-Haim, Y. (2006). *Info-gap decision theory: Decisions under severe uncertainty* (2nd ed.). London: Academic Press.

Bertsekas, D. P. (2007). *Dynamic programming and optimal control* (Vols. 1 and 2, 3rd ed.). Belmont: Athena Scientific.

Burgman, M. (2005). *Risks and decisions for conservation and environmental management*. Cambridge: Cambridge University Press.

Carpenter, S. R., Brock, W. A., & Ludwig, D. (2002). Collapse, learning and renewal. In L. H. Gunderson & C. S. Holling (Eds.), *Panarchy: understanding transformations in human and natural systems*. Washington, D.C.: Island Press.

Fischer, J., Peterson, G. D., Gardner, T. A., Gordon, L. J., Fazey, I., Elmqvist, T., Felton, A., Folke, C., & Dovers, S. (2009). Integrating resilience thinking and optimization for conservation. *Trends in Ecology and Evolution, 24,* 549–554.

Gregory, R. S., & Keeney, R. L. (2002). Making smarter environmental decisions. *Journal of the American Water Resources Association, 38,* 1601–1612.

Holling, C. S., & Meffe, G. K. (1996). Command and control and the pathology of natural resource management. *Conservation Biology, 10,* 328–337.

Hull, D. (2003). *Optimal control theory for applications*. New York: Springer-Verlag.

Kaelbling, L. P., Littman, M. L., & Cassandra, A. R. (1998). Planning and acting in partially observable stochastic domains. *Artificial Intelligence, 101,* 99–134.

Kangas, A., Kangas, J., & Kurttila, M. (2010). *Decision support for forest management*. New York: Springer.

Lempert, R. J., & Collins, M. T. (2007). Managing the risk of uncertain threshold response: Comparison of robust, optimum, and precautionary approaches. *Risk Analysis, 27,* 1009–1026.

McCarthy, M. A. (2007). Bayesian methods for ecology. Cambridge: Cambridge University Press.

Peterson, G. D., Carpenter, S. R., & Brock, W. A. (2003). Uncertainty and the management of multistate systems: An apparently rational route to collapse. *Ecology, 84,* 1403–1411.

Polasky, S., Carpenter, S. R., Folke, C., & Keeler, B. (2011). Decision-making under great uncertainty: Environmental management in an era of global change. *Trends in Ecology and Evolution, 26,* 398–404.

Regan, H. M., Ben-Haim, Y., Langford, B., Wilson, W. G., Lunberg, P., Andleman, S. J., & Burgman, M. A. (2005). Robust decision making under sever uncertainty for conservation management. *Ecological Applications, 15,* 1471–1477.

Runge, M. C. (2011). An introduction to adaptive management for threatened and endangered species. *Journal of Fish and Wildlife Management, 2,* 220–233.

Shenk, T., & Franklin, A. (Eds.). (2001). *Modeling in natural resource management: Development, interpretation, and application*. Washington, D.C.: Island Press.

Simonovic, S. P. (2009). *Managing water resources: Methods and tools for a systems approach*. Paris: UNESCO Publishing.

Sniedovich, M. (2011). *Dynamic programming: Foundations and principles* (2nd ed.) Boca Raton: CRC Press.

Stengel, R. F. 1994. *Optimal control and estimation*. New York: Dover Publications Inc.

Walker, B., & Salt, D. (2006). *Resilience thinking: Sustaining ecosystems and people in a changing world*. Washington, D.C.: Island Press.

Walker, B., Holling, C. S., Carpenter, S. R., & Kinzig, A. (2004). Resilience, adaptability and transformability in social-ecological systems. *Ecology and Society, 9*(2):5. http://www.ecologyandsociety.org/vol9/iss2/art5.

Walters, C. J. (1986). *Adaptive management of renewable resources*. New York: McGraw Hill.

Williams, B. K. (2001). Uncertainty, learning, and optimization in wildlife management. *Environmental and Ecological Statistics, 8,* 269–288.

Williams, B. K. (2009). Markov decision processes in natural resources management: Observability and uncertainty. *Ecological Modelling, 220,* 830–840.

Williams, B. K. (2011a). Adaptive resource management and the value of information. *Ecological Modelling, 222,* 3429–3436.

Williams, B. K. (2011b). Resolving structural uncertainty in natural resources management using POMDP approaches. *Ecological Modelling, 222,* 1092–1102.

Williams, B. K., Nichols, J. D., & Conroy, M. J. (2002). *Analysis and management of animal populations*. San Diego: Academic Press.

Chapter 13
Emerging Concepts in Adaptive Management

Derek Armitage, Steven Alexander, Mark Andrachuk, Samantha Berdej, Thomas Dyck, Prateep Kumar Nayak, Jeremy Pittman and Kaitlyn Rathwell

Keywords Adaptive management · Uncertainty · Natural resource management · Transdisciplinary · Governance

Introduction

Adaptive management is an elegant concept. Structure management interventions and policies as experiments, monitor feedback, and make necessary adjustments (Holling 1978, Walters 1986). Yet, the implementation of adaptive management has often been difficult, and the outcomes unclear. Lee (1993) offered a compelling account of the opportunities and challenges of adaptive management in the Columbia River Basin in the U.S. northwest in his book, 'Compass and Gyrscope'. He pointed not to matters of science as the primary stumbling block to adaptive management, but to a lack of enabling social and institutional conditions.

D. Armitage (✉) · S. Alexander · M. Andrachuk · S. Berdej · T. Dyck · P. K. Nayak · J. Pittman · K. Rathwell
Environmental Change and Governance Group, Faculty of Environment, University of Waterloo, Waterloo, ON N2L 3G1, Canada
e-mail: derek.armitage@uwaterloo.ca

S. Alexander
e-mail: s22alexa@uwaterloo.ca

M. Andrachuk
e-mail: mandrach@uwaterloo.ca

S. Berdej
e-mail: smberdej@uwaterloo.ca

J. Pittman
e-mail: jpittman@uwaterloo.ca

K. Rathwell
e-mail: kaitlyn.rathwell@uwaterloo.ca

C. R. Allen, A. S. Garmestani (eds.), *Adaptive Management of Social-Ecological Systems*,
DOI 10.1007/978-94-017-9682-8_13

McLain and Lee (1996) also documented the promise and pitfalls of scientific adaptive management. Their analysis highlighted adaptive management processes that did not effectively incorporate non-scientific forms of knowledge, recognize the social embeddedness of management decisions, engage with diverse stakeholders in cooperative ways, or orient adaptive management around more complex systems models. A special feature on adaptive management in Conservation Ecology (now Ecology and Society) in 1999 approached many of these same issues with a provocative question: "Adaptive management—scientifically sound, socially challenged?" Johnson's (1999) summary in this special feature outlined three main areas of concern: (1) the need to integrate stakeholders more effectively into adaptive management decision making; (2) the lack of institutional arrangements to support adaptive management; and (3) a failure to embrace management failure as a crucial part of learning for better outcomes. Two decades after Lee's (1993) book, many of these same constraints on adaptive management persist (Colfer 2005, Armitage et al. 2007, Westgate et al. 2013).

Examples of the successful application of adaptive management are few (Gunderson and Light 2006, Keith et al. 2011), and it remains more of an idealized concept than an empirically tested strategy to gain insights into the behavior of linked systems of people and nature (Lee 1999, Berkes et al. 2003). Westgate et al. (2013) documented the excess use of the term 'adaptive management' in a recent systematic literature review, and highlighted that only a small number of projects characterized as adaptive management effectively applied the concept to natural resource decision making (see also Gunderson and Light 2006). As Westgate et al. (2013) illustrate, further attention to the social context in which adaptive managers and scientists operate is crucial to achieving more credible and legitimate management outcomes.

We examine in this chapter six issues or concepts that emerge as central to ongoing efforts to advance the theory and practice of adaptive management of natural resources: (1) adopting a transdisciplinary perspective on adaptive management; (2) shifting from a natural resource management to social-ecological systems perspective; (3) situating adaptive management within a governance context; (4) surfacing the role of power in adaptive management processes; (5) engaging with knowledge co-production; and (6) exploring the role of adaptive management as a deliberative tool in support of social-ecological transformations.

Choices about what concepts to include here reflect our collective experiences and interest with adaptive management as it pertains to environmental change and governance. The concepts examined here further reflect recent directions in adap-

T. Dyck
Geography and Environmental Studies, Faculty of Arts, Wilfrid Laurier University, Waterloo, ON N2L 3CG, Canada
e-mail: dyck3730@mylaurier.ca

P. K. Nayak
School of Environment, Enterprise and Development, Faculty of Environment, University of Waterloo, Waterloo, ON N2L 3G1, Canada
e-mail: pnayak@uwaterloo.ca

tive management scholarship (broadly defined) and can help to build theory and situate the practice of natural resource management in a broader sustainability context. Others might choose to emphasize different concepts or issues, or depict the substantive concerns in alternative ways. In our view, however, the promise and elegance of adaptive management is more likely to emerge if practitioners and researchers situate their thinking in a transdisciplinary context of linked systems of people and nature, with reference to the issues of governance, power and knowledge, and as a strategy to encourage broader reflection on societies' interaction with natural resources. Consideration of these emergent concepts is likely to intensify the challenge of adaptive management. Yet without taking these issues into account, the promise of adaptive management is less likely to be realized.

A Transdisciplinary Turn in Adaptive Management

Science is based on mental models and social framings that influence the types of questions we ask, the data we collect and analyze, and ultimately our approaches to adaptive management (Peterson et al. 2003, Cumming and Collier 2005, Glaser 2006). Social framings of the natural world as predictable and controllable have been a mainstay of natural resource management for a century or more, and have served as the foundation for key management tools (e.g., maximum sustainable yield) (Gunderson et al. 1995). The emergence of adaptive management in recognition of ecological complexity and uncertainty was an important step forward in how managers framed natural resource management problems and solutions. However, first generation adaptive management has been driven largely by disciplinary science, and implemented in the context of segmented thinking and sector-based bureaucracies (Pinkerton 2007).

A transdisciplinary frame is crucial to meaningfully address complexity and uncertainty of natural resource management and foster a second generation of adaptive management. We follow Lang et al. (2012) in defining transdisciplinarity as an approach aimed to address the practical and conceptual dimensions of socially important issues, by integrating across diverse bodies of knowledge and explicitly involving stakeholders throughout a research and decision making process. In our view, a transdisciplinary frame for adaptive management should include: (1) defining research and management goals in terms of both socially and ecologically important issues—that is, recognizing people and the biosphere as a tightly coupled social-ecological system that is characterized by feedbacks across scales; (2) engaging in learning processes (formal and informal) through which knowledge about complex social-ecological systems is co-produced (e.g., science with the knowledge of resource users, researchers and practitioners); and (3) using that knowledge to transform societies' interactions with natural resources in ways that generate novel options for the maintenance of ecosystem services and human wellbeing.

Cooperation and reflexive practice (Ison et al. 2013) among multiple scientific domains and social groups is at the core of adaptive management, and those adaptive management processes that reflect the ideals of transdisciplinary practice are most likely to yield outcomes that are legitimate and salient. Strategies to foster greater transdisciplinarity in adaptive management practice often emerge in the context of specific places and problems. These practices include innovative ways to utilize knowledge from a diverse range of actors, recognition that science is a crucial but bounded component of the sustainability challenge, and institutionalization of the learning processes that are at the core of efforts to deal with uncertainty and change.

Natural Resource Management in a Social-Ecological Systems Perspective

Social-ecological systems are defined as linked and co-evolutionary systems of society and nature (Berkes et al. 2003). A social-ecological system lens helps to situate adaptive management in the complexities of linked social and ecological systems, and is an important shift away from a focus solely on the management of individual natural resources (e.g., forest stand productivity). Acknowledging or anticipating feedbacks beyond an immediate resource system (e.g., forest stand, fishery) has emerged as a crucial component of managing for uncertainty (Gunderson et al. 1995). Choices about focal areas of concern and units of analysis place logistical constraints on the extent to which a social-ecological system lens might be applied in an adaptive management setting. However, using a social-ecological lens to frame adaptive management problems encourages multi-level analysis, incorporation of multiple social framings and ways of understanding social-ecological system problems, and therefore, recognition of the 'wicked' nature of many resource management problems (Rittel and Webber 1973, Allen et al. 2011). Specifically, thinking about adaptive management in terms of social-ecological systems helps to highlight a number of inherent features of linked systems of people and nature, including feedback processes among drivers of change across scales, and the nestedness of systems and social and ecological sub-systems (Table 13.1).

Orienting adaptive management theory and practice around the main features of social-ecological systems has a number of implications, such as: (1) reinforcing the philosophical foundations of adaptive management which are to embrace uncertainty and complexity (Holling 1978, Berkes 2003); (2) appreciating the resource context as a complex adaptive system that involves multi-directional flows between people and their environments (Kates et al. 2001, Berkes 2003, 2011, Mahon et al. 2008, Levin and Clark 2010); (3) recognizing connections among adaptive management of natural resources and the livelihood, food security and social wellbeing concerns of people and communities (Chuenpagdee et al. 2005, MEA 2005, Weeratunge et al. 2014); and (4) illustrating that decision making arrangements must reflect how the social domain (e.g., distribution of power) intersects with the ecological through multiple feedback processes (Berkes 2010, Nayak and Berkes 2010).

A social-ecological perspective compels adaptive management practitioners and researchers to look beyond theoretical, methodological and disciplinary boundaries to offer an overarching framework—an inclusive lens—to study social-ecological

Table 13.1 Implications of a social-ecological lens for adaptive management

Social- ecological system features	Description
Linkages	Emphasizes that the two parts (human systems and environmental/biophysical systems) are only arbitrarily separable and equally important, and that they function as a coupled, interdependent, and co-evolutionary system (e.g., human actions affect biophysical systems, biophysical factors affect human well-being, and humans in turn respond to these factors) (Berkes 2011)
	Recognizes the role of humans in shaping ecosystem processes and dynamics thus valuing their capacity to influence and be influenced by ecological outcomes (Dale et al. 2000, Waltner-Toews and Kay 2005)
Feedback	Coupled systems exhibit nonlinear dynamics, thresholds, surprises, legacy effects and time lags (Liu et al. 2007) Extent and nature of coupling varies spatially, temporally and organizationally (Liu et al. 2007)
	Coupled systems have multiple drivers, an array of impacts, unpredictable ways in which drivers act, and multiple feedback interaction between human and biophysical systems (Nayak 2011)
	Interconnections and cross-scale dynamics among the social-ecological attributes become important factors that define the nature and extent of system complexity
Nestedness and sub-systems	Complex systems have a structural architecture characterized by hierarchical organization and interactions that take place between these nested systems (Simon 1962, Levin 1999)
	Focusing on sub-systems as distinct parts of the larger social-ecological system aids the development of understanding them, because they are valued as integral to each other, bound as a coupled system (Turner et al. 2003, Glaser 2006, Kotchen and Young 2007)
Scale	Observed dynamics and behavior of ecosystems and social-ecological systems are the result of the interplay of structures and processes that vary spatially and temporally (Levin 1999, Gunderson and Holling 2002, Cash et al. 2006)
	Allows us to think about complex multi-scale processes within the social-ecological system and determine appropriate scales of intervention for adaptive management

systems. Finally, by bringing the social context into a conventionally resource-focused approach, a social-ecological system lens helps to highlight the conflicts and distributive justice challenges of adaptive management, along with impacts on livelihoods and potential for inequity and problems of participation in decision-making processes (Berkes et al. 2006, Liu et al. 2007, Nayak and Berkes 2014). We address several of these emergent challenges in subsequent sections below.

Situating Adaptive Management in a Governance Context

Management and governance are neither synonymous, nor mutually exclusive. Management typically involves the operational decisions taken to achieve specific outcomes (e.g., increases in yield of a desired resource stock). Governance often refers to the broader processes and institutions through which societies make decisions that affect the environment (see Oakerson 1992). Biermann et al. (2009) define governance as "the interrelated and increasingly integrated system of formal and informal rules, rule-making systems, and actor-networks at all levels of human society (from local to global) that are set up to steer societies toward preventing, mitigating, and adapting to global and local environmental change." In this context, institutions are the formal and informal "working rules" and associated decisions (e.g., for monitoring and enforcement) that mediate interactions among people and their environments (Ostrom 1990). We use governance to refer to both an analytical lens to examine the broader set of rules and actor networks within which adaptive management actions and decisions take place, as well as specific arrangements for adaptive decision making about natural resources among government agencies, industry and resource user groups (see Armitage et al. 2012).

Gunderson and Light (2006) suggested that thinking in terms of adaptive governance can help "increase responsiveness and generate more diverse and versatile competencies that create options for the future and develop the adaptive capacity to improvise and adjust to recurring crises." This makes good sense given the social-ecological complexities of most natural resource management settings. However, working towards such a goal inevitably requires managers and other actors in an adaptive management process to consider more thoroughly the social and institutional constraints within which they operate, reflect on levels of power and authority among the actors involved in adaptive management, bridge diverse knowledge systems, and build adaptive capacity to support more fundamental transformations in how societies interact with natural resources. As Gunderson and Light (2006) noted, "adaptive governance deals with the complex human interactions that have been obstacles to the implementation of adaptive management," which include institutional constraints and contested and divergent values, goals, and objectives between actors.

Situating adaptive management in a governance context generates a number of useful insights for managers and resource users (Box 1). For example, government agencies with the mandate for adaptive management cannot be the only source of decision making, although they have a crucial role to play in that regard. As more actors (industry, user groups, civil society organizations) enter the adaptive management arena, different types and sources of knowledge will gain legitimacy. Indeed, our current understanding of social-ecological systems is incomplete and multiple types of knowledge are necessary to inform decisions (Brunner et al. 2005, Folke et al. 2005). A governance perspective (see Garmestani et al. 2009) helps managers to recognize the legitimacy of diverse and sometimes peripheral actors with new roles in resource and ecosystem management, and helps convey a "multi-objective reality when handling conflicts among diverse stakeholders" (Folke et al. 2005).

Box 1: Implications of a Governance Lens for Scientists and Managers

Consider emergent actors with new roles in resource and ecosystem management
State agencies are no longer the main actor or sole source of decision-making. Hybrid arrangements involving state and non-state actors have emerged, offering alternative and promising models, but have also created new challenges associated with accountability, legitimacy and scale.

Recognize that adaptive management occurs in contested and power-laden social contexts
Power underlies all adaptive management processes, and influences how trade-offs between multiple, competing objectives are made. Acknowledging and understanding the role of power encourages reflection on and recognition of the contested and divergent assumptions, values and goals amongst actors involved in decision-making.

Appreciate the need for engaging and bridging diverse knowledge systems for learning
Scientific knowledge of complex social-ecological systems is often incomplete, creating pitfalls when relying on it as the exclusive source of information for decision-making. Knowledge that is co-produced by bridging diverse sources and types is typically better suited for navigating complexity and uncertainty (Berkes 2009).

Embrace the challenge of adaptation
Adaptation to maintain or preserve existing features of social-ecological systems is necessary to address environmental change. Capacity to meet the challenge of adaptation is crucial, as is the need to recognize maladaptive practices and consider more fundamental system transformations. In light of ongoing processes of change in social-ecological systems, expectations of adaptive management need to be continually refined.

A governance lens may also encourage adaptive managers to reflect on the multiple domains (social, economic, ecological) in which their problems are nested (see Westley 2002, Garmestani et al. 2009). In other words, a governance perspective can facilitate an integrative or social-ecological view (as above), rather than a traditional regulatory or sectoral view. Similarly, a governance lens highlights the social structures and processes (i.e., networks) that link individuals, organizations, agencies, and institutions in a multi-level world (Olsson et al. 2004). Since actors interact vertically and horizontally within such networks, strong network arrangements are hypothesized to enhance the capacity for adaptive management by facilitating processes of learning and building legitimacy of decision outcomes (Armitage et al. 2012). However, such networked and/or multi-level arrangements also have disadvantages. They may require more time for decisions to be made, and exacerbate political, economic or livelihood conflict if not carefully facilitated. A governance lens thus highlights the need to strengthen capacity to manage adaptively across scales, but also to recognize that any management process is bounded by broader political, economic and institutional conditions that will ultimately define transitions towards sustainability.

Surfacing Power in Adaptive Management

The emergence of hybrid governance arrangements emphasizes a transition from the single state/agency actors in resolving management challenges, to network strategies involving combinations of actors from states, the private sector, and civil society (Lemos and Agrawal 2006). As the preceding discussion on governance highlights, adaptive managers are increasingly engaged with a broad array of actors outside formal (i.e., government) spheres that seek to influence the management of natural resources (Ansell and Gash 2008, Ali-Khan and Mulvihill 2008). Any decision regarding natural resources is inherently influenced by social relations of power (Bryant 1998, Brechin et al. 2003, Ansell and Gash 2008). Adaptive management processes are required to more effectively consider questions about actor inclusion (i.e., who participates in hypothesis generation, knowledge production, data analysis?), as well as questions about influence, the legitimacy of actor participation, and the distribution of power among actors (i.e., how effectively do different actors participate in various phases of adaptive management?).

We define power here as the application of action, knowledge and resources to resolve problems and further interests (Adger et al. 2005, Raik et al. 2008), and we identify four related arenas through which to consider power in adaptive management: (1) decision-making; (2) authority and control; (3) action; and (4) knowledge. These categories are not exclusive, and some social actors may span multiple categories (Table 13.2).

In adaptive management, a failure to address or consider differences in power among actors can have far-reaching implications for the legitimacy of decisions about natural resources (Borrini-Feyerabend et al. 2007, Larson and Soto 2008, Biermann and Gupta 2011). The differences in power among actors may be linked to capacity limitations (e.g., financial, technical), which may contribute to uneven representation in terms of the issues addressed and the interests considered (e.g., Stringer et al. 2006, Kallis et al. 2009). However, structurally embedded constraints related to institutions (e.g., rights, rules) and the marginalization of certain groups are more likely to be a foundational reason for uneven distribution of power among participants in an adaptive management process. In either case, unequal distributions

Table 13.2 Arenas of power in adaptive management

Arenas of power	Description	Roles and responsibilities	Examples	References
Decision-making	The power to meaningfully influence decisions	Participant, negotiator, discussant, persuader, advisor, consultant, communicator	Engagement of actors (e.g., tourism, government, NGO, community etc.) via a multi-stakeholder Management Advisory Board in Bunaken National Park, Indonesia (Erdman et al. 2004)	Mannigel 2008, Ferse et al. 2010
Authority and control	The power to coerce or constrain human action	Rule maker, decision maker, enforcement	Devolution of Brazil's water sector to local multi-stakeholder river basin councils generates varied levels of authority across states (Engle et al. 2011). Ongoing decentralization reforms across sub-Saharan Africa are transferring decision-making powers to local governments and organizations in the context of natural resource management (Ribot 2003)	Agrawal and Ribot 1999, Njaya et al. 2011, Campbell et al. 2013
Action	The power to execute	Implementer, monitor, adjudicator	Local enforcers (*kewang*) and traditional local leaders play a vital role in the functioning of customary *sasi* marine management systems in eastern Indonesia (Harkes 1998, Satria and Adhuri 2010)	Mappatoba 2004
Knowledge	The power to gather, learn, possess, and exclude knowledge	Knowledge holder, knowledge broker, knowledge (co)producer	Multi-stakeholder riparian management in the Sprucedale National Forest, southwestern USA results in competing discourses where differential power amongst actors is used to select knowledge sources and influence decision making (Arnold et al. 2012)	Natcher 2005, Nadasdy 2007, McGregor 2012

of power may lead to poor social, and ultimately ecological, outcomes (Lebel et al. 2005, Nadasdy 2007).

Explicit recognition of structural and agent-based dimensions of power and their interactions (see Raik et al. 2008) can prove crucial to successful adaptive management, particularly given the increasingly hybrid, networked and multi-level decision making arenas within which adaptive managers are situated. The sharing of authority and control among diverse actors is increasingly encouraged as an approach to management of a wide range of natural resources (Borrini-Feyerabend et al. 2007), and manifests in many ways, such as in the form of government-indigenous partnerships, community agreements on conservation, or collaborative management arrangements more generally (Press et al. 1995, Mappatoba 2004, Fox et al. 2008). In these contexts, the focus is less on active adaptive management, and more on the social process of learning by doing, monitoring and collaborative decision making in response to the uneven distribution of power among communities, conservation organizations and government agencies (Salafsky et al. 2001).

Knowledge Co-production in Principle and Practice

Knowledge systems are defined as interconnected symbols that create meaning about reality that humans co-construct and adapt over time (Dryzek 2005, Reid et al. 2006). Knowledge systems thus reflect a knowledge-practice-belief complex (Berkes 2012), where meaning emerges from actors co-constructing symbols, artifacts, competencies, and norms to enact 'what we know' and 'how we know it' (Midgley 2000). It is crucial for actors in adaptive management to recognize that knowledge is as much a social process (i.e., governance) as it is a set of outcomes (e.g., management plans).

Undertaking how to bridge knowledge systems in adaptive management is an area still in need of significant effort. Where efforts to bridge knowledge systems have been meaningfully attempted, they have often occurred in the context of collaborative and deliberative processes (Berkes and Davidson-Hunt 2008). Knowledge co-production can be defined as "the collaborative process of bringing a plurality of knowledge sources and types together to address a defined problem and build an integrated or systems-oriented understanding of that problem" (Armitage et al. 2011). Such processes encourage managers and other actors to: (1) examine different narratives of (or stories about) environmental change and uncertainty (Batterbury 1997, Dietz et al. 2003); (2) enhance their overall capacity to understand and accept uncertainty (Reidlinger and Berkes 2001); (3) allow actors to formulate shared visions to guide decision making (Peterson 2007); and (4) encourage a shift away from knowledge integration towards knowledge exchange (Fazey et al. 2013). A knowledge co-production approach seeks to maintain the integrity of participating knowledge systems and knowledge holders, while creating space for the development of novel and hybrid understandings needed to learn through uncertainty. This is a hallmark of adaptive management.

Evidence shows that bridging diverse knowledge systems improves the overall understanding of environmental phenomena among different groups (Reidlinger and Berkes 2001, Reid et al. 2006), and enhances the perceived salience, credibility and legitimacy of adaptive management (Cash et al. 2003, Mitchell et al. 2006, Reid et al. 2006). Knowledge of different types and from different sources (scientific, local, traditional) can improve the quality of decisions (Reid et al. 2006, Reed et al.

2013). For example, Laidler (2006) illustrates how Inuit and scientific knowledge holders use very different processes to understand and act upon changes in Arctic sea ice. Scientists use satellite imagery or local instruments to measure and verify changes with the aid of statistical modeling, whereas Inuit use their observations gained during hunting or other land-based practices and verify changes by sharing and discussing their experiences with other community members (Laidler 2006, Laidler et al. 2010). Hybrid knowledge, emerging from the contributions of different knowledge systems (e.g., western/scientific, local and indigenous), can create novel understandings of environment and natural resource management that are different from what either knowledge system could support on its own (Reidlinger and Berkes 2001, Armitage et al. 2011).

Knowledge emerging from artistic processes may also play an important role to bridge knowledge systems and thus can contribute to successful adaptive management (see Box 2). Art and artistic processes reflect a particular type of knowledge system, and they can also help to bridge different actors and enable reflection on how different knowledge systems can be used to make sense of environmental change and uncertainty (Vancouver Art Gallery 2006, Zurba and Berkes 2014). Further, art and artistic processes can help groups of individuals envision future changes at a number of scales—local to global (Elgin 2002, Davies and Sarpong 2013). Exploration of the role of art and artistic processes in adaptive management is warranted, and may include many different mediums (e.g., storytelling, digital media). In some cases, those engaged in artistic endeavors may be key resource users with direct connection to the decisions being made about resource systems. In other cases, art and artistic processes may be produced by independent actors but serve as a form of 'boundary object' around which dialogue and learning take place.

Box 2. A Role for Artistic Process in Adaptive Management?

Art and artistic processes can contribute to bridge different knowledge systems and may contribute in innovative ways to adaptive management. Artistic processes are similar to scenario planning, which is a reflective and forward-looking means of bridging knowledge systems in the adaptive management approach (Bennett and Zurek 2006). Artistic processes and mediums, such as music, theater and oral-history, offer a safe and culturally-embedded means of exploring and reflecting upon the human dimension of environmental change (Zurba and Berkes 2014). Creating space for these artistic forms in a governance setting is one way to demonstrate respect for diverse cultures, while providing opportunities for meaningful deliberation on key challenges. Take for example the Arctic Gnomes at Eden instillation project by Bullet Creative. This piece uses interactive instillation art to help individuals conceptualize changes in Arctic Sea ice (http://www.capefarewell.com/news/events/687-arctic-gnomes-at-eden.html). Shifting mental models about the way the world works is central to the process of adaptive management. Art and artistic process can enhance understanding of our own mental models and those of others, and may emerge as a key piece of the adaptive management puzzle.

Beyond the pragmatic benefits of bridging knowledge systems the inclusion of diverse knowledge systems in adaptive management has substantive political and ethical benefits (Bohensky and Maru 2011). And in contexts where uneven power distribution has undermined local and indigenous input in decision-making this need may be acute (see Tuhiwai Smith 1999, Mascarenhas 2007).

Adaptive Management as An Arena for Deliberative Transformations

Adaptive managers must define meaningful goals to achieve specific resource targets (e.g., annual allowable cut) or desired social and ecological outcomes (e.g., greater biodiversity, enhanced human wellbeing). Trade-offs among social and ecological outcomes or targets will incur additional uncertainties, and will generate conflict among managers, resources users and other civil society actors. In and of itself, the challenge of adaptive management is daunting. However, since adaptive management of natural resources is situated in a wider social-ecological system context (e.g., forest stand management in a wider regional planning process), defining goals and targets must also connect with broader debates about trajectories of desired change—not just natural resource management outcomes. A core tension in such debates often centers on the decision to adaptively manage natural resources in the context of uncertainty as opposed to fostering more deliberative transformations in situations where adaptive management may contribute to unsustainability in the first place.

Transformation refers to a fundamental shift in ecological, economic and social conditions when existing system trajectories (ecological, social, economic) are untenable (Walker et al. 2004, Chapin et al. 2009, Folke et al. 2010). With reference to adaptive management, we are interested primarily in the notion of deliberative or directional transformations, which are carried out with the intention of achieving particular (positive) outcomes (see O'Brien 2012). Transformations are different from adaptations because they typically challenge rather than seek adjustments to or maintain the current system or the *status quo* (Pelling 2011, O'Brien 2012). A key feature of deliberative transformations is recognition that fundamental shifts in some elements of a system or sub-system are needed to achieve desirable futures (see Miller et al. 2010).

Olsson et al. (2006) outlined a three-phase heuristic for thinking about transformations: (1) preparing for change; (2) navigating the transition from one regime to another; and (3) building resilience in the new regime. Strategies to operationalize such heuristics take us into the realm of adaptive management and governance. For example, preparation for change may occur through co-production of knowledge amongst diverse actors, which can help to identify undesirable or untenable regimes, possible alternatives, thresholds, and barriers to change (Hahn et al. 2006, Pahl-Wostl 2009, Chapin et al. 2009). Shadow or informal networks may be particularly important as these networks can facilitate experimentation and the identification of new approaches or governance arrangements (Olsson et al. 2006, Sendzimir et al. 2007, Moore and Westley 2011). By recognizing the broader governance context, adaptive managers may be better able to engage with some of these actors and processes that take place outside of formal adaptive management settings.

Navigating transitions is a highly unpredictable process, requiring significant flexibility and improvisation—key tenants of adaptive management. Transformative change implies significant uncertainty, and adaptive management provides an important strategy with which to monitor and assess specific interventions against long-term system goals. Adaptive management is thus a concrete way to encourage certain types of change in a deliberative manner. In the absence of careful thought about the broader context in which adaptive management occurs, there is a danger that unsustainable trajectories may be exacerbated or continued. In an effort to support the knowledge base required for deliberate transformations, however, adaptive management provides a setting to assess key variables (ecological and social) that contribute to social-ecological transformation.

Conclusions

Several decades of experience point to deeply embedded social and institutional constraints on the processes of adaptive management and resulting outcomes. We have outlined several concepts that are rooted in social and institutional processes and conditions that have emerged as fundamental to the adaptive management of natural resources (Table 13.3). These concepts provide an entrée to understand some of the pitfalls, but also the promises, of adaptive management, and they provide a frame through which to consider the theory and practice of adaptive management. Despite the challenges, adaptive management remains a set of concepts, principles and practices with significant potential to help societies navigate towards sustainability in an uncertain world (see Allen et al. 2011).

Table 13.3 Emerging concepts and the implications for adaptive management

Concept	Implication
Trans-disciplinarity	Supports adaptive managers and scientists to engage with alternative methodological approaches and knowledge systems
	Inherent uncertainty and unpredictability requires novel ways to understand social-ecological systems that are not bound by disciplinary traditions
	Encourages development of different hypotheses about, and analyses of, complex problems to better inform decisions and management interventions
Social-ecological systems approach	Situates adaptive management in the complexities of linked systems of people and nature and encourages consideration of their inherent features
	Challenges linear thinking, sectoral approaches, and the neglect of social drivers (positive, negative) of change and their feedbacks
	Encourages multiple-level analysis to determine appropriate scales of intervention for adaptive management, as well as linkages among natural resources, livelihoods, food security, social wellbeing, justice and power
	Embracing uncertainty and complexity may lead to fewer or less detrimental unintended consequences (i.e., surprises)
Governance	Encourages managers and scientists to consider more systematically the larger institutional frameworks and networks within which they operate, and linking a focus on outcome oriented operational decisions with societal processes and institutions that influence decisions about natural resources
	Helps adaptive management actors to recognize the institutions (rights, rules, norms), and their interplay across scales, at the core of decision-making
	Expands thinking about who is involved in and influences management processes (i.e., actors who may not be included in formal arrangements can have an important influence on adaptive management or perceptions of outcomes)
	Considers emergent actors and novel hybrid arrangements while situating adaptive management within a context of complex human interactions influenced by diverse values, goals and objectives
Knowledge co-production	Facilitates recognition that multiple types of knowledge are necessary to inform decisions
	Highlights availability of diverse strategies and processes to link different types and sources of knowledge (i.e., bridge knowledge systems)
	Co-producing knowledge about uncertain conditions can lead to robust understandings of the environment and for adaptive management, and novel hypotheses to be tested
Power	Encourages greater contextualization about natural resource ownership and control, and leads to enhanced credibility and legitimacy of decision outcomes
	Recognizes social power as a key driver in success and failure of adaptive management
	Facilitates increased attention to how power is distributed and its linkages to capacity and representation in adaptive management

Table 13.3 (continued)

Concept	Implication
Transformation	Expands perspectives and approaches of managers and scientists regarding pervasive problems contributing to the unsustainability of social-ecological systems
	Challenges core assumptions about the goal(s) of adaptive management by encouraging reflection on broader system trajectories and outcomes
	Highlights the potential role of adaptive management in fostering deliberative transformation through uncertainty.

References

Adger, W. N., Brown, K., & Tompkins, E. L. (2005). The political economy of cross-scale networks in resource co- management. *Ecology and Society, 10*(2), 9. http://www.ecologyandsociety.org/vol10/iss2/art9/.

Agrawal, A., & Ribot, J. (1999). Accountability in decentralization: A framework with South Asian and West African cases. *Journal of Developing Areas, 33,* 473–502.

Allen, C. R., Fontaine, J., Pope, K., & Garmestani, A. S. (2011). Adaptive management for a turbulent future. *Journal of Environmental Management, 92,* 1339–1345.

Ali-Khan, F., & Mulvihill, P. R. (2008). Exploring collaborative environmental governance: Perspectives on bridging and actor agency. *Geography and Compass, 2,* 1974–1994.

Ansell, C., & Gash, A. (2008). Collaborative governance in theory and practice. *Journal of Public Administration Research and Theory, 18,* 543–571.

Armitage, D., Berkes, F., & Doubleday, N. (Eds.). (2007). *Adaptive co-management: Collaboration, learning and multi-level governance.* Vancouver: UBC Press.

Armitage, D., Berkes, F., Dale, A., Kocho-Schellenberg, E., & Patton, E. (2011). Co-management and the co-production of knowledge: Learning to adapt in Canada's Arctic. *Global Environmental Change, 21,* 995–1004.

Armitage, D., de Loe, R., & Plummer, R. (2012). Environmental governance and its implications for conservation practice. *Conservation Letters, 5,* 245–255.

Arnold, J. S., Koro-Ljungberg, M., & Bartels, W. L. (2012). Power and conflict in adaptive management: Analyzing the discourse of riparian management on public lands. *Ecology and Society, 17*(1), 19. http://dx.doi.org/10.5751/ES-04636-170119.

Batterbury, S., Forsyth, T., & Thomson, K. (1997). Environmental transformations in developing countries: Hybrid research and democratic policy. *The Geographical Journal, 163,* 126–132.

Bennett, E., & Zurek, M. (2006). Integrating epistemologies through scenarios. In W. V. Reid, F. Berkes, & D. Capistrano (Eds.), *Bridging scales and knowledge systems: Concepts and applications in ecosystem assessment* (pp. 275–294). Washington, D.C.: Island Press.

Berkes, F. (2003). Alternatives to conventional management: Lessons from small-scale fisheries. *Environments, 31,* 5–19.

Berkes, F. (2009). Evolution of co-management: Role of knowledge generation, bridging organizations and social learning. *Journal of Environmental Management, 90,* 1692–1702.

Berkes, F. (2010). Shifting perspectives on resource management: Resilience and the reconceptualization of 'natural resources' and 'management'. *MAST Maritime Studies, 9,* 11–38.

Berkes, F. (2011). Restoring unity: The concept of marine social-ecological systems. In R. Ommer, R. Perry, K. Cochrane, & P. Cury (Eds.), *World fisheries. A social-ecological analysis* (pp. 9–28). London: Wiley-Blackwell.

Berkes, F. (2012). *Sacred Ecology* (3rd ed.). New York: Routledge.

Berkes, F., & Davidson-Hunt, I. (2008). The cultural basis for an ecosystem approach: Sharing across systems of knowledge. In D. Walter-Toews and E. Lister Nina-marie (Eds.), *The ecosys-*

tem approach: Complexity, uncertainty, and managing for sustainability (pp. 109–124). New York: Columbia University Press.

Berkes, F., Colding, J., & Folke, C. (Eds.). (2003). *Navigating social-ecological systems*. UK: Cambridge University Press.

Berkes, F., Hughes, T., Steneck, R., Wilson, J., Bellwood, D., Crona, B., & Worm, B. (2006). Globalization, roving bandits, and marine resources. *Science, 311,* 1557–1558.

Biermann, F., & Gupta, A. (2011). Accountability and legitimacy in earth system governance: A research framework. *Ecological Economics, 70,* 1856–1864.

Biermann, F., Betsill, M. M., Gupta, J., Kanie, N., Lebel, L., Liverman, D., & Zondervan, R. (2009). *Earth system governance: People, places and the planet. Science and implementation plan of the earth system governance project, ESG Report No. 1*. Bonn: The Earth System Governance Project.

Bohensky, E. L., & Maru, Y. (2011). Indigenous knowledge, science, and resilience: What have we learned from a decade of international literature on "Integration"? *Ecology and Society, 16*(4), 6. http://dx.doi.org/10.5751/ES-04342–160406.

Borrini-Feyerabend, G., Pimbert, M., Farvar, M. T., Kothari, A., & Renard, Y. (Eds.). (2007). *Sharing power: A global guide to collaborative management of natural resources*. London: Earthscan Publishing.

Brechin, S. R., Wilshusen, P. R., Fortwangler, C. L., & West, P. C. (Eds.). (2003). *Contested nature: Promoting international biodiversity with social justice in the twenty-first century*. Albany: State University of New York Press.

Brunner, R. D., Steelman, T. D., Coe-Juell, L., Cromley, C. M., Edwards, C. M., & Tucker, D. W. (2005). *Adaptive governance: Integrating science policy and decision making*. New York: Columbia University Press.

Bryant, R. L. (1998). Power, knowledge and political ecology in the third world: A review. *Progress in Physical Geography, 22,* 79–94.

Campbell, S. J., Kartawijaya, T., Yulianto, I., Prasetia, R., & Clifton, J. (2013). Co-management approaches and incentives improve management effectiveness in the Karimunjawa National Park, Indonesia. *Marine Policy, 41,* 72–79.

Cash, D. W., Clark, W. C., Alcock, F., Dickson, N. M., Eckley, N., Guston, D. H., Jäger, J., & Mitchell, R. B. (2003). Knowledge systems for sustainable development. *Proceedings of the National Academy of Sciences, 100,* 8086–8091.

Cash, D. W., Adger, W. N., Berkes, F., Garden, P., Lebel, L., Olsson, P., & Young, O. (2006). Scale and cross-scale dynamics: Governance and information in a multilevel world. *Ecology and Society, 11*(2), 8. http://www.ecologyandsociety.org/vol11/iss2/art8/.

Chapin, F. S., Kofinas, G. P., & Folke, C. (Eds.). (2009). *Principles of ecosystem stewardship: Resilience-based natural resource management in a changing world*. New York: Springer.

Chuenpagdee, R., Bundy, A., Charles, T., Christie, P., Fanning, L., Gonzales, P., & Zwanenburg, K. (2005). Creating a positive future for fisheries and coastal communities worldwide. In R. Chuenpagdee & A. Bundy (Eds.), *Innovation and outlook in fisheries: An assessment of the research presented at the 4th world fisheries congress* (pp. 77–88). Fisheries Centre Research Report, *13*(2), 77–88.

Colfer, C. (2005). *The complex forest: Communities, uncertainty, and adaptive collaborative management*. Washington, D.C.: RFF Press.

Cumming, G., & Collier, J. (2005). Change and identity in complex systems. *Ecology and Society, 10*(1): 29. http://www.ecologyandsociety.org/vol10/iss1/art29/.

Dale, V. H., Brown, S., Haeuber, R. A., Hobbs, N. T., & Huntly, N. (2000). Ecological principles and guidelines for managing the use of land. *Ecological Applications, 107,* 639–670.

Davies, C., & Sarpong, D. (2013). The epistemological relevance of the arts in foresight and futures studies. *Futures, 47,* 1–8.

Dietz, T., Ostrom, E., & Stern, P. C. (2003). The struggle to govern the commons. *Science, 302,* 1907–1912.

Dryzek, J. S. (2005). *The politics of the earth: Environmental discourses* (2nd ed.). New York: Oxford University Press.

Elgin, C. Z. (2002). Creation as reconfiguration: Art in the advancement of science. *International Studies in the Philosophy of Science, 16,* 13–25.

Engle, N. L., Johns, O. R., Lemos, M. C., & Nelson, D. R. (2011). Integrated and adaptive management of water resources: Tensions, legacies, and the next best thing. *Ecology and Society, 16*(1), 19. http://www.ecologyandsociety.org/vol16/iss1/art19/.

Erdman, M. V., Merrill, P. R., Mongdong, M., Arsyad, I., Harahap, Z., Pangalila, R., & Baworo, P. (2004). Building effective co-management systems for decentralized protected areas management in Indonesia: Bunaken National Park case study. http://www.nrm.or.id. Accessed 6 March 2013.

Fazey, I., Evely, A. C., Reed, M. R., Stringer, L. C., Kruijsen, J. H. J., White, P. C. L., Newsham, A., Jin, L., Cortazzi, M., Phillipson, J., Blackstock, K. L., Entwistle, N., Sheate, W. R., Armstrong, F., Blackmore, C., Fazey, J. A., Ingram, J., Gregson, J., Lowe, P., Morton, S., & Trevitt, C. (2013). Knowledge exchange: A review and research agenda for environmental management. *Environmental Conservation, 40,* 19–36.

Ferse, S. C., Máñez Costa, M., Schwerdtner Máñez, K., Adhuri, D. S., & Glaser, M. (2010). Allies, not aliens: Increasing the role of local communities in marine protected area implementation. *Environmental Conservation, 37,* 23–34.

Folke, C., Hahn, T., Olsson, P., & Norberg, J. (2005). Adaptive governance of social-ecological systems. *Annual Review of Environment and Resources, 30,* 441–473.

Folke, C., Carpenter, S., Walker, B., Scheffer, M., Chapin, T., & Rockström, J. (2010). Resilience thinking: Integrating resilience, adaptability and transformability. *Ecology and Society, 15*(4), 20. www.ecologyandsociety.org/vol15/iss4/art20/.

Fox, J., Bushley, B. R., Miles, E. B., & Quazi, S. A. (Eds.). (2008). *Connecting communities and conservation: Collaborative management of protected areas in Bangladesh.* Honolulu: East-West Centre.

Garmestani, A. S., Allen, C. R., & Cabezas, H. (2009). Panarchy, adaptive management and governance: Policy options for building resilience. *Nebraska Law Review, 87,* 1036–1054.

Glaser, M. (2006). The social dimensions in ecosystem management: Strengths and weaknesses of human-nature mind maps. *Research in Human Ecology, 13,* 122–142.

Gunderson, L., & Holling, C. S. (Eds.). (2002). *Panarchy.* Washington, D.C.: Island Press.

Gunderson, L., & Light, S. S. (2006). Adaptive management and adaptive governance in the everglades ecosystem. *Policy Science, 39,* 323–334.

Gunderson, L., Holling, C. S., & Light, S. S. (1995). *Barriers and bridges to the renewal of ecosystems and institutions.* New York: Columbia University Press.

Hahn, T., Olsson, P., Folke, C., & Johansson, K. (2006). Trust-building, knowledge generation and organizational innovations: The role of a bridging organization for adaptive co-management of a wetland landscape around Kristianstad Sweden. *Human Ecology, 34,* 573–592.

Harkes, I. (1998). The sasi laut system in Maluku Province, Indonesia. Fisheries Co-management Project Working Paper 33. ICLARM, Manila.

Holling, C. S. (1978). *Adaptive environmental assessment and management.* Caldwell: Blackburn Press.

Ison, R., Blackmore, C., & Iaquinto, B. (2013). Towards systemic and adaptive governance: Exploring the revealing and concealing aspects of contemporary social-learning metaphors. *Ecological Economics, 87,* 34–42.

Johnson, B. (1999). Introduction to the special feature: Adaptive management—scientifically sound, socially challenged? *Conservation Ecology, 3*(1): 10. http://www.consecol.org/vol3/iss1/art10/.

Kallis, G., Kiparsky, M., & Norgaad, R. (2009). Collaborative governance and adaptive management: Lessons from California's CALFED Water Program. *Environmental Science and Policy, 12,* 631–643.

Kates, R., Clark, W. C., Corell, R., Hall, J., Jaeger, C., Lowe, I., & Svedin, U. (2001). Sustainability science. *Science, 292,* 641–642.

Keith, D., Martin, T., McDonald-Madden, E., & Walters, C. (2011). Uncertainty and adaptive management for biodiversity conservation. *Biological Conservation, 144,* 1175–1178.

Kotchen, M., & Young, O. (2007). Meeting the challenges of the anthropocene: Towards a science of coupled human-biophysical systems. *Global Environmental Change, 17,* 149–151.
Laidler, G. (2006). Inuit and scientific perspectives on the relationship between sea ice and climate change: The ideal complement? *Climate Change, 78,* 407–444.
Laidler, G., Elee, P., Ikummaq, T., Joamie, E., & Aporta, C. (2010). Mapping Inuit sea ice knowledge, use, and change in Nunavut, Canada (Cape Dorset, Igloolik, Pangnirtung). In C. Aporta & S. Gearheard (Eds.), *SIKU: Knowing our ice: Documenting Inuit sea ice knowledge and use* (pp. 45–80). New York: Springer.
Lang, D., Wiek, A., Bergmann, M., Stauffacher, M., Martens, P., & Moll, P., (2012). Transdisciplinary research in sustainability science: Practice, principles, and challenges. *Sustainability Science,* (7 Supplement), 25–43.
Larson, A. M., & Soto, F. (2008). Decentralization of natural resource governance regimes. *Annual Review of Environment and Resources, 33,* 213–239.
Lebel, L., Garden, P., & Imamura, M. (2005). The politics of scale, position, and place in the governance of water resources in the Mekong region. *Ecology and Society, 10*(2), 18. http://www.ecologyandsociety.org/vol10/iss2/art18/.
Lee, K. (1993). *Compass and gyroscope: Integrating science and politics for the environment.* Washington, D.C.: Island Press.
Lee, K. (1999). Appraising adaptive management. *Conservation Ecology, 3*(2), 3. http://www.consecol.org/vol3/iss2/art3/.
Lemos, M. C., & Agrawal, A. (2006). Environmental governance. *Annual Review of Environment and Resources, 31,* 297–325.
Levin, S. (1999). *Fragile dominion: Complexity and the commons.* Reading: Perseus Books.
Levin, S. A., & Clark, W. C. (Eds.). (2010). *Toward a science of sustainability: Report from toward a science of sustainability conference. Airlie Center, Warrenton.* Virginia: November 29, 2009—December 2, 2009. CID Working Paper No. 196. Center for International Development at Harvard University.
Liu, J., Dietz, T., Carpenter, S., Alberti, M., Folke, C., Moran, E., & Taylor, W. (2007). Complexity of coupled human and natural systems. *Science, 317,* 1513–1516.
Mahon, R., McConney, P., & Roy, R. (2008). Governing fisheries as complex adaptive systems. *Marine Policy, 32,* 104–112.
Mannigel, E. (2008). Integrating parks and people: How does participation work in protected area management? *Society and Natural Resources, 21,* 498–511.
Mappatoba, M. (Ed.). (2004). *Co-management of protected areas: The case of community agreements on conservation in the Lore Lindu National Park, central Sulawesi—Indonesia.* Göttingen: Cuvillier Verlag.
Mascarenhas, M. (2007). Where the waters divide: First nations, tainted water and environmental justice in Canada. *Local Environment, 12,* 565–577.
McGregor, D. (2012). Traditional knowledge: Considerations for protecting water in Ontario. *The International Indigenous Policy Journal, 3*(3), Article 11. http://ir.lib.uwo.ca/iipj/vol3/iss3/11.
McLain, R., & Lee, R. (1996). Adaptive management: Promises and pitfalls. *Environmental Management, 20,* 437–448.
MEA (Millennium Ecosystem Assessment). (2005). *Ecosystems and human well-being: General synthesis.* Chicago: Island Press.
Midgley, G. (2000). *Systemic intervention philosophy, methodology, and practice.* New York: Kluwer Academic/Plenum Publishers.
Miller, F., Osbahr, H., Boyd, E., Thomalla, F., Bharwani, S., Ziervogel, G., & Nelson, D. (2010). Resilience and vulnerability: Complementary or conflicting concepts? *Ecology and Society, 15*(3), 11. http://www.ecologyandsociety.org/vol15/iss3/art11/.
Mitchell, R. B., Clark, W. C., Cash, D. W., & Dickson, N. M. (2006). *Global environmental assessments: Information and influence.* Cambridge: MIT Press.
Moore, M. L., & Westley, F. (2011). Surmountable chasms: Networks and social innovation for resilient systems. *Ecology and Society, 16*(1), 5. http://www.ecologyandsociety.org/vol16/iss1/art5/.

Nadasdy, P. (2007). Adaptive co-management and the gospel of resilience. In D. Armitage & F. Berkes (Eds.), *Adaptive co-management: Collaboration, learning and multi-level governance* (pp. 208–227). Vancouver: UBC Press.

Natcher, D. C., Davis, S., & Hickey, C. G. (2005). Co-management: Managing relationships, not resources. *Human Organization, 64,* 240–250.

Nayak, P. K. (2011). *Change and marginalisation: Livelihoods, commons institutions and environmental justice in Chilika Lagoon, India*. Dissertation, University of Manitoba.

Nayak, P. K., & Berkes, F. (2010). Whose marginalisation? Politics around environmental injustices in India's Chilika Lagoon. *Local Environment, 15,* 553–567.

Nayak, P. K., & Berkes, F. (2014). Linking global drivers with local and regional change: A social-ecological system approach in Chilika Lagoon, Bay of Bengal. *Regional Environmental Change, 14,* 2067–2078.

Njaya, F., Donda, S., & Béné, C. (2011). Analysis of power in fisheries co-management: Experiences from Malawi. *Society and Natural Resources, 25,* 652–666.

Oakerson, R. J. (1992). Analyzing the commons: A framework. In D. Bromley (Ed.), *Making the commons work: Theory, practice and policy*. San Francisco: ICS Press.

O'Brien, K. (2012). Global environmental change II: From adaptation to deliberate transformation. *Progress in Human Geography, 36,* 667–676.

Olsson, P., Folke, C., & Hahn, T. (2004). Social-ecological transformations for ecosystem management: The development of adaptive co-management of a wetland landscape in southern Sweden. *Ecology and Society, 9*(4), 2. http://www.ecologyandsociety.org/vol9/iss4/art2/.

Olsson, P., Gunderson, L., Carpenter, S. R., Ryan, P., Lebel, L., Folke, C., & Holling, C. S. (2006). Shooting the rapids: Navigating transitions to adaptive governance of social-ecological systems. *Ecology and Society, 11*(1), 18. http://www.ecologyandsociety.org/vol11/iss1/art18/.

Ostrom, E. (1990). *Governing the commons: The evolution of institutions for collective action*. UK: Cambridge University Press.

Pahl-Wostl, C. (2009). A conceptual framework for analysing adaptive capacity and multi-level learning processes in resource governance regimes. *Global Environmental Change, 19,* 354–365.

Pelling, M. (2011). *Adaptation to climate change: From resilience to transformation*. New York: Routledge.

Peterson, G. (2007). Using scenario planning to enable an adaptive co-management process in the Northern Highlands Lake District of Wisconsin. In D. Armitage, F. Berkes, & N. Doubleday (Eds.), *Adaptive co-management: Collaboration, learning and multi-level governance* (pp. 286–307). Vancouver: UBC.

Peterson, G., Carpenter, S., & Brock, W. (2003). Uncertainty and the management of multistate ecosystems: An apparently rational route to collapse. *Ecology, 84,* 1403–1411.

Pinkerton, E. (2007). Integrating holism and segmentalism: Overcoming barriers to adaptive co-management between management agencies and multi-sector bodies. In D. Armitage, F. Berkes, & N. Doubleday (Eds.), *Adaptive co-management: Collaboration, learning and multi-level governance* (pp. 151–171). Vancouver: UBC Press.

Press, T., Lea, D., Webb, A., & Graham, A. (Eds.). (1995). *Kakadu: Natural and cultural heritage management*. Darwin: Australian Nature Conservation Agency and North Australia Research Unit, Australian National University.

Raik, D. B., Wilson, A. L., & Decker, D. J. (2008). Power in natural resources management: An application of theory. *Society and Natural Resources, 21,* 729–39.

Reed, M. S., Fazey, I., Stringer, L. C., Raymond, C. M., Akhtar-Schuster, M., Begni, G., & Wagner, L. (2013). Knowledge management for land degradation monitoring and assessment: An analysis of contemporary thinking. *Land Degradation and Development, 24,* 307–322.

Reid, W., Berkes, F., Wilbanks, T., & Capistrano, D. (Eds.). (2006). *Bridging scales and knowledge systems: Concepts and applications in ecosystem assessment. Millenium Ecosystem Assessment*. Washington, D.C.: Island Press.

Ribot, J. (2003). Democratic decentralisation of natural resource: Institutional choice and discretionary power transfers in sub-Saharan Africa. *Public Administration Development, 23,* 53–65.

Rittel, H., & Webber, M. (1973). Dilemmas in a general theory of planning. *Policy Sciences, 4*, 155–169.

Salafsky, N., Margoluis, R., & Redford, K. H. (2001). *Adaptive management: A tool for conservation practitioners*. Washington, D.C.: Biodiversity Support Program.

Satria, A., & Adhuri, D. (2010). Pre-existing fisheries management systems in Indonesia, focusing on Lombok and Maluku. In K. Ruddle & A. Satria (Eds.), *Managing coastal and inland waters: Pre-existing aquatic management systems in southeast Asia* (pp. 31–55). New York: Springer.

Sendzimir, J., Magnuszewski, P., Flachner, Z., Balogh, P., Molnar, G., Sarvari, A., & Nagy Z. (2007). Assessing the resilience of a river management regime: Informal learning in a shadow network in the Tisza River Basin. *Ecology and Society, 13*(1), 11. http://www.ecologyandsociety.org/vol13/iss1/art11/.

Simon, H. (1962). The architecture of complexity. *Proceedings of the American Philosophical Society, 106*, 467–482.

Stringer, L. C., Dougill, A. J., Fraser, E., Hubacek, K., Prell, C., & Reed, M. S. (2006). Unpacking "participation" in the adaptive management of social– ecological systems: A critical review. *Ecology and Society, 11*(2), 39. http://www.ecologyandsociety.org/vol11/iss2/art39/.

Tuhiwai Smith, L. (1999). *Decolonizing methodologies: Research and indigenous peoples*. New York: Zed Books Ltd.

Turner, B. L., Matson, P. A., McCarthy, J. J., Corell, R. W., Christensen, L., Eckley, N., & Schiller, A. (2003). Illustrating the coupled human-environment systems for vulnerability analysis: Three case studies. *Proceedings of the National Academy of Science, 100*, 8080–8085.

Vancouver, A. G. (2006). *Raven travelling: Two centuries of Haida art*. Vancouver Art Gallery: Douglas and McIntyre.

Walker, B. H., Holling, C. S., Carpenter, S. R., & Kinzig, A. (2004). Resilience, adaptability and transformability in social–ecological systems. *Ecology and Society, 9*(2), 5. http://www.ecologyandsociety.org/vol9/iss2/art5.

Walters, C. J. (1986). *Adaptive management of renewable resources*. New York: McGraw Hill.

Waltner-Toews, D., & Kay, J. (2005). The evolution of an ecosystem approach: The diamond schematic and an adaptive methodology for ecosystem sustainability and health. *Ecology and Society, 10*(1), 38. http://www.ecologyandsociety.org/vol10/iss1/art38/.

Weeratunge, N., Bene, C., Siriwardane, R., Charles, A., Johnson, D., Allison, E. H., & Badjeck, M. (2014). Small-scale fisheries through the wellbeing lens. *Fish and Fisheries, 15*, 255–279.

Westgate, M., Likens, G., & Lindenmayer, D. (2013). Adaptive management of biological systems: A review. *Biological Conservation, 158*, 128–139.

Westley, F. (2002). *The devil in the dynamics*. In L. Gunderson & C. S. Holling (Eds.), *Panarchy* (pp. 333–360). Washington, D.C.: Island Press.

Zurba, M., & Berkes, F. (2014). Caring for country through participatory art: Creating a boundary object for communicating Indigenous knowledge and values. *Local Environment,* 19, 821–836.

Chapter 14
Adaptive Management of Social-Ecological Systems: The Path Forward

Ahjond S. Garmestani and Craig R. Allen

Keywords Adaptive management · Social-Ecological systems · Uncertainty · Resilience · Natural resource management

Introduction

Adaptive management is derived from resilience theory, and originally was developed as a way to explore the resilience of ecosystems without exceeding the resilience of the system of interest (Chap. 2, Holling 1973). Ecosystems are characterized by complexity and in most cases there is basic uncertainty regarding their dynamics. Uncertainty in the response of linked social-ecological systems to management interventions necessitates that an adaptive approach be utilized (Chap. 8, Bown et al. 2013). Adaptive management explicitly tests predictions against observations, which allows for iterative recalibration of the management process at pre-determined decision points as learning occurs (Williams 2011). This learning process allows for management actions to progress as uncertainty is reduced over time (Williams 2011). Adaptive management is not a panacea, but can be a powerful tool for environmental management when applied to appropriate problems in social-ecological systems.

This book is intended to present the state of the art of adaptive management by providing a historical perspective (Chaps. 2 and 3), highlighting bridges and barriers to its implementation (Chaps. 4, 10 and 11), and illuminating the evolution of

A. S. Garmestani (✉)
Office of Research and Development, U.S. Environmental Protection Agency,
26 W. Martin Luther King Drive, Cincinnati, OH 45268, USA
e-mail: garmestani.ahjond@epa.gov

C. R. Allen
U.S. Geological Survey, Nebraska Cooperative Fish and Wildlife Research Unit,
University of Nebraska-Lincoln, Lincoln, NE 68583, USA
e-mail: allencr@unl.edu

C. R. Allen, A. S. Garmestani (eds.), *Adaptive Management of Social-Ecological Systems*,
DOI 10.1007/978-94-017-9682-8_14

adaptive management since its development over the past 4 decades (Chaps. 5, 6, 7, 8, 9, 12 and 13). However, it is not prescriptive, and readers interested in "how to" should delve into the resources cited in chapter references. Here we discuss some of the recent themes recurring in the adaptive management literature, and discuss the different contexts of adaptive co-management, adaptive governance and resilience-based governance.

Adaptive Management: The Present and the Future

Present

Adaptive management has tremendous traction in the academic literature, demonstrating the persistence of the methodology (Westgate et al. 2013). There are several factors that act as "bridges" for successful adaptive management. These factors include: *collaboration* (Chap. 10, Reever Morghan et al. 2006, Stringer et al. 2006, Armitage et al. 2009, Johnson 2011, Moore et al. 2011, Williams 2011, Porzecanski et al. 2012, Susskind et al. 2012, Caves et al. 2013, Greig et al. 2013, LoSchiavo et al. 2013, Pratt Miles 2013, Westgate et al. 2013), *funding* (Chap. 4, Chap. 10, Armitage et al. 2009, Moore et al. 2011, Smith 2011, Caves et al. 2013, Greig et al. 2013, LoSchiavo et al. 2013, Rist et al. 2013, Westgate et al. 2013), *clear objectives* (Chap. 3, Chap. 5, Chap. 10, Moore et al. 2011, Williams 2011, Porzecanski et al. 2012, Susskind et al. 2012, Caves et al. 2013, Greig et al. 2013, LoSchiavo et al. 2013, Pratt Miles 2013), *leadership* (Chap. 3, Chap. 7, Chap. 10, Walters 2007, Munaretto and Huitema 2012, Caves et al. 2013, Greig et al. 2013), *presence of intermediaries* (Chap. 7, Stringer et al. 2006, Johnson 2011, Munaretto and Huitema 2012, Greig et al. 2013, Monroe et al. 2013, Pratt Miles 2013), *appropriate scale of project* (Chap. 10, Reever Morghan et al. 2006, Stringer et al. 2006), and a *favorable institutional, policy and social environment* (Chap. 10, Stringer et al. 2006, Armitage et al. 2009, Moore et al. 2011, Smith 2011, Porzecanski et al. 2012, Susskind et al. 2012, Caves et al. 2013, Greig et al. 2013, LoSchiavo et al. 2013). Some potential "barriers" to adaptive management are the lack of funding for project implementation and monitoring, and shifts in management policies, personnel and leadership (Conclusion, Jacobson et al. 2006, Westgate et al. 2013). In many cases where adaptive management was unsuccessful, the conditions necessary for success did not exist, whether those factors (and the interaction of those factors) are institutional, organizational or social (Chap. 3, Porzecanski et al. 2012). In some cases, adaptive assessments and experimentation have led to innovative environmental management and organizational learning (Chap. 3). Adaptive management isn't appropriate where there is little uncertainty and little controllability, thus excluding a large range of potential applications. Rather, other methods (e.g., scenario planning, building resilience and maximum sustained yield) may be better fits for the environmental problem to be managed (see Allen et al. 2011) where controllability is weak (i.e., management is largely not possible) or uncertainty is low.

In the United States the current legal framework is focused upon finality of process (e.g., National Environmental Policy Act), and not designed to accommodate iterative mechanisms, which are essential for adaptive management (Chap. 4, Benson and Garmestani 2011). The current focus upon finality in American law results from it being crafted around outdated scientific understanding about the dynamics of social-ecological systems (Garmestani et al. 2013). In essence, American law was built upon the understanding at the time that the world was characterized by a "balance of nature", which allowed natural resource managers to have a good sense about the manner in which the natural world will behave in the future (Garmestani et al. 2013). Thus, adaptive management is difficult to implement within the scope of current law. In a recent study, a majority of practitioners reported that implementation of adaptive management was hampered by legal constraints (Benson and Stone 2013). For example, most laws in the United States do not explicitly require monitoring, an essential component of adaptive management, and the lack of a regulatory "home" for adaptive management means agencies aren't typically bound by its requirements (Benson and Garmestani 2011). This means that adaptive management, as it is currently practiced in many of the most visible applications does not possess the legal grounding necessary for enforceability, which is essential to ensuring that the methodology is implemented as it was intended (Chap. 4, Holling 1978, Benson and Garmestani 2011). In addition, several legal scholars have concluded that conducting adaptive management is incompatible with current administrative law, and thus not possible without reform (Ruhl 1998, Karkkainen 2005, Garmestani et al. 2009).

Future

Adaptive management remains underutilized and poorly understood. A large part of this problem can be traced to its implementation through top-down authority or its highly visible but poorly functioning applications to large problems not well-suited to adaptive management purposes. An example of the latter case is the application of adaptive management to large river systems where endangered species recovery is the goal. In such cases, replicated experimentation is impossible and controllability is low. Here, structured decision making, which is closely related to adaptive management, is more appropriate. Top-down control is a problem in many cases where there are mandates to apply adaptive management, for example in some federal agencies, but with little guidance on implementation in the field. Adaptive management's promise is for a subset of mesoscale environmental problems. These mesoscales—larger and longer than typical graduate student-driven academic research but smaller and shorter than continental watersheds or most climate-driven change—remain poorly understood, but are amenable to replicated experimental manipulations that can yield tangible results in reasonable time frames. Examples include projects such as testing of green infrastructure impacts on water quality and quantity in urban settings, techniques for invasive species removal, and methods of ecological restoration.

Certainly, there is room for improvement in the process of adaptive management. Integrating adaptive management with law will likely require some reforms (see Chap. 4). For example, Benson and Garmestani (2011) have suggested that an American law such as the National Environmental Protection Act (NEPA) could be reconfigured to accommodate adaptive management. In particular, they advocate for a new process within NEPA that is iterative rather than linear, requiring monitoring, and reform that adds substance and structure to NEPA's mandate. The impediment of "final agency action" that serves to foment a linear administrative law process presents a substantial obstacle to adaptive management and has been the subject of recent discourse (Karkkainen 2005, Garmestani et al. 2009, Benson and Garmestani 2011). Karkkainen (2005) has argued for an "adaptive management track" that would be implemented if an agency could demonstrate that such a variation to current administrative law was warranted. Answering this call, Craig and Ruhl (2014) proposed and drafted model legislation that could create an "adaptive management track" for specific agency decisions, within the context of administrative law. This proposed new law, the Model Adaptive Management Procedure Act (MAMPA) could be a leap forward in our pursuit of operationalizing adaptive management (*sensu* Holling) for linked social-ecological systems. MAMPA balances the foundations of administrative law, while accounting for social-ecological resilience, and offers great promise for sound environmental governance (see Craig and Ruhl 2014).

In addition to legal reform, there are other mechanisms for improving the adaptive management process and environmental outcomes. For example, stakeholder evaluations (i.e., "what should be" vs. "what is") of an adaptive management project could be used as one metric for measuring the progress of a project (Berkley 2013). Assessing the context of adaptive management via metrics at each phase of the adaptive management process is another possible mechanism for improving environmental outcomes (Chap. 6). In particular, by assessing each phase of the adaptive management cycle, system-wide learning will occur, even if there are issues with a phase or the entire process (Chap. 6). While adaptive management is supposed to create the conditions for iterative management in response to system feedback, explicitly linking adaptive management to thresholds that require management intervention should be considered (Chap. 5, Garmestani and Allen 2014). Linking adaptive management to ecological and legal thresholds, with the capacity for recalibrating thresholds in light of new information, could be one aspect of the path forward for adaptive management.

Adaptive management is considered to be the best existing approach for dealing with the unpredictability of social-ecological systems (Westgate et al. 2013). While we (Allen et al. 2011) have stated that adaptive management is only appropriate under certain circumstances (e.g., when uncertainty and controllability are high), Rist et al. (2013) argue that there are no boundaries to the application of adaptive management (but see Chap. 10 and Chap. 11). Rather, when to apply adaptive management can be defined by the problem of interest, and the resources available to managers. In making this assertion, Rist et al. (2013) argue that adaptive management should be seen simply as a methodology to reduce uncertainty in environmental management, separate from the institutional, policy and social environment where management occurs. Their argument turns on the proposition that all environmental management is subject to institutional, policy and social constraints, and these factors are not endemic to adaptive management. Thus, according to Rist et al. (2013), adaptive management needs to be placed within an institutional and governance framework (e.g., adaptive governance) that facilitates its core purpose. In contrast, separating adaptive management from political and social processes is not possible for many adaptive management projects (Chap. 13, Gunderson and Light 2006, Cosens and Williams 2012) and furthermore, whether a project succeeds or fails is dependent upon human and social capital, regardless of the quality of the science (Chap. 7, Cundill et al. 2011). As a result, adaptive management has been integrated with collaborative management (adaptive co-management), which ultimately sets the stage for adaptive governance (Chap. 9, Folke et al. 2005, Munaretto and Huitema 2012). Adaptive governance attempts to take into account formal and informal institutions, and is at the intersection integrating adaptive co-management and governance (Garmestani et al. 2009, Huitema et al. 2009). While adaptive governance has been touted as the manner by which to implement resilience thinking, it is lacking in significant legal grounding that would allow for it to be incorporated into rules and regulations (Ruhl 2012, Garmestani and Benson 2013). Building upon the lack of legal grounding for adaptive governance, Garmestani and Benson (2013) offered a framework for resilience-based governance that integrates resilience theory (i.e., panarchy, adaptive management, and adaptive governance) with reflexive law. Cumming (2013) asserts that this framework has great potential for resilience-based governance, as it explicitly accounts for scale and governance mismatches, but would require major legal reform (Garmestani and Benson 2013).

Conclusion

Adaptive management remains at the forefront of environmental management nearly 40 years after its original conception, largely because we have yet to develop other methodologies that offer the same promise (Allen et al. 2011). Despite the criticisms of adaptive management and the numerous failed attempts to implement it, adaptive management has yet to be replaced. The concept persists because it is seen as critical to managing for resilience, and therefore an essential aspect of social-ecological resilience (Garmestani and Allen 2014). Moving forward, adaptive management of social-ecological systems provides policymakers, managers and scientists a powerful tool for managing for resilience in the face of uncertainty. The methodology has been developing for nearly half a century, and continues to resonate for environmental management, even though there are numerous barriers to its implementation. Over time, we have come to learn that "barriers" to adaptive management include: *lack of collaboration* (Plummer and Armitage 2007, Allen and Gunderson 2011, Allen et al. 2011, Johnson 2011, Keith et al. 2011, Williams

2011, Munaretto and Huitema 2012, Porzecanski et al. 2012, Susskind et al. 2012, Westgate et al. 2013), *lack of funding* (Plummer and Armitage 2007, Walters 2007, Allen and Gunderson 2011, Greig et al. 2013, LoSchiavo et al. 2013,Westgate et al. 2013), *lack of clear objectives* (Allen and Gunderson 2011, Porzecanski et al. 2012, Susskind et al. 2012, Greig et al. 2013, Pratt Miles 2013, Rist et al. 2013), *lack of leadership* (Gunderson and Light 2006, Walters 2007, Allen and Gunderson 2011, Munaretto and Huitema 2012,Westgate et al. 2013), *lack of intermediaries* (Stringer et al. 2006, Munaretto and Huitema 2012, Bown et al. 2013, Greig et al. 2013), *inappropriate scale of projects* (Chap. 10, Reever Morghan et al. 2006, Stringer et al. 2006), and *lack of a favorable institutional, policy and social environment* (Chap. 3, Plummer and Armitage 2007, Armitage et al. 2009, Allen and Gunderson 2011, Keith et al. 2011, Porzecanski et al. 2012, Susskind et al. 2012, Bown et al. 2013, LoSchiavo et al. 2013). Adaptive management can be successful under the right circumstances, and at the right scale. Large-scale, river basin projects (e.g., Everglades) have been extensively treated in the literature and unfortunately, many have met with limited success. This has led some commentators to claim that adaptive management is a failed management strategy. However, the limitations of adaptive management simply illuminate that adaptive management is not an appropriate strategy for large-scale social-ecological systems, with a host of complicating factors ranging from the ecosystem to the institutional, organizational and policy environment. Any of these factors individually, or in combination, likely dooms these large-scale adaptive management projects from their inception. Rather, adaptive management can be successful (e.g., waterfowl harvests, green infrastructure) at an appropriate scale and under appropriate conditions, especially in cases when there is a favorable institutional, organizational and policy environment.

Importantly, adaptive management is not a solution for every context, and should not be viewed as such. Rather, adaptive management should be viewed as flowing from social-ecological resilience and a critical component of adaptive governance, and therefore resilience-based governance (Garmestani and Benson 2013). Ultimately this means that adaptive management is a very useful tool for sound environmental management and governance (Chap. 12).

References

Allen, C. R., & Gunderson, L. H. (2011). Pathology and failure in the design and implementation of adaptive management. *Journal of Environmental Management, 92,* 1379–1384.

Allen, C. R., Fontaine, J. J., Pope, K. L., & Garmestani, A. S. (2011). Adaptive management for a turbulent future. *Journal of Environmental Management, 92,* 1339–1345.

Armitage, D. R., Plummer, R., Berkes, F., Arthur, R. I., Charles, A. T., Davidson-Hunt, I. J., Diduck, A. P., Doubleday, N. C., Johnson, D. S., Marschke, M., McConney, P., Pinkerton, E. W., & Wollenberg, E. K. (2009). Adaptive con-management for social-ecological complexity. *Frontiers in Ecology and the Environment, 7,* 95–102.

Benson, M. H., & Garmestani, A. S. (2011). Embracing panarchy, building resilience and integrating adaptive management through a rebirth of the National Environmental Policy Act. *Journal of Environmental Management, 92,* 1420–1427.

Benson, M. H., & Stone, A. B. (2013). Practitioner perceptions of adaptive management implementation in the United States. *Ecology and Society, 18*(3), 32. http://dx.doi.org/10.5751/ES-05613-180332.

Berkley, J. (2013). Opportunities for collaborative adaptive management progress: integrating Stakeholder assessments into progress measurement. *Ecology and Society, 18*(4), 69. http://dx.doi.org/10.5751/ES-05988-180469.

Bown, N. K., Gray, T. S., & Stead, S. M. (2013). Co-management and adaptive co-management: Two modes of governance in a Honduran marine protected area. *Marine Policy, 39,* 128–134.

Caves, J. K., Bodner, G. S., Simms, K., Fisher, L. A., & Robertson, T. (2013). Integrating collaboration, adaptive management, and scenario-planning: experiences at Las Cienegas National Conservation Area. *Ecology and Society, 18*(3), 43. http://dx.doi.org/10.5751/ES-05749-180343.

Cosens, B. A., & Williams, M. K. (2012). Resilience and water governance: Adaptive governance in the Columbia River basin. *Ecology and Society, 17*(4), 3. http://dx.doi.org/10.5751/ES-04986-170403.

Craig, R. K., & Ruhl, J. B. (2014). Designing administrative law for adaptive management. *Vanderbilt Law Review, 67,* 1–87.

Cumming, G. S. (2013). Scale mismatches and reflexive law. *Ecology and Society, 18*(1), 15. http://dx.doi.org/10.5751/ES-05407-180115.

Cundill, G., Cumming, G. S., Biggs, D., & Fabricius, C. (2011). Soft systems thinking and social learning for adaptive management. *Conservation Biology, 26,* 13–20.

Folke, C., Hahn, T., Olsson, P., & Norberg, J. (2005). Adaptive governance of social-ecological systems. *Annual Review of Environment and Resources, 30,* 441–473.

Garmestani, A. S., & Benson, M. H. (2013). A framework for resilience-based governance of social-ecological systems. *Ecology and Society, 18*(1), 9. http://www.ecologyandsociety.org/vol18/iss1/art9/

Garmestani, A. S., & Allen, C. R. (2014). *Social-ecological resilience and law*. New York: Columbia University Press.

Garmestani, A. S., Allen, C. R., & Cabezas, H. (2009). Panarchy, adaptive management and governance: Policy options for building resilience. *Nebraska Law Review, 87,* 1036–1054.

Garmestani, A. S., Allen, C. R., & Benson, M. H. 2013. Can law foster social-ecological resilience? *Ecology and Society, 18*(2): 37. http://www.ecologyandsociety.org/vol18/iss2/art37/

Green, O. O., & Garmestani, A. S. (2012). Adaptive management to protect biodiversity, best available science and the Endangered Species Act. *Diversity, 4,* 164–178.

Greig, L. A., Marmorek, D. R., Murray, C., & Robinson, D. C. E. (2013). Insight into enabling adaptive management. *Ecology and Society, 18*(3), 24. http://dx.doi.org/10.5751/ES-05686-180324.

Gunderson, L. H., & Light, S. S. (2006). Adaptive management and adaptive governance in the Everglades. *Policy Sciences, 39,* 323–334.

Holling, C. S. (1973). Resilience and stability of ecological systems. *Annual Review of Ecological Systems, 4,* 1–23.

Holling, C. S. (1978). *Adaptive environmental assessment and management*. Chichester: Wiley.

Huitema, D., Mostert, E., Egas, W., Moellenkamp, S., Pahl-Wostl, C., & Yalcin, R. (2009). Adaptive water governance: Assessing the institutional prescriptions of adaptive (co-) management from a governance perspective and defining a research agenda. *Ecology and Society, 14*(1), 26. http://www.ecologyandsociety.org/vol14/iss1/art26/.

Jacobson, S. K., Morris, J. K., Sanders, J. S., Wiley, E. N., Brooks, M., Bennetts, R. E., Percival, H. F., & Marynowski, S. (2006). Understanding barriers to implementation of an adaptive land management program. *Conservation Biology, 20,* 1516–1527.

Johnson, F. A. (2011). Learning and adaptation in the management of waterfowl harvests. *Journal of Environmental Management, 92,* 1385–1394.

Karkkainen, B. C. (2005). Panarchy and adaptive change: Around the loop and back again. *Minnesota Journal of Law, Science & Technology, 7,* 59–77.

Keith, D. A., Martin, T. G., McDonald-Madden, E., & Walters, C. (2011). Uncertainty and adaptive management for biodiversity conservation. *Biological Conservation, 144,* 1175–1178.

LoSchiavo, A. J., Best, R. G., Burns, R. E., Gray, S., Harwell, M. C., Hines, E. B., McLean, A. R., Clair, T. St., Traxler, S., & Vearil, J. W. (2013). Lessons learned from the first decade of adaptive management in comprehensive Everglades restoration. *Ecology and Society, 18*(4), 70. http://dx.doi.org/10.5751/ES-06065-180470.

Monroe, M. C., Plate, R., & Oxarart, A. (2013). Intermediate collaborative adaptive management strategies build stakeholder capacity. *Ecology and Society, 18*(2), 24. http://dx.doi.org/10.5751/ES-05444-180224.

Moore, C. T., Lonsdorf, E. V., Knutson, M. G., Laskowski, H. P., & Lor, S. K. (2011). Adaptive management in the U.S. National Wildlife Refuge System: Science-management partnerships for conservation delivery. *Journal of Environmental Management, 92,* 1395–1402.

Munaretto, S., & Huitema, D. (2012). Adaptive comanagement in the Venice lagoon? An analysis of current water and environmental management practices and prospects for change. *Ecology and Society, 17*(2), 19. http://dx.doi.org/10.5751/ES-04772-170219.

Plummer, R., & Armitage, D. R. (2007). Charting the new territory of adaptive co-management: A Delphi study. *Ecology and Society, 12*(2), 10. http://www.ecologyandsociety.org/vol12/iss2/art10/.

Porzecanski, I., Saunders, L. V., & Brown, M. T. (2012). Adaptive management fitness of watersheds. *Ecology and Society, 17*(3), 29. http://dx.doi.org/10.5751/ES-05061-170329.

Pratt Miles, J. D. (2013). Designing collaborative processes for adaptive management: Four structures for multistakeholder collaboration. *Ecology and Society, 18*(4), 5. http://dx.doi.org/10.5751/ES-05709-180405.

Reever Morghan, K. J., Sheley, R. L., & Svejcar, T. J. (2006). Successful adaptive management—the integration of research and management. *Rangeland Ecology and Management, 59,* 216–219.

Rist, L., Felton, A., Samuelsson, L., Sandström, C., & Rosvall, O. (2013). A new paradigm for adaptive management. *Ecology and Society, 18*(4), 63. http://dx.doi.org/10.5751/ES-06183-180463.

Ruhl, J. B. (1998). The Endangered Species Act and private property: A matter of timing and location. *Cornell Journal of Law and Public Policy, 8,* 37–53.

Ruhl, J. B. (2012). Panarchy and the law. *Ecology and Society, 17*(3), 31. http://dx.doi.org/10.5751/ES-05109-170331.

Smith, C. B. (2011). Adaptive management on the central Platte River—Science, engineering, and decision analysis to assist in the recovery of four species. *Journal of Environmental Management, 92,* 1414–1419.

Stringer, L. C., Dougill, A. J., Fraser, E., Hubacek, K., Prell, C., & Reed, M. S. (2006). Unpacking "participation" in the adaptive management of social–ecological systems: A critical review. *Ecology and Society, 11*(2), 39. http://www.ecologyandsociety.org/vol11/iss2/art39/.

Susskind, L., Camacho, A. E., & Schenk, T. (2012). A critical assessment of collaborative adaptive management in practice. *Journal of Applied Ecology, 49,* 47–51.

Walters, C. (2007). Is adaptive management helping to solve fisheries problems? *Ambio, 35,* 304–307.

Westgate, M. J., Likens, G. E., & Lindenmayer, D. B. (2013). Adaptive management of biological systems: A review. *Biological Conservation, 158,* 128–139.

Williams, B. K. (2011). Passive and active adaptive management: Approaches and an example. *Journal of Environmental Management, 92,* 1371–1378.

Index

C. R. Allen, A. S. Garmestani (eds.), *Adaptive Management of Social-Ecological Systems*,
DOI 10.1007/978-94-017-9682-8

MIX
Papier aus verantwortungsvollen Quellen
Paper from responsible sources
FSC® C105338

Printed by Books on Demand, Germany